Plant Cell Monographs
Volume 13

Series Editor: David G. Robinson
Heidelberg, Germany

Plant Cell Monographs
Recently Published Titles

The Chloroplast
Interactions with the Environment
Volume Editors: Sandelius, A. S., Aronsson, H.
Vol. 13, 2009

Root Hairs
Volume Editors: Emons, A.M.C., Ketelaar, T.
Vol. 12, 2009

Plant Microtubules
Development and Flexibility
2nd Edition
Volume Editor: Nick, P.
Vol. 11, 2008

Plant Growth Signalling
Volume Editors: Bögre, L., Beemster, G.
Vol. 10, 2008

Cell Division Control in Plants
Volume Editors: Verma, D. P. S., Hong, Z.
Vol. 9, 2008

Endosperm
Volume Editor: Olsen, O.-A.
Vol. 8, 2007

Viral Transport in Plants
Volume Editors: Waigmann, E., Heinlein, M.
Vol. 7, 2007

**Nitric Oxide in Plant Growth,
Development and Stress Physiology**
Volume Editors: Lamattina, L., Polacco, J.
Vol. 6, 2007

The Expanding Cell
Volume Editors: Verbelen, J.-P., Vissenberg, K.
Vol. 5, 2007

The Plant Endoplasmic Reticulum
Volume Editor: Robinson, D. G.
Vol. 4, 2006

The Pollen Tube
A Cellular andMolecular Perspective
Volume Editor: Malhó, R.
Vol. 3, 2006

Somatic Embryogenesis
Volume Editors: Mujib, A., Šamaj, J.
Vol. 2, 2006

Plant Endocytosis
Volume Editors:
Šamaj, J., Baluška, F., Menzel, D.
Vol. 1, 2005

Anna Stina Sandelius • Henrik Aronsson
Editors

The Chloroplast

Interactions with the Environment

Editors
Prof. Dr. Anna Stina Sandelius
and
Dr. Henrik Aronsson

Department of Plant and Environmental Sciences
University of Gothenburg
P.O. Box 461, 405 30 Gothenburg
Sweden

Series Editor
Prof. Dr. David G. Robinson
Ruprecht-Karls-University of Heidelberg
Heidelberger Institute for Plant Sciences (HIP)
Department Cell Biology
Im Neuenheimer Feld 230
D-69120 Heidelberg
Germany

ISBN: 978-3-540-68692-7   e-ISBN: 978-3-540-68696-5
DOI: 10.1007/978-3-540-68696-5

Library of Congress Control Number: 2008939144

© 2009 Springer-Verlag Berlin Heidelberg

This work is subject to copyright. All rights are reserved, whether the whole or part of the material is concerned, specifically the rights of translation, reprinting, reuse of illustrations, recitation, broadcasting, reproduction on microfilm or in any other way, and storage in data banks. Duplication of this publication or parts thereof is permitted only under the provisions of the German Copyright Law of September 9, 1965, in its current version, and permission for use must always be obtained from Springer. Violations are liable to prosecution under the German Copyright Law.

The use of general descriptive names, registered names, trademarks, etc. in this publication does not imply, even in the absence of a specific statement, that such names are exempt from the relevant protective laws and regulations and therefore free for general use.

Cover design: WMX Design GmbH, Heidelberg, Germany

Printed on acid-free paper

9 8 7 6 5 4 3 2 1 0

springer.com

Editors

Anna Stina Sandelius pursued her PhD degree in Plant Physiology with Prof Conny Liljenberg at the University of Gothenburg. She graduated in 1983 and spent the following year and a half with Prof D. James Morré at Purdue University in West Lafayette, Indiana, USA. She returned to the University of Gothenburg in 1985, where she attained full professorship in 1999. She is presently vice dean of the science faculty. As a graduate student, she studied the role of galactolipids during chloroplast development. She then switched to the plasma membrane and its lipid metabolism and supply, initially focusing on inositolphospholipids. The interest in lipid trafficking brought back plastids as an object of study and, thanks to her group's recent discovery that the plasma membrane uses plastid-synthesized galactolipids to replace phospholipids during phosphate-limiting cultivation, the two objects of study, along with the endoplasmic reticulum, have been brought together in efforts to elucidate lipid trafficking between plastids and the plasma membrane.

Henrik Aronsson pursued his PhD degree in Plant Physiology with Dr Clas Dahlin at the University of Gothenburg. He graduated in 2001 and spent the following year and a half as a postdoctoral student in Dr Paul Jarvis' group at Leicester University. The next year he spent at Gotland University and Skövde University as senior lecturer. He then returned to the University of Gothenburg in 2004, where he attained associate professorship in 2007. As a graduate student, he studied plastid protein targeting of the light-dependent enzyme NADPH: protochlorophyllide reductase (POR) both to the envelope and the internal membrane system. He then switched to studying the chloroplast protein import machinery with a focus on the components that make up the machinery. His group has recently started studying the plastid vesicular transport system between the envelope and the internal membrane system with emphasis on putative proteins involved in the process.

Preface

A complete book on chloroplast would contain a vast number of chapters! We chose to focus on interactions between the chloroplast and its immediate as well as distant environments, with a first chapter on plastid evolution. When we received the manuscripts, also the chapters related to communication and/or physical interactions between chloroplasts and their surroundings maintained this temporal interaction as a background theme. Communication, physical interactions, evolution – but hardly anything on photosynthesis or pigments. The latter topics are probably the most clearly obvious ones for a chloroplast book, but here also lies our rationale behind the choice of chapter subjects; we want to present chloroplasts in a different perspective. The recent rapid evolvement of the presented research areas, largely made possible by the development of molecular techniques and genetic screens of an increasing number of plant model systems, makes the interactive theme timely. We are truly grateful to all the contributing authors for providing exciting chapters!

The first two chapters set the stage: in the first, the evolution of plastids is presented and the structural, functional and genomic variations among plastids of land plants and algae are described in an evolutionary context. Double membrane-bound plastids, which are believed to derive directly from a cyanobacterial endosymbiont, as well as plastids of a more complex ancestry with more than two delimiting membranes are covered. The former kind, well studied in land plants and green algae, is the main object in the following chapters. The second chapter defines the borderline, the chloroplast envelope. A current state-of-the-art list of chloroplast envelope proteins is presented, which, together with the lipid setup of this membrane system, reflects the prokaryotic origin of the chloroplast as well as its integration into the host cell.

Three chapters focus on transport across the envelope. The reduced genome of the chloroplast, compared to its ancestors, necessitates import of nuclear-encoded proteins from the cytoplasm. Chapter three presents the protein import machinery and its constituents in the two envelope membranes and how import is regulated and the chloroplast protein level maintained. Several of the imported proteins are involved in chloroplast lipid metabolism and the fourth chapter presents the interdependence of the chloroplast and the rest of the cell in providing lipid constituents to all membranes, within or outside plastids, during various environmental conditions.

This and the following chapter, on metabolite transporters, share an evolutionary feature, that plastids have acquired the role of sole provider of certain compounds once synthesized in the host cell, such as fatty acids and certain amino acids. The metabolite transport chapter uses an evolutionary perspective to present the vast array of metabolite transporters that connect chloroplast metabolism with that of the surrounding cell and also address the specific features of apicoplast transporter systems.

Transport of proteins, lipids and metabolites between plastids and the surrounding cell is regulated by feedback controls at several levels, stemming from intracellular as well as external conditions. The last four chapters cover different aspects of communication between chloroplasts and their surroundings. Plastid-nucleus signalling is the topic of the sixth chapter, with focus on the retrograde information flow, from the plastid to the nucleus, through different pathways. The next chapter presents plastid division at the molecular and cellular level with emphasis on the integration of the host and former endosymbiont and the roles of environmental and endogenous signals in controlling the process. Chapter eight also deals with communication and the chloroplast as a physical entity. The focus is on chloroplast movement and positioning in relation to light quality and quantity, where cytosolic components are involved in motility changes to optimize chloroplast function and survival. Sensing the environment and communication are also central themes of the final chapter. Here the focus switches to the involvement of chloroplasts as providers of metabolites that benefit plant individuals and communities, and the examples include defence coordination compounds and attractants for pollinators and seed dispersal.

Again, we are greatly indebted to all authors! We also extend our thanks to all colleagues in Göteborg and elsewhere who helped us with the review process, to the series editor David G. Robinson for trust and encouragement and last but not least to Anette Lindqvist, Christina Eckey and Elumalai Balamurugan of Springer-Verlag for continuous and patient support and help.

August 2008

Anna Stina Sandelius
Henrik Aronsson

Contents

Diversity and Evolution of Plastids and Their Genomes 1
E. Kim and J.M. Archibald

The Chloroplast Envelope Proteome and Lipidome 41
N. Rolland, M. Ferro, D. Seigneurin-Berny,
J. Garin, M. Block, and J. Joyard

**The Chloroplast Protein Import Apparatus,
Its Components, and Their Roles** 89
H. Aronsson and P. Jarvis

Chloroplast Membrane Lipid Biosynthesis and Transport 125
M.X. Andersson and P. Dörmann

**The Role of Metabolite Transporters in Integrating
Chloroplasts with the Metabolic Network of Plant Cells** 159
A.P.M. Weber and K. Fischer

Retrograde Signalling 181
L. Dietzel, S. Steiner, Y. Schröter, and T. Pfannschmidt

**Plastid Division Regulation and Interactions
with the Environment** 207
J. Maple, A. Mateo, and S.G. Møller

Chloroplast Photorelocation Movement 235
N. Suetsugu and M. Wada

A Sentinel Role for Plastids 267
F. Bouvier, A.S. Mialoundama, and B. Camara

Index 293

Contributors

Mats X. Andersson
Plant and Soil Science Laboratory, Department of Agricultural Sciences, Faculty of Life Sciences, The University of Copenhagen, Thorvaldsensvej 40,1871 Frederiksberg C, Denmark

John M. Archibald
Department of Biochemistry and Molecular Biology, Canadian Institute for Advanced Research, Dalhousie University, Halifax, NS, Canada, B3H 1X5
jmarchib@dal.ca

Henrik Aronsson
Department of Plant and Environmental Sciences, University of Gothenburg, P.O. Box 461, 405 30 Gothenburg, Sweden
henrik.aronsson@dpes.gu.se

Maryse Block
Laboratoire de Physiologie Cellulaire Végétale; CEA; CNRS; INRA; Université Joseph Fourier, iRTSV, CEA-Grenoble, 38054 Grenoble-cedex 9, France

Florence Bouvier
Institut de Biologie Moléculaire des Plantes, Centre National de la Recherche Scientifique and Université Louis Pasteur, 67084 Strasbourg Cedex, France

Bilal Camara
Institut de Biologie Moléculaire des Plantes, Centre National de la Recherche Scientifique and Université Louis Pasteur, 67084 Strasbourg Cedex, France
bilal.camara@ibmp-ulp.u-strasbg.fr

Lars Dietzel
Department for Plant Physiology, Friedrich-Schiller-University Jena, Dornburger Str. 159, 07743 Jena, Germany

Peter Dörmann
Institute of Molecuar Physiology and Biotechnology of Plants, University of Bonn, Karlrobert-Kreiten-Str. 13, 53115 Bonn, Germany
doermann@uni-bonn.de

Myriam Ferro
Laboratoire Etude de la Dynamique des Protéomes; CEA; INSERM; Université Joseph Fourier, iRTSV, CEA-Grenoble, 38054 Grenoble-cedex 9, France

Karsten Fischer
Institutt for Biologi, University of Tromsø, 9037 Tromsø, Norway
karsten.fischer@ib.uit.no

Jérôme Garin
Laboratoire Etude de la Dynamique des Protéomes; CEA; INSERM; Université Joseph Fourier, iRTSV, CEA-Grenoble, 38054 Grenoble-cedex 9, France

Paul Jarvis
Department of Biology, University of Leicester, University Road, Leicester LE1 7RH, UK
rpj3@le.ac.uk

Jacques Joyard
Laboratoire de Physiologie Cellulaire Végétale; CEA; CNRS; INRA; Université Joseph Fourier, iRTSV, CEA-Grenoble, 38054 Grenoble Cedex 9, France
jacques.joyard@cea.fr

Eunsoo Kim
Department of Biochemistry and Molecular Biology, Canadian Institute for Advanced Research, Dalhousie University, Halifax, NS, Canada B3H 1X5

Jodi Maple
Centre for Organelle Research, Faculty of Science and Technology, University of Stavanger, 4036 Stavanger, Norway

Alfonso Mateo
Centre for Organelle Research, Faculty of Science and Technology, University of Stavanger, 4036 Stavanger, Norway

Alexis Samba Mialoundama
Institut de Biologie Moléculaire des Plantes, Centre National de la Recherche Scientifique and Université Louis Pasteur, 67084 Strasbourg Cedex, France

Simon Geir Møller
Centre for Organelle Research, Faculty of Science and Technology, University of Stavanger, 4036 Stavanger, Norway
simon.g.moller@uis.no

Thomas Pfannschmidt
Department for Plant Physiology, Friedrich-Schiller-University Jena, Dornburger Str. 159, 07743 Jena, Germany
thomas.pfannschmidt@uni-jena.de

Contributors

Norbert Rolland
Laboratoire de Physiologie Cellulaire Végétale; CEA; CNRS; INRA; Université Joseph Fourier, iRTSV, CEA-Grenoble, 38054 Grenoble Cedex 9, France

Yvonne Schröter
Department for Plant Physiology, Friedrich-Schiller-University Jena, Dornburger Str. 159, 07743 Jena, Germany

Daphné Seigneurin-Berny
Laboratoire de Physiologie Cellulaire Végétale; CEA; CNRS; INRA; Université Joseph Fourier, iRTSV, CEA-Grenoble, 38054 Grenoble Cedex 9, France

Sebastian Steiner
Department for Plant Physiology, Friedrich-Schiller-University Jena, Dornburger Str. 159, 07743 Jena, Germany

Noriyuki Suetsugu
Department of Biology, Faculty of Science, Kyushu University, Fukuoka, 812-8581, Japan

Masamitsu Wada
Department of Biology, Faculty of Science, Kyushu University, Fukuoka, 812-8581, Japan
wada@nibb.ac.jb

Andreas P.M. Weber
Institut für Biochemie der Pflanzen, Heinrich-Heine-Universität, Universitätsstraße 1, 40225 Düsseldorf, Germany
andreas.weber@uni-duesseldorf.de

Diversity and Evolution of Plastids and Their Genomes

E. Kim and J. M. Archibald (✉)

Abstract Plastids, the light-harvesting organelles of plants and algae, are the descendants of cyanobacterial endosymbionts that became permanent fixtures inside nonphotosynthetic eukaryotic host cells. This chapter provides an overview of the structural, functional and molecular diversity of plastids in the context of current views on the evolutionary relationships among the eukaryotic hosts in which they reside. Green algae, land plants, red algae and glaucophyte algae harbor double-membrane-bound plastids whose ancestry is generally believed to trace directly to the original cyanobacterial endosymbiont. In contrast, the plastids of many other algae, such as dinoflagellates, diatoms and euglenids, are usually bound by more than two membranes, suggesting that these were acquired indirectly via endosymbiotic mergers between nonphotosynthetic eukaryotic hosts and eukaryotic algal endosymbionts. An increasing amount of genomic data from diverse photosynthetic taxa has made it possible to test specific hypotheses about the evolution of photosynthesis in eukaryotes and, consequently, improve our understanding of the genomic and biochemical diversity of modern-day eukaryotic phototrophs.

1 Introduction

The origin and evolution of plastids,[1] the light-gathering organelles of photosynthetic eukaryotes, is a subject that has intrigued biologists for more than a century. Since the original musings of Schimper (1885) and Mereschkowsky (1905), a wealth of structural, biochemical and, most recently, molecular sequence data

---

J.M. Archibald
Department of Biochemistry and Molecular Biology, Canadian Institute for Advanced Research, Dalhousie University, Halifax, NS, Canada, B3H 1X5
e-mail: jmarchib@dal.ca

---

[1] The term "chloroplast" is sometimes used to refer to all photosynthetic plastids among eukaryotic phototrophs and, occasionally, only in reference to the photosynthetic organelle of green algae and land plants.

has accumulated and shown convincingly that these important cellular structures are of endosymbiotic origin (Gray and Spencer 1996). The exact timing of the first plastid-generating endosymbiosis, and the ecological and physiological conditions that facilitated such an event is not known, but it is commonly thought that plastids evolved from once free-living cyanobacteria that were originally ingested as food by a heterotrophic eukaryote. Under this scenario, rather than being digested, these prokaryotic cells escaped the confines of their phagocytic vacuole and gradually became fully integrated components of their eukaryotic hosts. As an obvious consequence of the cyanobacterial ancestry of their plastids, eukaryotic phototrophs perform oxygenic photosynthesis and have contributed greatly to the burial of organic carbon and oxygenation of Earth's atmosphere (Katz et al. 2004).

While the notion that plastids are derived from endosymbiotic cyanobacteria is now widely accepted, many important questions about the origin and diversification of plastids remain. The challenges associated with inferring the evolutionary history of plastids are in large part due to the exceptional structural, biochemical and molecular diversity seen in modern-day photosynthetic organisms. In this chapter, we provide an overview of the distribution of plastids across the known spectrum of eukaryotic life, summarize the diversity of photosynthetic pigments and storage carbon biochemistry seen in plants and algae, and present recent advances in our understanding of the tempo and mode of plastid diversification. Finally, we discuss the evolution of the plastid genome and proteome, providing recent insight into the significant role of lateral (or horizontal) gene transfer (LGT). As we shall see, the evolutionary history of plastids is exceedingly complex and while the use of molecular phylogenetics has improved our understanding of plastid evolution significantly, it has also raised as many questions as it has answered.

2 Distribution of Plastids

From an evolutionary perspective, the most fundamental distinction between different plastid types is between those whose ancestry can be traced directly to the original cyanobacterial endosymbiont (i.e., "primary" plastids) and those that were acquired indirectly as a result of an endosymbiosis between a plastid-bearing eukaryote and an unrelated eukaryotic host (i.e., "secondary" or "tertiary" plastids). The plastids of glaucophytes, rhodophytes (red algae) and Viridiplantae (green algae and land plants) are bound by two envelope membranes, which are thought to be derived from the inner and outer membranes of the original cyanobacterial endosymbiont (Jarvis and Soll 2001) (Fig. 1a–c). The lack of additional plastid membranes, a feature of the organelle in many other organisms, led to the notion that these plastids arose through a primary endosymbiotic event involving a cyanobacterial endosymbiont (Gibbs 1981). The evidence for and against a single origin for all primary plastids will be discussed in Sect. 4.1.

In contrast to the plastids of glaucophytes, rhodophytes and Viridiplantae, the euglenids and chlorarachniophytes are believed to have acquired their plastids from green

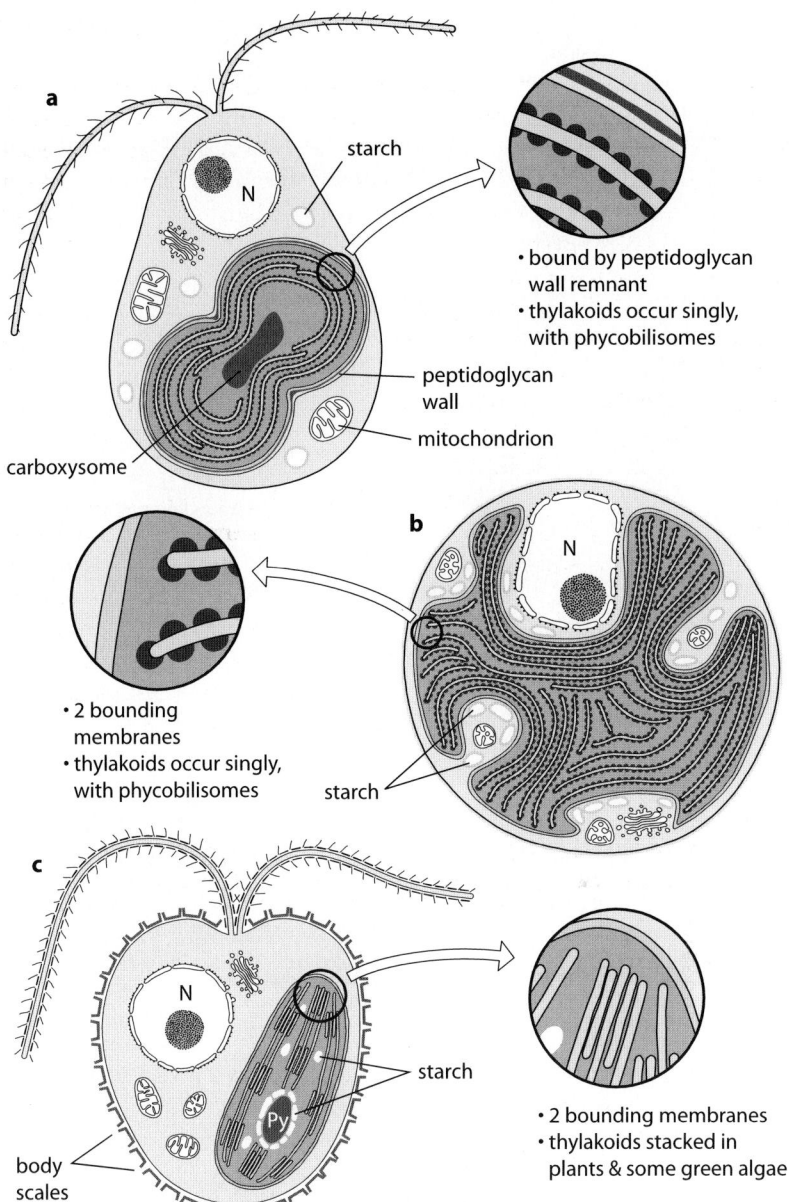

Fig. 1 Diversity of photosynthetic eukaryotes and their plastids. The plastids of glaucophytes (**a**), rhodophytes (**b**) and Viridiplantae (**c**) are bound by two membranes. Note the presence of phycobilisomes attached to the lumen side of the thylakoid membrane in **a** and **b**. The euglenids (**d**) and chlorarachniophytes (**e**) both possess plastids of green algal ancestry, which are surrounded by three and four membranes, respectively. The plastids of haptophytes (**f**), cryptophytes (**g**) and stramenopiles (**h**) are of red algal origin and are each surrounded by four membranes. Note the continuity of the nuclear envelope and the outermost plastid membrane, and the presence of ribosomes on the outer plastid membrane. Cells shown in **i** and **j** correspond to peridinin-containing dinoflagellates and apicomplexans, respectively. *N* nucleus, *Py* pyrenoid. (Art by L. Wilcox)

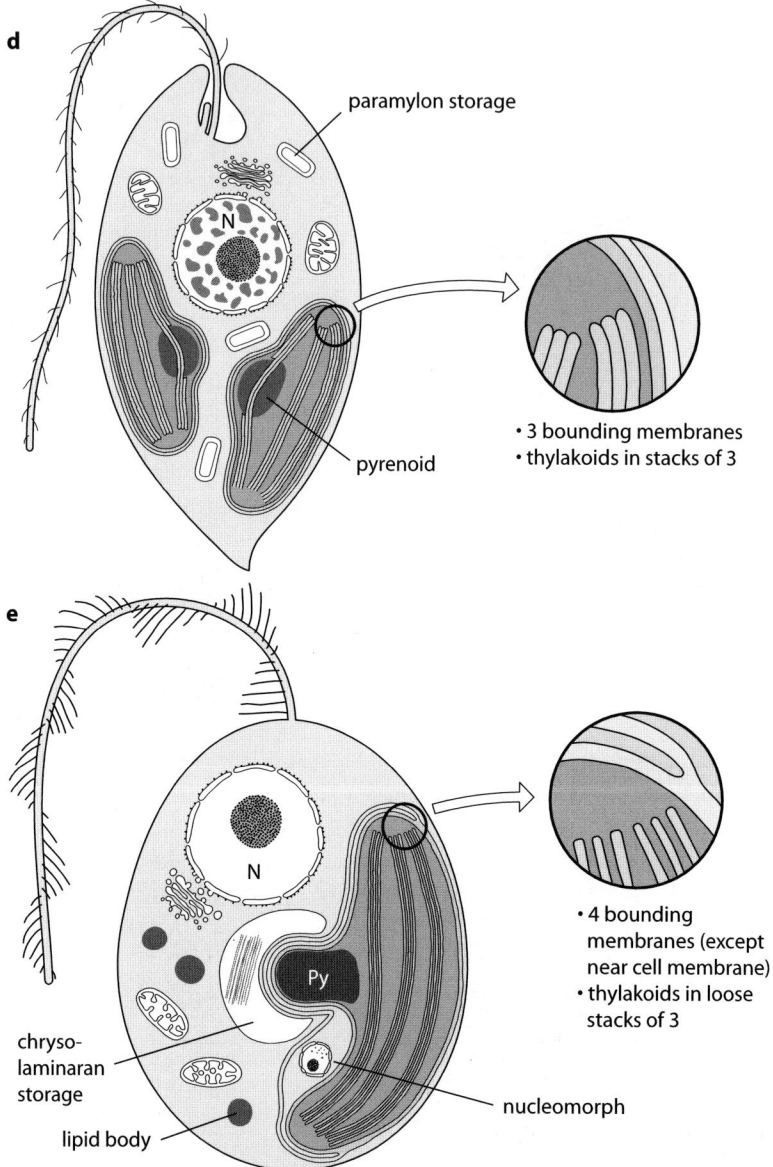

Fig. 1 (continued)

algae via secondary endosymbiosis (Table 1, Fig. 1d, e). Secondary or tertiary origins are proposed for the red-algae-derived plastids of cryptophytes, haptophytes, stramenopiles, most photosynthetic dinoflagellates, and apicomplexans (Table 1, Fig. 1f–j), although the origin of the apicomplexan plastid remains controversial (see below). With

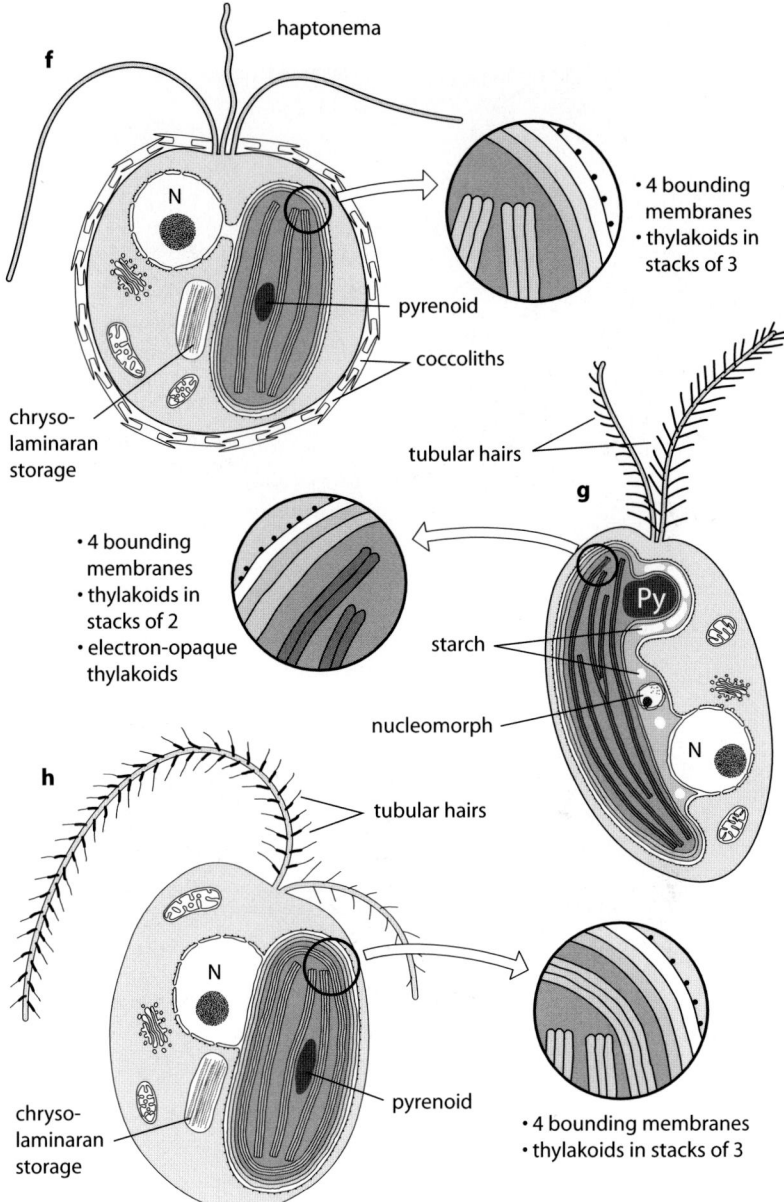

Fig. 1 (continued)

respect to the abundance of plastid-containing species in each of these groups, all known members of the glaucophytes, rhodophytes, Viridiplantae and haptophytes possess photosynthetic or nonphotosynthetic plastids, indicating that plastid acquisition

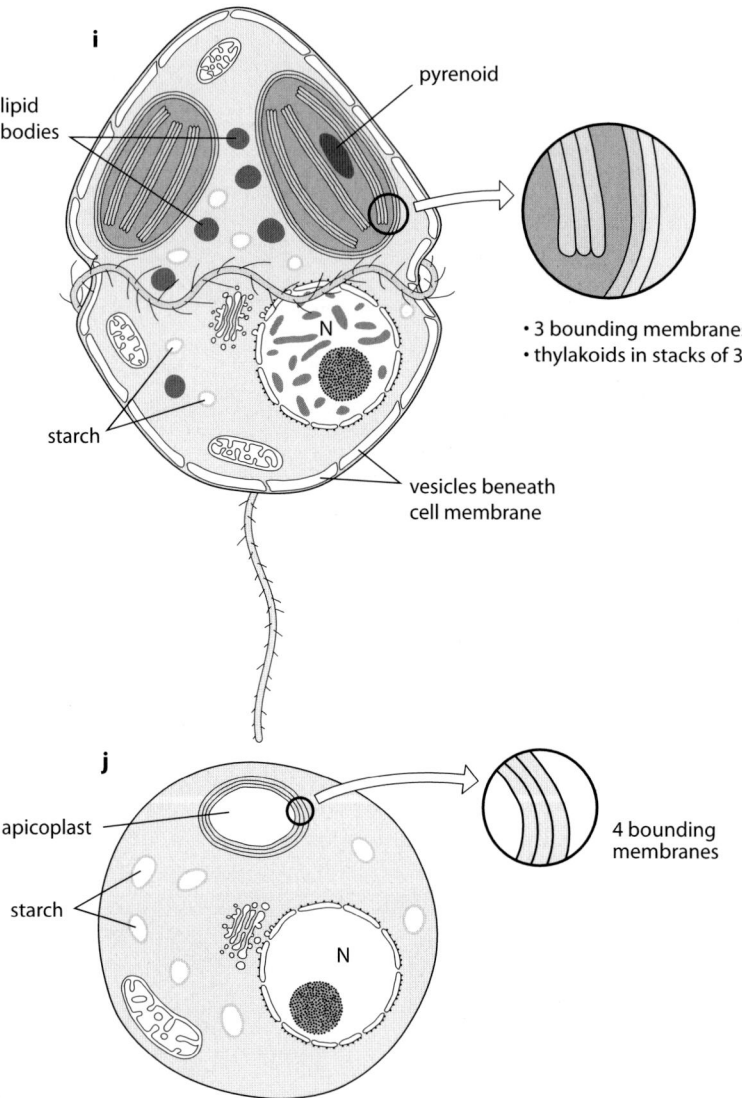

Fig. 1 (continued)

occurred prior to the diversification within each of these lineages. While the chlorarachniophytes are an exclusively photosynthetic lineage, they belong to the eukaryotic "supergroup" Rhizaria, which is composed mostly of plastid-less taxa (Archibald and Keeling 2004; Nikolaev et al. 2004). Similarly, alveolates, cryptophytes, euglenids and stramenopiles each contain plastid-less members that are closely related to plastid-containing taxa. As will be elaborated upon below, the study of nonphotosynthetic

Table 1 General features of plastids

Taxonomic classification	Chlorophyll	Phycobilisomes	Number of plastid membranes	Putative plastid origin(s)	Nucleomorph?	Storage carbon (location)	Plastid-nuclear envelope continuity	Number of thylakoid stack
Glaucophyta	a	Yes	2	Primary	No	Starch (cytoplasm)	No	1
Rhodophyta	a	Yes	2	Primary	No	Starch (cytoplasm)	No	1
Viridiplantae	a, b	No	2	Primary	No	Starch (plastid stroma)	No	Multiple
Cryptophyta	a, c	No	4	Secondary (red)	Yes	Starch (periplastid space)	Yes	2
Haptophyta	a, c	No[a]	4	Secondary (red)	No	Chrysolaminaran (cytoplasm)	Yes	3
Stramenopiles	a, c[b]	No	4	Secondary (red)	No	Chrysolaminaran (cytoplasm)	Yes (not always)	3 (usually)
Alveolata — Dinophyta — Peridinin type	a, c	No	3 (occasionally 2°)	Secondary (red)?	No	Starch (cytoplasm)	No	3
Alveolata — Dinophyta — Green algal origin	a, b	No	4	Secondary (green)	?[d]	Starch (cytoplasm)	No	2–3
Alveolata — Dinophyta — Diatom origin	a, c	No	3	Tertiary (diatom)	?[d]	Starch (cytoplasm)	No	3
Alveolata — Dinophyta — Haptophyte origin	a, c	No	2–4[e]?	Tertiary (haptophyte)	No	Starch (cytoplasm)	No	3
Alveolata — Apicomplexa	None	No	2–4[f]?	Secondary (red or green?)	No	Starch (cytoplasm)	No	NA
Euglenida	a, b	No	3	Secondary (green)	No	Paramylon (cytoplasm)	No	3
Chlorarachniophyta	a, b	No	4[g]	Secondary (green)	Yes	Chrysolaminaran (cytoplasm)	No	3

NA not available

[a] Phycobiliproteins – phycoerythrin or phycocyanin – occur in the plastid lumen
[b] Members of Eustigomatophyceae lack chlorophyll c (Andersen 2004)
[c] Schnepf and Elbrachter (1999)
[d] Although the nuclear genomes of the algal endosymbionts seem to be retained, the extent of genome reduction, characteristic of a "genuine" nucleomorph, is not known in these plastids
[e] Bergholtz et al. (2006) and Schnepf and Elbrachter (1999)
[f] Hopkins et al. (1999), Kohler (2005), McFadden and Roos (1999) and Tomova et al. (2006)
[g] Near the plasmalemma, the chloroplast envelope is commonly incomplete and usually composed of only two membranes (Moestrup and Sengco 2001)

relatives of plastid-containing lineages has the potential to improve our understanding of the timing of plastid acquisition and loss in eukaryotic evolution.

The Alveolata is comprised of three major subclades, the apicomplexans, dinoflagellates and ciliates, of which the first two groups include plastid-containing members. About half of known dinoflagellates harbor plastids obtained from diverse algal sources (see below for details), while the rest are plastid-less (Schnepf and Elbrachter 1999). Although the "early-diverging" dinoflagellate genera *Perkinsus* and *Oxyrrhis* were originally thought to lack plastids (Leander and Keeling 2003; Saldarriaga et al. 2003), a recent study identified a four-membrane-bound plastid-like organelle in *Perkinsus atlanticus* (Teles-Grilo et al. 2007). Apicomplexans such as *Plasmodium* and *Toxoplasma* contain two to four membrane-bound nonphotosynthetic plastids known as apicoplasts (Kohler et al. 1997; Hopkins et al. 1999; Kohler 2005; Tomova et al. 2006). On the other hand, other apicomplexan parasites, including *Cryptosporidium parvum* and the gregarines, are apparently devoid of such organelles (Toso and Omoto 2007). In addition, plastids have thus far not been identified in colpodellids, predatory, free-living heterotrophs that are closely related to Apicomplexa (Brugerolle 2002; Leander et al. 2003). Nevertheless, the recent discovery of *Chromera velia*, a photosynthetic relative of apicomplexans, suggests that these plastid-less apicomplexans and colpodellids might have lost their plastids secondarily (Moore et al. 2008).

In Cryptophyta, most genera possess four membrane-bound plastids and are noteworthy in that the relic nucleus of their red algal endosymbiont, known as the nucleomorph, still persists between the second and the third plastid membranes (Hoef-Emden et al. 2002; Archibald 2007). Molecular phylogenies have shown that the single plastid-less cryptophyte genus *Goniomonas* is sister to the plastid-containing cryptophytes and includes three species, which show substantial genetic diversity comparable to that of all other cryptophytes combined (Deane et al. 2002; Von der Heyden et al. 2004). The cryptophytes are sister to the katablepharids, an enigmatic lineage comprising plastid-less, free-living biflagellates common in aquatic environments (Lee and Kugrens 1991; Okamoto and Inouye 2005; Kim et al. 2006).

The Euglenida include plastid-containing members such as *Euglena gracilis* and paraphyletic plastid-less taxa such as the primary osmotroph *Distigma* and phagotrophs such as *Petalomonas* and *Entosiphon* (Busse et al. 2003; Breglia et al. 2007). The plastids of euglenids are surrounded by three membranes (Fig. 1d), and unlike the plastids of cryptophytes are not directly associated with the nuclear envelope (van Dooren et al. 2001). The Euglenida are closely related to Diplonemida and Kinetoplastida, which include free-living (e.g., *Bodo*) and parasitic (e.g., *Trypanosoma*) plastid-less heterotrophs (Busse and Preisfeld 2002; Leander 2004).

Stramenopiles (also known as heterokonts) include morphologically diverse forms of eukaryotes ranging from pico-sized (less than 3 μm) flagellates to giant kelps. Molecular phylogenies using the small subunit ribosomal RNA (rRNA) gene suggest that plastid-containing stramenopiles form a clade (i.e., Ochrophyta; Cavalier-Smith and Chao 1996) to the exclusion of paraphyletic plastid-less subgroups such as Bicosoecida, Developayella, Hyphochytriales, labyrinthulomycetes, Opalinata, peronospromycetes, Placididea and *Pirsonia* (Guillou et al. 1999;

Moriya et al. 2000, 2002; Karpov et al. 2001; Andersen 2004; Kuhn et al. 2004). Within Ochrophyta, loss of photosynthesis has occurred multiple times, especially in the Chrysophyceae, but plastids usually persist as in the common freshwater flagellate *Paraphysomonas* (Preisig and Hibberd 1983). Although a few ochrophyte taxa were once thought to have completely lost their plastids (Cavalier-Smith et al. 1995/1996), bona fide organelles were subsequently identified or plastid-derived *rbcL* genes were successfully PCR-amplified, supporting the presence of plastids in the genera *Pteridomonas* and *Ciliophrys* (Sekiguchi et al. 2002). Comparable investigations are needed to test for the presence or absence of plastids in *Oikomonas* and *Picophagus*, two additional nonphotosynthetic ochrophyte taxa (Cavalier-Smith et al. 1995/1996; Guillou et al. 1999).

Several additional examples of photosynthetic eukaryotes are worthy of mention, two of which are somewhat controversial. First, a putative plastid-like organelle has been identified in the "picobiliphytes," a newly discovered eukaryotic lineage that has yet to be cultured in the laboratory (Not et al. 2007). These cells appear to harbor a photosynthetic body that emits orange autofluorescence under blue light, suggesting the presence of phycobiliproteins (Not et al. 2007). A small DNA-containing region was identified in close association with the picobiliphyte "plastid" and proposed to be a nucleomorph, though this has not been proven. Second, the thecate amoeba *Paulinella chromatophora* (Rhizaria) possesses two to four elongate blue-green photosynthetic bodies that recent molecular investigations have shown to be very closely related to cyanobacteria of the *Prochlorococcus/Synechococcus* clade (Lukavsky and Cepak 1992; Marin et al. 2005, 2007; Yoon et al. 2006; Nowack et al. 2008). Finally, the diatom *Rhopalodia gibba* (Stramenopiles) harbors *Cyanothece*-like, cyanobacterium-derived spheroid bodies that fix nitrogen using ATP and/or photosynthate derived from the plastids of its host (Prechtl et al. 2004). While the cyanobacterial-derived entities of both *P. chromatophora* and *R. gibba* appear to be "permanent" cellular inclusions that cannot be cultured in isolation, there is considerable debate as to whether the term "endosymbiont" or "organelle" is most appropriate (see Archibald (2006), Bhattacharya and Archibald (2006), Theissen and Martin (2006) and Bodyl et al. (2007) for recent discussion).

3 Biochemical Diversity of Plastids

3.1 *Photosynthetic Pigments*

In photosynthetic plants and algae, the harvesting of light energy involves three major types of pigments: chlorophylls (Chl), carotenoids and phycobilins (Graham and Wilcox 2000). Chl *a* is an essential component of the core complexes of photosystems I and II and is universally distributed in photosynthetic plastids and cyanobacteria, whereas Chl *b*, *c* and *d* are regarded as accessory pigments, which absorb and transfer excitation energy to Chl *a* (Falkowski and Raven 2007). Chl *b* occurs in the plastids of Viridiplantae and their secondary derivatives (chlorarachniophytes, euglenids and the dinoflagellate *Lepidodinium*) and in three cyanobacterial genera, *Prochlorococcus*, *Prochloron* and *Prochlorothrix*, which are collectively referred to as prochlorophytes even though they do not form a monophyletic assemblage (Green and Durnford 1996; Griffiths 2006). Biosynthesis of Chl *b* requires the enzyme Chl *a* oxygenase (CAO), which converts Chl *a* into Chl *b* (Tomitani et al. 1999). Molecular sequence analyses suggest that the CAO gene sequences of Viridiplantae and two prochlorophytes, *Prochloron* and *Prochlorothrix*, share a common evolutionary origin (Tomitani et al. 1999), while that of *Prochlorococcus* may be of a separate origin (Hess et al. 2001). On the other hand, Chl *a*/*b*-binding proteins of green plastids belong to the eukaryotic light-harvesting complex (LHC) family, and are not related to the prochlorophyte functional equivalent, prochlorophyte-like Chl binding protein (Pcb) (Green and Durnford 1996; La Roche et al. 1996). The eukaryotic LHC family is thought to be derived from the high-light-inducible protein (HLIP) through successive gene duplication events, whereas prochlorophyte Pcb is related to the iron stress-induced protein (IsiA) or the photosystem II protein PsbC (Chen et al. 2005; Green 2005).

Chl *d* occurs in the cyanobacterium *Acaryochloris marina* as a major pigment (Miyashita et al. 2003). Although several red algae were thought to produce a small amount of Chl *d*, its detection seems to be due to the presence of epiphytic *Acaryochloris* on them (Murakami et al. 2004). While the molecular structure suggests that Chl *d* is likely synthesized directly from Chl *a*, its biosynthetic pathway is poorly understood (Beale 1999). Chl *c* occurs in red algal-derived plastids (although not in red plastids themselves) as a major pigment fraction, as well as in some phycobilisome-lacking cyanobacteria and prasinophycean green algae (Wilhelm 1987; Larkum et al. 1994; Green and Durnford 1996; Miyashita et al. 2003; Six et al. 2005). Unlike Chl *b* and *d*, which are structurally chlorin-based like Chl *a*, Chl *c* (c_1, c_2, c_3) is structurally more similar to a Chl *a* precursor, protochlorophyllide (porphyrin), and generally does not possess a hydrophobic phytol tail (Zapata and Garrido 1997). The existence of Chl *c* in red algal-derived plastids has been argued as evidence for their common origin (Cavalier-Smith 1999; see below). However, the biosynthetic pathway underlying the synthesis of Chl *c* is essentially unknown (Beale 1999), and, hence, its utility as a phylogenetic marker is unclear.

Carotenoids, which are structurally related to tetraterpenoids, exist as two major types, the hydrocarbon carotenes and their oxygenated derivatives, xanthophylls such as alloxanthin, peridinin and fucoxanthin (Cunningham and Gantt 1998). They are a vital component of the thylakoid membrane and play an important role in energy transfer as well as photoprotection by dissipating excess energy (Cunningham and Gantt 1998). With respect to carotenoid distribution, β-carotene occurs universally in chloroplasts, whereas other carotenoids show a more restricted distribution and are recognized as potentially valuable phylogenetic markers. Alloxanthin, for example, is uniquely found among cryptophyte algae (Reid et al. 1990). Peridinin is another unique pigment, which is found in the majority of plastid-containing dinoflagellates and occurs within the lumen as water-soluble peridinin–Chl *a*–protein complexes (Green and Durnford 1996).

Phycobilin pigments – phycourobilin, phycoerythrobilin and phycocyanobilin – are linear tetrapyrroles that bind to proteins to form phycobiliproteins, which can be further assembled into a large hemispherical structure, about 40 nm in diameter, known as a phycobilisome (Falkowski and Raven 2007). Anchored into the stromal side of the thylakoid membrane, phycobilisomes are visible under the electron microscope and appear to be responsible for the spatial separation of thylakoid membranes. Prochlorophytes and plastids that are devoid of phycobilisomes apparently have paired or stacked thylakoid membranes (Walsby 1986) (Fig. 1c–i). While cryptophytes and the cyanobacterium *Prochlorococcus* do not possess typical phycobilisomes, they nevertheless possess phycoerythrin and/or phycocyanin in the lumen part of the thylakoid (Hess et al. 1999; Griffiths 2006; Dammeyer et al. 2007).

It has been suggested that the ancestral cyanobacterium that gave rise to the plastid possessed both Chl *b* and phycobilisomes and during plastid evolution Chl *b* was lost in glaucophytes and rhodophytes, whereas phycobilisomes were lost in the lineage leading to green algae and land plants (Tomitani et al. 1999). This hypothesis, however, is not supported by pigment composition patterns seen in extant cyanobacteria. To date, no cyanobacterium possessing both Chl *b* and phycobilisomes is known. Even in the case where Chl *b* and phycobiliproteins co-occur as in the prochlorophyte *Prochlorococcus*, these phycobiliproteins do not form highly organized phycobilisomes. This suggests that the Chl *b* biosynthetic capacity of green plastids was likely acquired after the divergence of green plastids from red plastids through LGT of the CAO gene and potentially other gene compliments from *Prochlorothrix* or *Prochloron*-like cyanobacteria or other vectors such as cyanophage.

3.2 *Storage Carbon Biochemistry*

The end products of photosynthesis in algae and plants are stored as polysaccharides comprising d-glucose monomers, linked via either α-1,4-glycosidic bonds with α-1,6-branches (starch), or β-1,3-glycosidic bonds with occasional β-1,6-branches (chrysolaminaran and paramylon). The storage polysaccharides of most algal groups are found in the cytoplasm, with cryptophytes and members of the green algae and land plants being interesting exceptions. In cryptophytes, starch accumulates in the space between the second and third plastid envelopes (i.e., the periplastidal compartment; Fig. 1g), which corresponds to the cytoplasm of the red algal endosymbiont (McFadden et al. 1994). In green algae and land plants, starch accumulates within the plastid stroma (Ball and Morell 2003).

Starch is found in photosynthetic members of the cryptophytes, dinoflagellates, glaucophytes, rhodophytes, Viridiplantae and their nonphotosynthetic derivatives (Raven 2005). Apicomplexans such as *Toxoplasma gondii* and even *C. parvum*, which lacks a plastid, also produce starch granules in the cytoplasm (Harris et al. 2004). Unlike the glycogen seen in animals, fungi and prokaryotes, which is a water-soluble and highly branched (10–12%) polymer of less than 50 nm in diameter, starch is a large (0.1 to over 50 µm) and complex semicrystalline polymer generally made of amylopectin and amylose (Ball and Morell 2003). Amylose is a linear chain of α-1,4-linked glucose containing less than 1% α-1,6-branches, whereas amylopectin is a much larger molecule with frequent α-1,6-branches (5–6%) (Buleon et al. 1998). In land plants, where starch has been extensively studied, the starch granule is organized as concentric rings of alternating semicrystalline and amorphous layering patterns resulting from regularly branching amylopectin molecules (Buleon et al. 1998; Buleon et al. 2007). Like green algae and land plants, the starch granules of cryptophytes and dinoflagellates are composed of amylose (up to 40% for cryptophytes) and amylopectin and stain blue-black with iodine solution (Vogel and Meeuse 1968; Antia et al. 1979; McFadden et al. 1994; Coppin et al. 2005). In contrast, the storage carbohydrate of apicomplexans appears to lack amylose and consist only of amylopectin (Coppin et al. 2005). Red algae show more variations in storage polysaccharides, containing either glycogen (*Cyanidium caldarium*) or amylopectin (florideophycean red algae) alone, or a mixture of amylose and semi-amylopectin (*Porphyridium purpureum*) (Yu et al. 2002; Shimonaga et al. 2007). Interestingly, some cyanobacteria produce semi-amylopectin instead of glycogen, which shows a chain length distribution similar to that of the red alga *P. purpureum* (Nakamura et al. 2005; Shimonaga et al. 2007); however, unlike *P. purpureum*, these cyanobacteria do not synthesize amylose and it is not clear whether these organisms share a similar biochemical machinery to produce semi-amylopectin (Nakamura et al. 2005).

Paramylon is the β-1,3-glucose linked storage carbohydrate of euglenids, and does not stain with iodine solution. A paramylon granule, bound by a single membrane, is composed of triangular and rectangular segments, each segment made of several layers (Kiss et al. 1987). Similar to cellulose, paramylon is organized into microfibrils (of 4.0 nm in diameter) composed of triple helices of β-1,3-glucose chains, which are further bundled into thicker fibers (Marchessault and Deslandes 1979; Kiss et al. 1987). The higher-order assembly of microfibrils and their interactions with water molecules contribute to the highly crystalline nature of paramylon (Marchessault and Deslandes 1979; Kiss et al. 1988). The paramylon granules occur in the cytoplasm and generally near the pyrenoid region of photosynthetically active plastids (Kiss et al. 1986). However, heterotrophically grown *Euglena* cells and even some plastid-less "primitive" euglenids such as certain *Petalomonas* species also contain abundant paramylon granules (Kiss et al. 1986; Lee et al. 2000). This suggests that when the colorless euglenid ancestor acquired its plastid through the secondary endosymbiotic engulfment of a green algal cell, the carbon storage system of the host was utilized, as opposed to that of the algal endosymbiont.

Chlorarachniophytes, haptophytes and stramenopiles also store photosynthetic end products as β-1,3-polyglucans and are referred to as chrysolaminaran (also known as leucosin or laminaran) (McFadden et al. 1997; Granum and Myklestad 2001; Chiovitti et al. 2006; Hirokawa et al. 2007). However, unlike paramylon, which is insoluble, chrysolaminaran is water-soluble and generally consists of only 20–60 glucose units (Janse et al. 1996). In chlorarachniophytes, the cap-shaped chrysolaminaran vesicle is tightly associated with the pyrenoid, albeit separated from it by the four membranes surrounding the plastid (Fig. 1e) (McFadden et al. 1997; Moestrup and Sengco 2001). Some brown seaweeds (Phaeophyceae) have manitol groups attached to the ends of the β-1,3-polyglucan chain (Chizhov et al. 1998). The haptophyte *Pavlova* is unusual in that it produces water-insoluble, crystalline β-1,3-polyglucans like the paramylon of euglenoids, although the two crystalline granules have different structures and likely arose independently (Kiss and Triemer 1988).

Storage polyglucans are synthesized from either ADP glucose- or UDP glucose-based pathways (Ball and Morell 2003). Green algae and plants are unique among eukaryotic algae in that their starch synthesis pathway utilizes ADP glucose as a donor, as in bacteria, where the pathway is likely to have originated (Ball and Morell 2003). In contrast, other eukaryotic algae utilize UDP glucose as precursors to synthesize glucose polymers, similar to glycogen synthesis in animals and fungi (Viola et al. 2001; Ball and Morell 2003; Barbier et al. 2005; Deschamps et al. 2006). During the acquisition of plastids, the majority of algal groups other than Viridiplantae and cryptophytes have transferred the location of their storage carbohydrate to the host cytoplasm, which suggests that the endosymbiont's photosynthetic carbon metabolism has been amalgamated into the host carbohydrate biochemistry. Indeed, analyses of starch synthesis pathway genes have revealed that red algae and even land plants integrated the endosymbiont and host carbohydrate pathways (Patron and Keeling 2005).

4 Origin of Plastids

4.1 Origin of Primary Plastids

The question of single or multiple origins for the primary plastids of glaucophytes, rhodophytes and Viridiplantae has been extensively debated (Nozaki et al. 2003; Palmer 2003; Stiller et al. 2003; Larkum et al. 2007; Stiller 2007). Several plastid-related characters support the hypothesis that plastids evolved from a single type of cyanobacterial ancestor. First, the plastids of Rhodophyta and Viridiplantae (and their derivatives) share eukaryote-specific LHCs that are not present in cyanobacteria (Durnford et al. 1999). LHC homologs, however, have not been identified in the plastids of Glaucophyta (Rissler and Durnford 2005). Second, Tic110, an important component of the protein import apparatus, is present in all three primary plastids and their descendants, but is absent in cyanobacteria (McFadden and van Dooren 2004). This suggests that the nucleus-encoded Tic110 may represent a postendosymbiotic innovation (McFadden and van Dooren 2004). Third, the organization of the *atpA* gene cluster is another line of evidence in support of the common origin of the three types of primary plastids (Stoebe and Kowallik 1999). Plastid-encoded gene phylogenies also support the notion that the three primary plastids are closely related to each other, suggesting their common origin (Rodriguez-Ezpeleta et al. 2005), although the acquisition of similar cyanobacterial endosymbionts on multiple occasions could also produce such a topology. Currently available data thus favor a single origin for all primary plastids. Nevertheless, it should be emphasized that plastid-related characters do not directly address whether or not the three primary plastids share a single endosymbiotic origin. This is because the number of plastid membranes, upon which the concept of "primary" plastids depends, does not necessarily reflect the primary, secondary or tertiary origin of the organelle because of possible losses of membranes during or after endosymbiosis (Stiller and Hall 1997). For example, the plastids of some peridinin-containing dinoflagellates have lost one membrane and are now bound by only two plastid membranes, although these are not likely to be of primary origin (Schnepf and Elbrachter 1999). The relationships among the nucleocytoplasmic component of Glaucophyta, Rhodophyta and Viridiplantae are also unsettled. While the monophyly of the three groups was initially strongly supported in a combined nuclear-encoded gene phylogeny by Rodriguez-Ezpeleta et al. (2005), addition of sequences from cryptophytes and haptophytes resulted in a reduction in statistical support or the monophyly of the three groups was no longer inferred (Burki et al. 2007; Patron et al. 2007).

4.2 Evolution of Green Algal-Derived Secondary Plastids

A wealth of biochemical, ultrastructural and molecular data has shown that euglenids and chlorarachniophytes (Fig. 1 d, e) acquired photosynthesis secondarily through the uptake of green algal endosymbionts (McFadden 2001; Archibald and Keeling 2002). Cavalier-Smith (1999) hypothesized that the plastids in these two groups are specifically related to one another, i.e., that the secondary endosymbiosis occurred in their common ancestor. This is parsimonious in the sense that it requires only a single endosymbiotic event, but is problematic when one considers that their respective host cells are not obviously related to one another: euglenids belong to the Euglenozoa, while the chlorarachniophytes reside within an entirely different eukaryotic supergroup, the Rhizaria (Adl et al. 2005; Keeling et al. 2005). Indeed, recent phylogenies of the plastid-targeted psbO protein and concatenated plastid-encoded proteins support the hypothesis that the euglenid and chlorarachniophyte plastids are of independent origin (Rogers et al. 2007; Takahashi et al. 2007). Finally, it is worth noting that some dinoflagellates have plastids of green algal origin (Hansen et al. 2007), having replaced their red algal secondary plastids (or reacquired a plastid after having lost it) (see below).

4.3 Evolution of Red Algal-Derived Secondary Plastids

A number of eukaryotes acquired their plastids through secondary or tertiary endosymbioses involving red algal endosymbionts. These include the cryptophytes, haptophytes and stramenopiles, whose plastids are generally bound by four membranes (Fig. 1f–h, Table 1). An interesting feature of these plastids, albeit with some exceptions in stramenopiles (Andersen 2004), is the confluence of the outer plastid membrane with the nuclear envelope either directly, or via endoplasmic reticulum (Fig. 1f–h; Cavalier-Smith 1999). Consequently, the outer membrane of these plastids is typically studded with cytoplasmic ribosomes. Apicoplasts and the peridinin-containing plastids of dinoflagellates may also have originated from red algal endosymbionts (see below), although their outer membranes are not continuous with the host cell endomembrane system (Schnepf and Elbrachter 1999). The origin and evolution of red algal secondary plastids is hotly debated, and there is currently no generally agreed upon consensus on the pattern of plastid gain and/or loss that best describes the current distribution of these organelles (Falkowski et al. 2004; Grzebyk et al. 2004; Keeling et al. 2004).

Cavalier-Smith (1999) hypothesized that the red algal-derived plastids of cryptophytes, haptophytes, stramenopiles (i.e., "chromists") and plastid-bearing alveolates (apicomplexans and dinoflagellates; see below) are the product of an ancient secondary endosymbiosis in a common ancestor each of these lineages shared to the exclusion of all other eukaryotic groups. As was the case for the origin of green algal secondary plastids, the so-called chromalveolate hypothesis was postulated in order to minimize the number of secondary endosymbioses needed to explain the observed distribution of red secondary plastids. A wide range of data has been brought to bear on the question of chromalveolate monophyly, most notably, phylogenies of plastid and host nuclear gene sequences. Host gene phylogenies and the presence of a number of plastid-less sister taxa, for example, conflict with the chromalveolate hypothesis, or require additional underlying assumptions (e.g., extensive plastid loss; Kim et al. 2006; Hackett et al. 2007; Patron et al. 2007). Analyses of plastid-encoded genes also do not consistently support the chromalveolate hypothesis; the relationships among red algal-derived plastids vary depending on taxonomic sampling, analytical methods, and types and the number of genes included (Martin et al. 2002; Yoon et al. 2002b; Ohta et al. 2003; Sanchez-Puerta et al. 2007). Furthermore, the monophyly of the plastid component of cryptophytes, haptophytes, stramenopiles and alveolates can also be explained by the "serial hypothesis," which proposes serial transfers of red algal-derived plastids among different host lineages (Bachvaroff et al. 2005; Bodyl 2005; Sanchez-Puerta et al. 2007). The acquisition of similar red algal endosymbionts by different host eukaryotes is also predicted to produce a gene topology where their plastid genes are clustered together (Grzebyk et al. 2004).

The most widely cited molecular marker in support of the chromalveolate hypothesis is glyceraldehyde-3-phosphate dehydrogenase (GAPDH). Photosynthetic eukaryotes possess cytosolic and plastid isoforms of this protein and unlike Viridiplantae and rhodophytes, in which the plastid-targeted GAPDH likely derived from a cyanobacterial donor, red plastid-containing eukaryotes – apicomplexans, cryptophytes, dinoflagellates, haptophytes and stramenopiles – possess a plastid-targeted GAPDH that arose through duplication of the eukaryotic cytosolic isoform (Fast et al. 2001; Harper and Keeling 2003); notable exceptions exist in some dinoflagellates (Fagan and Hastings 2002; Takishita et al. 2003). Although a close relationship among these plastid-targeted GAPDH copies was interpreted as evidence for a single origin for red algal-derived plastids (Harper and Keeling 2003), these results have been called into question (Bodyl 2005; Bodyl and Moszczynski 2006). One prominent reason is the apparent discrepancy between the branching pattern of the cytosolic and the plastid-targeted GAPDH subtrees. Although the plastid-targeted GAPDH sequences of "chromalveolate" taxa cluster together with strong bootstrap support (more than 95%), their cytosolic GAPDH sequences do not form a clade (Fast et al. 2001; Harper and Keeling 2003). The observed disparities between the two homologs would seem to refute the underlying assumption of Fast et al. (2001) that the two homologs have coevolved since the endosymbiotic common origin. Secondly, plastid-targeted GAPDH protein phylogenies are not always consistent with accepted organismal relationships. For example, in the plastid-targeted GAPDH subtree the apicomplexan *T. gondii* is strongly associated with haptophytes to the exclusion of peridinin-containing dinoflagellates (Harper and Keeling 2003; Takishita et al. 2004). This suggests that *T. gondii* and peridinin-containing dinoflagellates obtained the genes for their plastid-targeted GAPDH proteins independently.

The phylogeny of several other nucleus-encoded plastid-targeted proteins, such as sedoheptulose bisphosphatase, fructose bisphosphatase, phosphoribulokinase and fructose bisphosphate aldolase, has also been explored in an attempt to elucidate the origin of red algal-derived plastids. Unfortunately, these analyses are largely inconclusive because of the complex evolutionary history of such genes (Kroth et al. 2005; Petersen et al. 2006; Teich et al. 2007). With the growing recognition of the impact of LGT in eukaryotic evolution, especially in phagotrophs (Andersson 2005), the complex evolutionary patterns of plastid-targeted proteins may have arisen during the early stages of endosymbiosis when the host–endosymbiont relationship had not been permanently established. If ancient host eukaryotes were exposed to, and "experimented" with, diverse kinds of algae in the context of transient endosymbioses, analogous to modern sea-slug-plastid symbioses (Rumpho et al. 2000), nuclear genes for plastid-targeted proteins could have originated from multiple sources, and thus display a mosaic evolutionary pattern. Additionally, the mixotrophic life style of many plastid-containing eukaryotes could provide a continuous source of "foreign" genes to the host. Seen in this light, the analysis and interpretation of plastid-targeted proteins requires caution.

4.4 Alveolate Plastids

The red or green algal secondary endosymbiotic origin of the apicoplast has been intensely debated (Funes et al. 2002, 2003; Waller et al. 2003). While sequence analyses of *tufA*, *rpoB*, *rpoC1* and *rpoC2* genes suggest that apicoplasts are related to green plastids (Kohler et al. 1997; Cai et al. 2003), rRNA, transfer RNA (tRNA) and ribosomal protein gene trees tend to support (when long-branching sequences of euglenids are excluded) the alternative hypothesis that apicoplasts are of red algal origin (Blanchard and Hicks 1999). Consistent with the results of phylogenetic analyses, the ribosomal protein gene cluster of apicoplasts shares structural similarity with that of red algal plastids (Blanchard and Hicks 1999). As noted above, a photosynthetic relative of apicomplexans, *C. velia*, has been discovered recently (Moore et al. 2008), the plastid of which appears to be related to apicoplasts on the basis of plastid rRNA gene phylogeny and to be of red algal origin. This result suggests that the hypothesis of green algal ancestry of the apicoplast is unlikely, although it is still possible that the apicoplast is of chimeric origin, i.e., derived from both red and green plastids through multiple endosymbiotic events, as has been suggested (Funes et al. 2004). Comprehensive phylogenetic analysis of plastid-encoded genes in *C. velia* will be necessary to test whether the apicoplast is of red algal origin or of chimeric origin.

Dinoflagellates are remarkable in that they have acquired plastids from diverse algal sources (Schnepf and Elbrachter 1999). Approximately half of known dinoflagellate species possess "genuine" plastids, while other photosynthetic dinoflagellates harbor transient cyanobacterial endosymbionts or plastids (i.e., kleptoplastids) borrowed from haptophytes (e.g., *Dinophysis mitra*), cryptophytes (e.g., *Dinophysis acuminata*, *Gymnodinium acidoum*) or green algae (e.g., *Noctiluca scintillans*) (Wilcox and Wedemayer 1984; Schnepf and Elbrachter 1999; Takishita et al. 2002; Koike et al. 2005; Minnhagen and Janson 2006).

Among the plastid-containing dinoflagellates, most are characterized by the presence of the pigment peridinin as a major carotenoid fraction, as well as a nuclear-encoded form II ribulose-1,5-bisphosphate carboxylase/oxygenase (RuBisCO), which otherwise only occurs in some proteobacteria (Morse et al. 1995). In contrast, the plastids of the dinoflagellate genera *Karenia*, *Karlonidium* and *Takayama* contain fucoxanthin and its derivatives (19′-hexanoyloxyfucoxanthin and 19′-butanoyloxyfucoxanthin) instead of peridinin, reminiscent of certain haptophytes (Takishita et al. 2004). Yet other dinoflagellates harbor plastids taken from prasinophyte green algae (*Lepidodinium viride*, *L. chlorophorum*) or diatoms (*Kryptoperidinium foliaceum*, *Durinskia baltica*, *Galeidinium rugatum* and *Peridinium quinoquecorne*), some of which seem to retain the relic nucleus and, in some cases, even the mitochondria of their algal endosymbionts (Schnepf and Elbrachter 1999; Horiguchi and Takano 2006; Hansen et al. 2007; Imanian and Keeling 2007). The thecate dinoflagellate *Podolampas bipes* harbors plastids that originate from a dictyophyte (Stramenopiles) (Schnepf and Elbrachter 1999; Schweiker and Elbrachter 2004), but whether these plastids are permanent or transient remains to be demonstrated.

The origins of peridinin-containing and fucoxanthin-containing (19′-hexanoyloxyfucoxanthin and 19′-butanoyloxyfucoxanthin) plastids of dinoflagellates have been difficult to discern (Ishida and Green 2002; Yoon et al. 2002a; Bodyl and Moszczynski 2006). Growing evidence suggests that the two types of plastids arose separately and that the fucoxanthin-containing plastids originated from a haptophyte ancestor (Ishida and Green 2002; Yoon et al. 2002a; Bodyl and Moszczynski 2006). This hypothesis is supported by the phylogenies of two nuclear-encoded and plastid-targeted proteins, psbO and GAPDH, and the plastid-encoded psbC (Ishida and Green 2002; Takishita et al. 2004, 2005). Although combined DNA sequence analysis of plastid-encoded *psaA* and *psbA* genes suggested a strong affiliation between the two types of plastids and consequently a common haptophyte origin (Yoon et al. 2002a), this relationship has been suggested to be a phylogenetic artifact caused by codon usage heterogeneity (Inagaki et al. 2004). The true origin of peridinin-containing plastids is thus presently unclear (Cavalier-Smith 1999; Bodyl and Moszczynski 2006). Unfortunately, the plastid genome of peridinin-containing dinoflagellates is fragmented into multiple minicircles (see Sect. 1.3) (Zhang et al. 2002) and as a result, information on gene order cannot be used as a phylogenetic character. In addition, only a handful of plastid-encoded genes remain in the peridinin plastid and those that do are highly derived in sequence (Sanchez-Puerta et al. 2007) and thus susceptible to phylogenetic artifacts.

5 Plastid Genome Evolution

5.1 Plastid Genome Structure

As of January 2008, approximately 200 plastid genomes have been completely sequenced. The vast majority of these are from land plants and green algae, with only a single plastid genome sequence currently available from members of the Chlorarachniophyta, Glaucophyta and Haptophyta. Consequently, our views on the "typical" features of plastid genomes are significantly biased. Nevertheless, it is possible to identify a number of near-universal features of plastid genomes and in some cases these features are shared with the genomes of modern-day cyanobacteria.

One of the most widely distributed features of plastid genomes is the presence of ribosomal DNA (rDNA) containing repeats that form a quadripartite structure consisting of two inverted repeats and small and large single-copy regions (Stirewalt et al. 1995; Oudot-Le Secq et al. 2007). The rDNA repeat unit typically contains three rRNA genes (*rns*, *rnl*, *rrn5*) and two tRNAs (*trnA*, *trnI*), but can harbor as few as four genes or up to 161 genes through contraction or expansion of this region (Chumley et al. 2006) (Table 2). Although the repeats are rarely 100% identical to one another, they are always highly similar, and apparently evolve by concerted evolution. The presence of rDNA-containing repeats in all three plastid types and in some cyanobacteria such as *Synechococcus* sp. WH8102 suggests that the repeats likely predate the origin of plastids (Glockner et al. 2000). Some plastids, however, have apparently lost rDNA-containing repeats, or have rearranged the repeats in tandem (Ohta et al. 2003; Turmel et al. 2005; De Koning and Keeling 2006). In the case of the red alga *Porphyra*, the repeats are directly oriented (Reith and Munholland 1993).

Far and away the most unusual plastid genomes belong to the peridinin-containing dinoflagellates, which are composed of minicircles 2–10 kbp in size. Individual minicircles usually carry one to three genes, although "empty" minicircles have also been detected (Barbrook et al. 2006). The noncoding region of the

Table 2 Plastid genome features

Taxonomic classification	Genome size (nt)	No. of ORFs/structural RNAs[a]	Size of rDNA repeat unit (nt)	Genes in rDNA-containing repeats[b]	LSC/SSC ratio	Intron[a]
Glaucophyta						
Cyanophora paradoxa	135,599	145/37	11,285	(SSC)-*rns-trnI(gau)-trnA(ugc)-rnl-r rn5-dnaK-trnC(gca)-groEL-groES-clpP*-(LSC)	5.3	1 (group I)
Rhodophyta and red-plastid-containing eukaryotes						
Cyanidium caldarium (Rhodophyta)	164,921	197/33	Not present	NA	NA	0
Cyanidioschyzon merolae (Rhodophyta)	149,987	207/35	Not present	NA	NA	0
Gracilaria tenuistipitata (Rhodophyta)	183,883	203(1)/33	Not present	NA	NA	0
Porphyra yezoensis (Rhodophyta)	191,952	209/50	4,829	DR: *rns-trnI(gau)-trnA(ugc)-rnl-rrn5*	4.4	0
Porphyra purpurea (Rhodophyta)	191,028	209/38	4,826	DR: *rns-trnI(gau)-trnA(ugc)-rnl-rrn5*	4.4	0
Emiliania huxleyi (Haptophyta)	105,309	119/31	4,841	(LSC)-*rns-trnI(gau)-trnA(ugc)-rnl-rrn5*-(SSC)	7.6	0
Guillardia theta (Cryptophyta)	121,524	147/31	4,967	(LSC)-*rns-trnI(gau)-trnA(ugc)-rnl-rrn5*-(SSC)	6.2	0
Rhodomonas salina (Cryptophyta)	135,854	146(3)/32	4,927	(LSC)-*rns-trnI(gau)-trnA(ugc)-rnl-rrn5*-(SSC)	6.6	2 (group II)
Odontella sinensis (diatom, Stramenopiles)	119,704	137/29	7,725	(LSC)-*trnP(ugg)-orf355-rns-trnI(gau)-trnA(ugc)-rnl-rrn5-ycf32-rpl32*-(SSC)	1.7	0
Phaeodactylum tricornutum (diatom, Stramenopiles)	117,369	130/30	6,912	(LSC)-*trnP(ugg)-ycf89-rns-trnI(gau)-trnA(ugc)-rnl-rrn5-psbY*-(SSC)	1.6	0
Thalassiosira pseudonana (diatom, Stramenopiles)	128,814	127/30	18,337	(LSC)-*trnP(ugg)-ycf89-rns-trnI(gau)-trnA(ugc)-rnl-rrn5-psbY-rpl32-trnL(uag)-rbcR-rpl21-rpl27-secA-rpl34-ycf46-ccs1-psbA-ycf35-psaC-ccsA*-(SSC)	2.4	0

(continued)

Table 2 (continued)

Taxonomic classification	Genome size (nt)	No. of ORFs/structural RNAs[a]	Size of rDNA repeat unit (nt)	Genes in rDNA-containing repeats[b]	LSC/SSC ratio	Intron[a]
Eimeria tenella NP (apicomplexan, Alveolata)	34,750	28/26	5,361	*trnI(gau)-rns-trnA(ugc)-trnN(guu)-trnL(uag)-trnR(ucu)-trnV(uac)-trnR(acg)-trnM(cau)-rnl-trnT(ugu)*-(LSC)	NA (SSC = 0)	0
Theileria parva NP (apicomplexan, Alveolata)	39,579	44/26	Not present	NA	NA	0
Toxoplasma gondii NP (apicomplexan, Alveolata)	34,996	26/26	5,316	*trnI(gau)-rns-trnA(ugc)-trnN(guu)-trnL(uag)-trnR(ucu)-trnV(uac)-trnR(acg)-trnM(cau)-rnl-trnT(ugu)*-(LSC)	NA (SSC = 0)	1 (group I)
Viridiplantae and green-plastid-containing eukaryotes						
Chlamydomonas reinhardtii (chlorophyte, Viridiplantae)	203,827	68/33	22,211	(SSC)-*rns-trnI(gau)-trnA(ugc)-rrn7-rrn3-rnl-rrn5-psbA*-(LSC)	1.0	7 (group I, II)
Stigeoclonium helveticum (chlorophyte, Viridiplantae)	223,902	79/31	Not present	NA	NA	21 (group I, II)
Mesostigma viride (streptophyte, Viridiplantae)	118,360	105/36	6,057	(LSC)-*trnT(ugu)-rns-trnI(gau)-trnA(ugc)-rnl-rrn5-trnR(acg)*-(SSC)	3.7	0
Zygnema circumcarinatum (streptophyte, Viridiplantae)	165,372	103/37	Not present	NA	NA	13 (group I, II)
Nicotiana tabacum (embryophyte, Viridiplantae)	155,943	85(1)/34	25,343	(LSC)-*rpl2-rpl23-trnI(cau)-ycf2-trnL(caa)-orf79-ndhB-rps7-3'rps12-orf131-orf70B-trnV(gac)-rns-trnI(gau)-trnA(ugc)-rnl-rrn4.5-rrn5-trnR(acg)-orf75-trnN(guu)*-(SSC)	4.7	21 (group I, II)

Diversity and Evolution of Plastids and Their Genomes

t1.60	*Epifagus virginiana* NP (embryophyte, Viridiplantae)	70,028	22(9)/21(5)	22,735	(LSC)-*rpl2*-ψ*rpl23*-*trnI(caa)*-*ycf2*-*trnL(caa)*-ψ*ndhB*-*rps7*-3'*rps12*-*rns*-ψ*trnI(gau)*-ψ*trnA(ugc)*-*rnl*-*rrn4.5*-*rrn5*-*trnR(acg)*-*trnN(guu)*-(SSC)	4.2	6 (group II)
t1.64	*Euglena gracilis* (euglenid, Kinetoplastida)	143,171	66/31(6)	5,918	DR (x3):ψ*trnI*-ψ*trnW*-*rns*-*trnI(gau)*-*trnA(ugc)*-*rnl*-*rrn5*	NA	155[c] (group II)
t1.66	*Euglena longa* NP (euglenid, Kinetoplastida)	73,345	46/31	~5,400	DR (x3): *rns*-*rnl*-*rrn5*-*trnV(uac)*	NA	Numerous (group II)
t1.69	*Bigelowiella natans* (chlorarachniophyte, Rhizaria)	69,166	57/30	9,380	(SSC)-*psbM*-*rns*-*trnE(uac)*-*trnL(caa)*-*trnA(ugc)*-*trnI(gau)*-*rrn5*-*rnl*-*petD*-*petB*-*psaA*-(LSC)	11.2	0

t1.71 Genes predicted to be nonfunctional are preceded by ψ. *ORF* open reading frame, *nt* nucleotide, *DR* direct repeats, *LSC* large single copy region, *SSC* small single copy region, *NP* nonphotosynthetic plastid

t1.73 [a]Genes or introns within the rDNA repeats are counted once. The numbers of predicted pseudogenes, which were separately counted from predicted functional genes, are indicated in *parentheses*. In this table, structural RNAs include only rRNAs, tRNAs, and *rnpB*

t1.75 [b]Unless otherwise noted, rDNA-containing repeats are inversely oriented

t1.76 [c]Thompson et al. (1995)

minicircle usually harbors two to four "cores," which are highly conserved within species and show evidence of concerted evolution (Zhang et al. 2002). Thus far, only about a dozen protein-coding genes, ribosomal small and large subunit rRNA genes, and several tRNAs have been found to reside on minicircles, with most of the plastid genome apparently having been transferred to the nucleus (Bachvaroff et al. 2004; Barbrook et al. 2006; Nelson et al. 2007).

Plastid genome density varies greatly when the full breadth of eukaryotic diversity is considered. The percentage coding sequence ranges from 50.1% in the green alga *Chlamydomonas reinhardtii* (Maul et al. 2002) to 93.5% in the red alga *Cyanidioschyzon merolae* (Ohta et al. 2003). The *C. reinhardtii* chloroplast genome is more than 200 kbp in size, yet harbors only 99 genes, with more than 20% of the genome being occupied by randomly dispersed repetitive elements (Maul et al. 2002). Such repeat sequences, which are especially abundant in chlorophyte chloroplast DNA (cpDNA), are thought to promote genome arrangements (Palmer et al. 1987; Maul et al. 2002; Turmel et al. 2005; De Cambiaire et al. 2006). In contrast, the plastid genome of *C. merolae*, which is approximately 150 kbp in size, encodes 243 genes and has numerous instances of overlapping genes (Ohta et al. 2003).

With respect to physical structure, plastid genomes have traditionally been assumed to exist as individual genome-sized circles. However, growing evidence suggests that plastid genomes may in fact exist predominantly as linear forms of multiple genome units (Maul et al. 2002; Bendich 2004; Rogers et al. 2007). For example, more than 95% of the cpDNA in maize has been shown to exist in a branched linear form, visible in high-resolution fluorescence microscopy (Oldenburg and Bendich 2004). The size and complexity of cpDNA does not seem to be uniform and appear to be developmentally regulated (Bendich 2004).

5.2 Plastid Genome Content

Completely sequenced plastid genomes are between 35 and 220 kbp in size, with the genomes of nonphotosynthetic plastids residing at the bottom of this distribution. The apicoplast genome of *Eimeria tenella* is approximately 35 kbp in size, and is the smallest described thus far, encoding 54 unique genes (Cai et al. 2003). Among photosynthetic plastids, the chlorarachniophyte *Bigelowiella natans* possesses the smallest plastid genome, at approximately 70 kbp, the result of genome compaction and numerous gene losses (Rogers et al. 2007). The green alga *Stigeoclonium helvetiucum* contains the largest plastid genome (approximately 220 kbp), although the approximately 190 kbp genome of the red alga *Porphyra yezoensis* possess the largest known gene repertoire, with a total of 259 unique genes (Belanger et al. 2006).

During the evolutionary transition from endosymbiont to fully integrated eukaryotic organelle, extensive plastid genome reduction has occurred through the combined effects of gene loss and gene transfer to the nucleus (Martin et al. 2002). This is readily apparent when one considers that the smallest cyanobacterial genome, that of the prochlorophyte *Prochlorococcus marinus* MIT 9301, is still significantly larger than the largest plastid genome (approximately 1,640 kbp vs. approximately 220 kbp) and encodes approximately 8 times more genes than the most gene-rich plastid DNA (2,005 vs. 259 genes). This suggests that more than 85% of the genes present in the cyanobacterial progenitor of plastids have been lost or transferred to the nucleus. Plastid DNA fragments have also been transferred to the mitochondrial genome in some land plants, but only tRNAs seem to remain functional (Joyce and Gray 1989; Notsu et al. 2002); all known protein-coding "mtpts" (mitochondrial plastid DNAs) appear to have become pseudogenes (Wang et al. 2007).

Plastid-encoded genes can be classified into two categories, protein-coding genes and structural RNAs. With respect to protein-coding genes, RNA polymerase subunits, ribosomal proteins and proteins directly involved in photosynthesis commonly occur among the three types of primary plastids, whereas many of the genes involved in biosynthesis are restricted to those of red algae (Table 3). Almost all plastid genomes encode rRNAs and tRNAs, although some plastids are predicted to import at least some nucleus-encoded tRNAs from the cytoplasm to make up the deficit (Lohan and Wolfe 1998). Small RNAs, including transfer-messenger RNA (*ssra*), P RNA (the RNA component of RNase P, *rnpB*) and chloroplast signal recognition particle RNA (*ffs*) are encoded in some plastid genomes, but the genes are not easy to identify and, consequently, are poorly annotated (De Novoa and Williams 2004; Rosenblad and Samuelsson 2004; Oudot-Le Secq et al. 2007).

Some plastids have completely lost photosynthetic capacity, a prominent example being the apicoplast of apicomplexans. Multiple instances of conversion into nonphotosynthetic plastids have also occurred within cryptophytes, dinoflagellates, euglenids, stramenopiles, red algae and Viridiplantae (Goff et al. 1997; Hoef-Emden and Melkonian 2003; Nudelman et al. 2003; Bungard 2004; De Koning and Keeling 2006). Retention of nonphotosynthetic plastids and their genomes is due to the fact that these organelles carry out a variety of important metabolic processes in addition to photosynthesis, including amino acid and fatty acid biosynthesis (Mazumdar et al. 2006). In general, the genomes of nonphotosynthetic plastids are significantly reduced compared with those of their closest photosynthetic relatives and as expected are devoid of photosynthesis-related genes (Wolfe et al. 1992; Gockel and Hachtel 2000). However, the large subunit RuBisCO gene (*rbcL*) still remains in a number of parasitic land plants and the nonphotosynthetic euglenid *Euglena longa* (formerly known as *Astasia longa*), suggesting an alternative function(s) (Gockel and Hachtel 2000).

5.3 *Lateral Gene Transfer and the Plastid Genome and Proteome*

Unlike mitochondria, whose genomes appear to have been significantly impacted by LGT (Bergthorsson et al. 2004), current evidence suggests that plastids have

Table 3 Functional classification of plastid-encoded genes from analysis of the completely sequenced genomes of 24 green plastids, 14 red plastids and one glaucophyte plastid

Functional categories			Genes
Genetic system	Replication, transcription, recombination, and repair		*cfxQ(cbbX) dnaB dnaX*[a] *ntcA(ycf28) ompR(ycf27) rbcR(ycf30) recO rne rpoA(ycf67) rpoB rpoC1 rpoC2 rpoZ(ycf61) tctD(ycf29)*
	Translation		*infA infB infC tsf tufA*
	Ribosomal proteins	Small subunit	*rps1 rps2 rps3 rps4 rps5 rps6 rps7 rps8 rps9 rps10 rps11 rps12 rps13 rps14 rps15 rps16 rps17 rps18 rps19 rps20*
		Large subunit	*rpl1 rpl2 rpl3 rpl4 rpl5 rpl6 rpl9 rpl11 rpl12 rpl13 rpl14 rpl16 rpl18 rpl19 rpl20 rpl21 rpl22 rpl23 rpl24 rpl27 rpl28 rpl29 rpl31 rpl32 rpl33 rpl34 rpl35 rpl36*
	Protein quality control		*clpC clpP dnaK groEL groES*
	ORFs within introns (group I, II)		*I-CuvI*[b] *matA(ycf13)*[c] *matK(ycf14)*[c] *rdpO*[d]
Photosynthetic apparatus	Phycobiliproteins		*apcA apcB apcD apcE apcF cpcA cpcB cpcG cpeA cpeB nblA(ycf18) pbsA*
	Chlorophyll biosynthesis		*chlB chlI chlL chlN hemA*
	Photosystem I		*psaA psaB psaC psaD psaE psaF psaI psaJ psaK psaL psaM ycf3 ycf4*
	Photosystem II		*psbA psbB psbC psbD psbE psbF psbH psbI psbJ psbK psbL psbM psbN psbT(ycf8) psbV(petK) psbX psbY(ycf32) psbZ(ycf9) psb28(psbW, ycf79)*
	Cytochrome b₆/f complex		*ccsA(ycf5) ccs1(ycf44) dsbD petA petB petD petG petJ petL(ycf7) petM(ycf31) petN(ycf6)*
	Redox system		*bas1(ycf42) fdx ftrB ndhA ndhB ndhC ndhD ndhE ndhF ndhG ndhH ndhI ndhJ ndhK(psbG) petF trxA(trxM)*
	ATP synthase		*atpA atpB atpD atpE atpF atpG atpH atpI*
Membrane transport			*cemA(ycf10, hbp) cysA cysT cysW mntA mntB secA secG(ycf47) secY sufB(ycf24) sufC(ycf16) tatC(ycf43, yigU, ycbT) ycf84 ycf85*

Biosynthesis	Carbohydrates	*odpA odpB pgmA rbcLg rbcLr rbcSg rbcSr*
	Lipids	*accA accB accD(ycf11, zfpA) acpP crtE crtR(desA) fabH*
	Nucleotides	*carA upp*
	Amino acids	*argB glnB gltB hisH ilvB ilvH leuC leuD trpA trpG*
	Complex sugars	*glmS lpxA lpxC*
	Cofactors	*bioY cobA lipB menA menB menC menD menE menF moeB nadA preA thiG thiS(ycf40)*
	tRNA synthesis/maturation	*syfB(pheT) syh(hisS) trmE(thdF)*
Other cellular processes	Cell division	*minD minE* *ftsH(ycf2, ycf25) ftsI ftsW*
	Signal transduction	*dfr(ycf26)*
	DNA organization	*hlpA(hupA)*
Hypothetical plastid ORFs		*ycf1 ycf12 ycf15 ycf17*[CAB/ELIP/HILP family] *ycf19 ycf20 ycf21 ycf22 ycf23 ycf33 ycf34 ycf35 ycf36 ycf37 ycf38 ycf39 ycf41 ycf45 ycf46 ycf48 ycf49 ycf50 ycf51 ycf52 ycf53 ycf54 ycf55 ycf56 ycf57(ycf83) ycf58 ycf59 ycf60 ycf62 ycf63 ycf64 ycf65 ycf66 ycf68 ycf69 ycf70 ycf71 ycf72 ycf73 ycf74 ycf75 ycf76 ycf77 ycf78 ycf80 ycf81 ycf82 ycf86ᵍ ycf88 ycf89 ycf90 ycxr*
RNA genes	rRNAs	*rns rnl rrn5*
	tRNAs	*trnA(ggc) trnA(uge) trnC(gca) trnD(guc) trnE(uuc) trnF(gaa) trnG(gcc) trnG(ucc) trnH(gug) trnI(cau) trnI(gau) trnK(uuu) trnL(caa) trnL(gag) trnL(uaa) trnL(uag) trnM(cau) trnfM(cau) trnN(guu) trnP(ggg) trnP(ugg) trnQ(uug) trnR(acg) trnR(ccg) trnR(ccu) trnR(ucu) trnS(cga) trnS(gcu) trnS(gga) trnS(uga) trnT(ggu) trnT(ugu) trnV(gac) trnV(aac) trnV(uac) trnW(cca) trnY(gua)*
	Miscellaneous	*ffs*ʰ *ssrA rnpB*

Unique ORFs of unknown functions are not included. Synonyms are indicated in *parentheses*. Genes labeled in *blue*, *green* and *red* are those that show restricted distribution to plastids of glaucophytes, green lineage, and red lineage, respectively. Genes present in all three types of plastids are shown in **bold**. *Underlined* genes do not show significant sequence homology to any bacterial genes in BLAST searches. Nomenclature follows Stoebe et al. (1998), Douglas and Penny (1999) and De Las Rivas et al. (2002), and the following updates:

(continued)

Table 3 (continued)

1. *ycf67* of *Euglena* plastid genomes has been identified as *rpoA* (Sheveleva et al. 2002).
2. *rpoZ* (*ycf61*), *secG* (*ycf47*), *sufB* (*ycf24*), *sufC* (*ycf40*) and *thiS* (*ycf16*), according to Oudot-Le Secq et al. (2007).
3. *tatC* (*ycf43*), according to Ohta et al. (2003) and Hagopian et al. (2004).
4. Plastid genomes of *Cyanidioschyzon merolae*, *Cyanidium caldarium* and *Galdieria sulphuraria* encode a *crtR* (or *desA*) like gene, but its function remains unclear (Cunningham et al. 2007).
5. New plastid-encoded genes added since Stoebe et al. (1998): *crtR, trmE* (*thdF*), *hlpA* (*hupA*), *ycf81–90*.
6. CypaCp027 (orf206) of *Cyanophora paradoxa* is annotated as *recO*, as is curated by GenBank.

Plastid genomes summarized in this table are as follows:

Green plastids: Streptophyta, Viridiplantae (*Acorus calamus, Adiantum capillus-veneris, Agrostis stolonifera, Amborella trichopoda, Anthoceros formosae, Chaetosphaeridium globosum, Chara vulgaris, Chlorokybus atmophyticus, Marchantia polymorpha, Mesostigma viride, Physcomitrella patens, Staurastrum punctulatum, Zygnema circumcarinatum*); Chlorophyta, Viridiplantae (*Chlamydomonas reinhardtii, Chlorella vulgaris, Helicosporidium sp., Nephroselmis olivacea, Oltmannsiellopsis viridis, Ostreococcus tauri, Scenedesmus obliquus, Stigeoclonium helveticum*); Chlorarachniophyta, Rhizaria (*Bigelwiella natans*); Euglenida, Kinetoplastida (*Euglena gracilis, Euglena longa*)

Red plastids: Rhodophyta (*Cyanidium caldarium, Cyanidioschyzon merolae, Gracilaria tenuistipitata, Porphyra purpurea, Porphyra yezoensis*); diatoms, Stramenopiles (*Odontella sinensis, Phaeodactylum tricornutum, Thalassiosira pseudonana*); Haptophyta (*Emiliania huxleyi*); Cryptophyta (*Guillardia theta, Rhodomonas salina*); Apicomplexa, Alveolata (*Eimeria tenella, Theileria parva, Toxoplasma gondii*)

Glaucophyte plastid (*Cyanophora paradoxa*)

[a]*dnaX*, detected in the plastid genome of *Rhodomonas salina*, was acquired by lateral gene transfer (LGT) (Khan et al. 2007)

[b]A homing endonuclease gene within ribosomal DNA group I intron of plastid genomes. It has been identified in a number of green plastid genomes, but its evolutionary origin is poorly understood (Haugen et al. 2007)

[c]*matA* and *matK* are ORFs found within group II introns, which appear to have invaded plastid genomes through LGT from an unknown source (Hausner et al. 2006; Khan and Archibald 2008). *matA* occurs in the plastid genomes of *Rhodomonas salina*, *Euglena gracilis* and *Euglena longa*, while *matK* is found in the plastid genomes of streptophytes, except for the "earliest-diverging" streptophyte *Mesostigma viride*

[d]Refers to maturase-like gene within group II introns of unknown origin

[e]*Chlorella vulgaris* is reported to have a *minE*-like gene in the plastid genome (Wakasugi et al. 1997), but it shows only marginal sequence similarity to bacterial or other plastid-encoded *minE* gene sequences

[f]The *ycf49* gene sequence of *Cyanophora paradoxa* shows only marginal similarities to those of *Cyanidioschyzon merolae* and *Cyanidium caldarium*

[g]*ycf86* is annotated in the plastid genomes of *Cyanidioschyzon merolae* and *Cyanidium caldarium*; however, the two copies do not show sequence homology or positional conservation. Until further studies reveal that the two copies indeed encode proteins of similar function, we suggest that the two copies be annotated as unique ORFs

[h]*ffs* has been identified in green and red algal plastid genomes using the conserved signal recognition particle RNA model, and cannot be easily detected on the basis of sequence homology searches (Rosenblad and Samuelsson 2004). Its presence or absence, however, has not been examined for the glaucophyte *Cyanophora paradoxa* plastid genome

been influenced only minimally in this regard. Nevertheless, several examples of LGT involving the plastid genome have been documented, including the RuBisCO small and large subunit genes (*rbcS* and *rbcL*), ribosomal protein gene *rpl36* and *dnaX*, and perhaps group I and group II introns (Sheveleva and Hallick 2004; Hausner et al. 2006; Rice and Palmer 2006; Khan et al. 2007). Unlike green plastids and those of glaucophytes, red algal plastids (and their derivatives) encode *rbcS* and *rbcL* genes that are not of cyanobacterial origin, but rather of α-proteobacterial ancestry, and were presumably acquired by LGT early in the evolution of rhodophytes (Delwiche and Palmer 1996; Rice and Palmer 2006). The *cbbx* gene found in red algal plastid genomes has also been suggested to be the product of LGT involving an α-proteobacterial donor (Maier et al. 2000). However, with the accumulation and analysis of additional bacterial sequences it now appears that this gene is most closely related to cyanobacterial homologs, as would be predicted under a model of vertical evolution. The shared presence of a noncyanobacterial *rpl36* gene in the plastid genomes of cryptophytes and haptophytes has been cited as another example of plastid LGT and used to support the notion that these two lineages may be each other's closest relatives (Rice and Palmer 2006). The plastid genome of the cryptophyte *Rhodomonas salina* was found to contain a gene with significant similarity to the bacterial DNA polymerase subunit gene *dnaX*, the first described case of putative DNA replication machinery encoded in plastid DNA (Khan et al. 2007). Phylogenetic analysis suggests that the *Rhodomonas dnaX* gene was acquired by LGT from a firmicute-like bacterium.

Self-splicing introns are found in a number of plastid genomes, most notably those of euglenids and land plants, and are thought to have invaded the plastid genome on multiple occasions (Simon et al. 2003; Hausner et al. 2006; Haugen et al. 2007). Plastid-encoded group I introns, which can be found in rRNA, tRNA or protein-coding genes, have been reported in apicomplexans, glaucophytes, stramenopiles and Viridiplantae, with some of these containing a homing endonuclease-coding region, which promotes intron mobility (Blanchard and Hicks 1999; Haugen et al. 2005, 2007). Group II introns are found in the plastids of cryptophytes, euglenids and Viridiplantae and usually inhabit tRNA or protein-coding genes (Ems et al. 1995; Zimmerly et al. 2001; Sheveleva and Hallick 2004; Hausner et al. 2006; Khan et al. 2007). Many plastid group II introns have highly derived RNA secondary structures and have often lost their open reading frames (Hausner et al. 2006). The most extreme examples of group II intron derivatives reside within the plastid genome of euglenids. These include "group III" introns, which have been reduced to 73–119 nucleotides, and "twintrons," which are group II introns nested within another group II intron (Copertino and Hallick 1993; Thompson et al. 1995; Lambowitz and Zimmerly 2004; Sheveleva and Hallick 2004). While introns are common in the plastid genomes of euglenids and Viridiplantae, the chlorarachniophyte *B. natans* plastid, as well as those of diatoms, haptophytes, rhodophytes and at least some cryptophytes (e.g., *Guillardia theta*) are completely devoid of group I or group II introns (Douglas and Penny 1999; Hagopian et al. 2004; Puerta et al. 2005; Oudot-Le Secq et al. 2007) (Table 3). Interestingly, members of the cryptophyte genus *Rhodomonas* possess highly unusual group II introns that appear to have been acquired by LGT from a euglenid-like ancestor (Khan and Archibald 2008).

While a number of other plastid genes display anomalous branching patterns in phylogenetic analyses, current evidence is insufficiently strong to refute the null hypothesis of vertical inheritance (Rice and Palmer 2006). One enigmatic example is *psbA*, which encodes the D1 protein of photosystem II. This gene has been noted as a "problematic" phylogenetic marker (Inagaki et al. 2004; Bachvaroff et al. 2005; Rice and Palmer 2006), as it seems to be in conflict with other plastid-encoded genes in that it strongly refutes the monophyly of "chromalveolate" plastids (Bachvaroff et al. 2005). Interestingly, the *psbA* gene products in green algae share a seven amino acid C-terminal deletion with their homologs in the prochlorophyte *Prochlorothrix*, but not with any other cyanobacteria (including *Prochlorococcus* and *Prochloron*) or nongreen plastids (Morden et al. 1992; Griffiths 2006). This is intriguing given that green plastids and *Prochlorothrix* also share similar pigment composition, which originally led to the idea that they share a common ancestry (Morden and Golden 1989). However, the *psbA* gene sequences of green plastids and *Prochlorothrix* do not branch together in phylogenetic analyses (Scherer et al. 1991; Zeidner et al. 2003), suggesting that the shared seven amino acid deletion is either the result of convergent evolution or the product of partial recombination involving only the portion of the gene encoding the C-terminal end of the protein. These observations suggest that *psbA* and possibly other plastid-encoded genes might have an even more complex evolutionary history than currently appreciated.

Finally, LGT has also played a role in the evolution of nucleus-encoded, plastid-targeted proteins, in some organisms much more so than appears to be the case for the plastid genome itself. An expressed sequence tag based study of the chlorarachniophyte *B. natans* (Archibald et al. 2003) showed that while the majority of plastid-targeted proteins in this organism appear to be green algal in origin, as would be predicted on the basis of what is known about the origin of the chlorarachniophyte plastid, a number of these genes appear most closely related to red algae (or their red algal derivatives) and prokaryotes other than cyanobacteria. This result is significant given that chlorarachniophyte algae have been shown to ingest a wide range of algal species (Hibberd and Norris 1984). LGT also appears to have contributed to the complement of plastid-targeted proteins in other algae, including dinoflagellates (Hackett et al. 2004), a lineage well known for its mixotrophic life style. However, as noted above, the dinoflagellates have repeatedly replaced their ancestral peridinin-containing plastid with those derived from other algae, so it is possible that genes displaying atypical phylogenetic patterns are in fact remnants of past endosymbioses.

Acknowledgments We thank Linda Graham and Hameed Khan for helpful comments on an earlier version of this chapter and Lee Wilcox for producing the drawings presented in Fig. 1. Research in the Archibald laboratory is supported by the Natural Sciences and Engineering Research Council of Canada, the Canadian Institutes of Health Research and the Dalhousie Medical Research Foundation. E.K. is supported by a postdoctoral fellowship from the Tula Foundation and J.M.A. is a Scholar of the Canadian Institute for Advanced Research, Program in Integrated Microbial Biodiversity.

References

Adl SM, Simpson AGB, Farmer MA, Andersen RA, Anderson OR, Barta JR, Bowser SS, Brugerolle G, Fensome RA, Fredericq S, James TY, Karpov S, Kugrens P, Krug J, Lane CE, Lewis LA, Lodge J, Lynn DH, Mann DG, McCourt RM, Mendoza L, Moestrup O, Mozley-Standridge SE, Nerad TA, Shearer CA, Smirnov AV, Spiegel FW, Taylor MFJR (2005) The new higher level classification of eukaryotes with emphasis on the taxonomy of protists. *J Eukaryot Microbiol* 52:399–451

Andersen RA (2004) Biology and systematics of heterokont and haptophyte algae. *Am J Bot* 91:1508–1522

Andersson JO (2005) Lateral gene transfer in eukaryotes. *Cell Mol Life Sc*i 62:1182–1197

Antia NJ, Cheng JY, Foyle RAJ, Percival E (1979) Marine cryptomonad starch from autolysis of glycerol-grown *Chroomonas salina*. *J Phycol* 15:57–62

Archibald JM (2006) Endosymbiosis: double-take on plastid origins. *Curr Biol* 16:R690–R692

Archibald JM (2007) Nucleomorph genomes: structure, function, origin and evolution. *Bioessays* 29:392–402

Archibald JM, Keeling PJ (2002) Recycled plastids: a 'green movement' in eukaryotic evolution. *Trends Genet* 18:577–584

Archibald JM, Keeling PJ (2004) Actin and ubiquitin protein sequences support a cercozoan/foraminiferan ancestry for the plasmodiophorid plant pathogens. *J Eukaryot Microbiol* 51:113–118

Archibald JM, Rogers MB, Toop M, Ishida K, Keeling PJ (2003) Lateral gene transfer and the evolution of plastid-targeted proteins in the secondary plastid-containing alga *Bigelowiella natans*. *Proc Natl Acad Sci USA* 100:7678–7683

Bachvaroff TR, Concepcion GT, Rogers CR, Herman EM, Delwiche CF (2004) Dinoflagellate expressed sequence tag data indicate massive transfer of chloroplast genes to the nuclear genome. *Protist* 155:65–78

Bachvaroff TR, Puerta MVS, Delwiche CF (2005) Chlorophyll *c*-containing plastid relationships based on analyses of a multigene data set with all four chromalveolate lineages. *Mol Biol Evol* 22:1772–1782

Ball SG, Morell MK (2003) From bacterial glycogen to starch: understanding the biogenesis of the plant starch granule. *Annu Rev Plant Biol* 54:207–233

Barbier G, Oesterhelt C, Larson MD, Halgren RG, Wilkerson C, Garavito RM, Benning C, Weber APM (2005) Comparative genomics of two closely related unicellular thermo-acidophilic red algae, *Galdieria sulphuraria* and *Cyanidioschyzon merolae*, reveals the molecular basis of the metabolic flexibility of *Galdieria sulphuraria* and significant differences in carbohydrate metabolism of both algae. *Plant Physiol* 137:460–474

Barbrook AC, Santucci N, Plenderleith LJ, Hiller RG, Howe CJ (2006) Comparative analysis of dinoflagellate chloroplast genomes reveals rRNA and tRNA genes. *BMC Genomics* 7:297

Beale SI (1999) Enzymes of chlorophyll biosynthesis. *Photosynth Res* 60:43–73

Belanger AS, Brouard JS, Charlebois P, Otis C, Lemieux C, Turmel M (2006) Distinctive architecture of the chloroplast genome in the chlorophycean green alga *Stigeoclonium helveticum*. *Mol Genet Genomics* 276:464–477

Bendich AJ (2004) Circular chloroplast chromosomes: The grand illusion. *Plant Cell* 16:1661–1666

Bergholtz T, Daugbjerg N, Moestrup O, Fernandez-Tejedor M (2006) On the identity of *Karlodinium veneficum* and description of *Karlodinium armiger* sp nov (Dinophyceae), based on light and electron microscopy, nuclear-encoded LSU rDNA, and pigment composition. *J Phycol* 42:170–193

Bergthorsson U, Richardson AO, Young GJ, Goertzen LR, Palmer JD (2004) Massive horizontal transfer of mitochondrial genes from diverse land plant donors to the basal angiosperm *Amborella*. *Proc Natl Acad Sci USA* 101:17747–17752

Bhattacharya D, Archibald JM (2006) Response to Theissen and Martin. *Curr Biol* 16:R1017–R1018

Blanchard JL, Hicks JS (1999) The non-photosynthetic plastid in malarial parasites and other apicomplexans is derived from outside the green plastid lineage. *J Eukaryot Microbiol* 46:367–375

Bodyl A (2005) Do plastid-related characters support the chromalveolate hypothesis? *J Phycol* 41:712–719

Bodyl A, Moszczynski K (2006) Did the peridinin plastid evolve through tertiary endosymbiosis? A hypothesis. *Eur J Phycol* 41:435–448

Bodyl A, Mackiewicz P, Stiller JW (2007) The intracellular cyanobacteria of *Paulinelia chromatophora*: endosymbionts or organelles? *Trends Microbiol* 15:295–296

Breglia SA, Slamovits CH, Leander BS (2007) Phylogeny of phagotrophic euglenids (Euglenozoa) as inferred from hsp90 gene sequences. *J Eukaryot Microbiol* 54:86–92

Brugerolle G (2002) *Colpodella vorax*: Ultrastructure, predation, life-cycle mitosis, and phylogenetic relationships. *Eur J Protistol* 38:113–125

Buleon A, Colonna P, Planchot V, Ball S (1998) Starch granules: structure and biosynthesis. *Int J Biol Macromol* 23:85–112

Buleon A, Veronese G, Putaux JL (2007) Self-association and crystallization of amylose. *Aust J Chem* 60:706–718

Bungard RA (2004) Photosynthetic evolution in parasitic plants: Insight from the chloroplast genome. *Bioessays* 26:235–247

Burki F, Shalchian-Tabrizi K, Minge M, Skjaeveland A, Nikolaev SI, Jakobsen KS, Pawlowski J (2007) Phylogenomics reshuffles the eukaryotic supergroups. PLoS ONE 2:e790

Busse I, Preisfeld A (2002) Phylogenetic position of *Rhynchopus* sp. and *Diplonema ambulator* as indicated by analyses of euglenozoan small subunit ribosomal DNA. *Gene* 284:83–91

Busse I, Patterson DJ, Preisfeld A (2003) Phylogeny of phagotrophic euglenids (Euglenozoa): a molecular approach based on culture material and environmental samples. *J Phycol* 39:828–836

Cai XM, Fuller AL, McDougald LR, Zhu G (2003) Apicoplast genome of the coccidian *Eimeria tenella*. *Gene* 321:39–46

Cavalier-Smith T (1999) Principles of protein and lipid targeting in secondary symbiogenesis: euglenoid, dinoflagellate, and sporozoan plastid origins and the eukaryote family tree. *J Eukaryot Microbiol* 46:347–366

Cavalier-Smith T, Chao EE (1996) 18S rRNA sequence of *Heterosigma carterae* (Raphidophyceae), and the phylogeny of heterokont algae (Ochrophyta). *Phycologia* 35:500–510

Cavalier-Smith T, Chao EE, Thompson CE, Hourihane SL (1995/1996) *Oikomonas*, a distinctive zooflagellate related to chrysomonads. *Arch Protistenkd* 146:273–279

Chen M, Hiller RG, Howe CJ, Larkum AWD (2005) Unique origin and lateral transfer of prokaryotic chlorophyll *b* and chlorophyll *d* light-harvesting systems. *Mol Biol Evol* 22:21–28

Chiovitti A, Ngoh JE, Wetherbee R (2006) 1,3-Beta-d-glucans from *Haramonas dimorpha* (Raphidophyceae). *Bot Mar* 49:360–362

Chizhov AO, Dell A, Morris HR, Reason AJ, Haslam SM, McDowell RA, Chizhov OS, Usov AI (1998) Structural analysis of laminarans by MALDI and FAB mass spectrometry. *Carbohydr Res* 310:203–210

Chumley TW, Palmer JD, Mower JP, Fourcade HM, Calie PJ, Boore JL, Jansen RK (2006) The complete chloroplast genome sequence of *Pelargonium* x *hortorum*: organization and evolution of the largest and most highly rearranged chloroplast genome of land plants. *Mol Biol Evol* 23:2175–2190

Copertino DW, Hallick RB (1993) Group II and group III introns of twintrons: Potential relationships with nuclear premessenger RNA introns. *Trends Biochem Sci* 18:467–471

Coppin A, Varre JS, Lienard L, Dauvillee D, Guerardel Y, Soyer-Gobillard MO, Buleon A, Ball S, Tomavo S (2005) Evolution of plant-like crystalline storage polysaccharide in the protozoan parasite *Toxoplasma gondii* argues for a red alga ancestry. *J Mol Evol* 60:257–267

Cunningham FX, Gantt E (1998) Genes and enzymes of carotenoid biosynthesis in plants. *Annu Rev Plant Physiol Plant Mol Biol* 49:557–583

Cunningham FX, Lee H, Gantt E (2007) Carotenoid biosynthesis in the primitive red alga *Cyanidioschyzon merolae*. *Eukaryot Cell* 6:533–545

Dammeyer T, Michaelsen K, Frankenberg-Dinkel N (2007) Biosynthesis of open-chain tetrapyrroles in *Prochlorococcus marinus*. *FEMS Microbiol Lett* 271:251–257

Deane JA, Strachan IM, Saunders GW, Hill DRA, McFadden GI (2002) Cryptomonad evolution: nuclear 18S rDNA phylogeny versus cell morphology and pigmentation. *J Phycol* 38:1236–1244

De Cambiaire JC, Otis C, Lemieux C, Turmel M (2006) The complete chloroplast genome sequence of the chlorophycean green alga *Scenedesmus obliquus* reveals a compact gene organization and a biased distribution of genes on the two DNA strands. *BMC Evol Biol* 6:37

De Koning AP, Keeling PJ (2006) The complete plastid genome sequence of the parasitic green alga *Helicosporidium* sp is highly reduced and structured. *BMC Biol* 4:12

De Las Rivas J, Lozano JJ, Ortiz AR (2002) Comparative analysis of chloroplast genomes: functional annotation, genome-based phylogeny, and deduced evolutionary patterns. *Genome Res* 12:567–583

Delwiche CF, Palmer JD (1996) Rampant horizontal transfer and duplication of rubisco genes in eubacteria and plastids. *Mol Biol Evol* 13:873–882

De Novoa PG, Williams KP (2004) The tmRNA website: Reductive evolution of tmRNA in plastids and other endosymbionts. *Nucleic Acids Res* 32:D104–D108

Deschamps P, Haferkamp I, Dauvillee D, Haebel S, Steup M, Buleon A, Putaux JL, Colleoni C, d'Hulst C, Plancke C, Gould S, Maier U, Neuhaus HE, Ball S (2006) Nature of the periplastidial pathway of starch synthesis in the cryptophyte *Guillardia theta*. *Eukaryot Cell* 5:954–963

Douglas SE, Penny SL (1999) The plastid genome of the cryptophyte alga, *Guillardia theta*: complete sequence and conserved synteny groups confirm its common ancestry with red algae. *J Mol Evol* 48:236–244

Durnford DG, Deane JA, Tan S, McFadden GI, Gantt E, Green BR (1999) A phylogenetic assessment of the eukaryotic light-harvesting antenna proteins, with implications for plastid evolution. *J Mol Evol* 48:59–68

Ems SC, Morden CW, Dixon CK, Wolfe KH, dePamphilis CW, Palmer JD (1995) Transcription, splicing and editing of plastid RNAs in the nonphotosynthetic plant *Epifagus virginiana*. *Plant Mol Biol* 29:721–733

Fagan TF, Hastings JW (2002) Phylogenetic analysis indicates multiple origins of chloroplast glyceraldehyde-3-phosphate dehydrogenase genes in dinoflagellates. *Mol Biol Evol* 19:1203–1207

Falkowski PG, Raven JA (2007) Aquatic photosynthesis. Princeton University Press, Princeton

Falkowski PG, Katz ME, Knoll AH, Quigg A, Raven JA, Schofield O, Taylor FJR (2004) The evolution of modern eukaryotic phytoplankton. *Science* 305:354–360

Fast NM, Kissinger JC, Roos DS, Keeling PJ (2001) Nuclear-encoded, plastid-targeted genes suggest a single common origin for apicomplexan and dinoflagellate plastids. *Mol Biol Evol* 18:418–426

Funes S, Davidson E, Reyes-Prieto A, Magallon S, Herion P, King MP, Gonzalez-Halphen D (2002) A green algal apicoplast ancestor. *Science* 298:2155–2155

Funes S, Davidson E, Reyes-Prieto A, Magallon S, Herion P, King MP, Gonzalez-Halphen D (2003) Response to comment on "A green algal apicoplast ancestor". *Science* 301:49b

Funes S, Reyes-Prieto A, Perez-Martinez X, Gonzalez-Halphen D (2004) On the evolutionary origins of apicoplasts: revisiting the rhodophyte vs. chlorophyte controversy. *Microbes Infect* 6:305–311

Gibbs SP (1981) The chloroplasts of some algal groups may have evolved from endosymbiotic eukaryotic algae. *Ann N Y Acad Sci* 361:193–208

Glockner G, Rosenthal A, Valentin K (2000) The structure and gene repertoire of an ancient red algal plastid genome. *J Mol Evol* 51:382–390

Gockel G, Hachtel W (2000) Complete gene map of the plastid genome of the nonphotosynthetic euglenoid flagellate *Astasia longa*. *Protist* 151:347–351

Goff LJ, Ashen J, Moon D (1997) The evolution of parasites from their hosts: a case study in the parasitic red algae. *Evolution* 51:1068–1078

Graham LE, Wilcox LW (2000) Algae. Prentice Hall, Upper Saddle River

Granum E, Myklestad SM (2001) Mobilization of β-1,3-glucan and biosynthesis of amino acids induced by NH_4^+ addition to N-limited cells of the marine diatom *Skeletonema costatum* (Bacillariophyceae). *J Phycol* 37:772–782

Gray MW, Spencer DF (1996) Organellar evolution. In: Roberts DM, Sharp P, Alderson G Collins M (eds) Evolution of microbial life (Society for General Microbiology Symposium 54). Cambridge University Press, Cambridge, pp 109–126

Green BR (2005) Lateral gene transfer in the cyanobacteria: Chlorophylls, proteins, and scraps of ribosomal RNA. *J Phycol* 41:449–452

Green BR, Durnford DG (1996) The chlorophyll-carotenoid proteins of oxygenic photosynthesis. *Annu Rev Plant Physiol Plant Mol Biol* 47:685–714

Griffiths DJ (2006) Chlorophyll b-containing oxygenic photosynthetic prokaryotes: Oxychlorobacteria (prochlorophytes). *Bot Rev* 72:330–366

Grzebyk D, Katz ME, Knoll AH, Quigg A, Raven JA, Schofield O, Taylor FJR, Falkowski PG (2004) Response to comment on "The evolution of modern eukaryotic phytoplankton". *Science* 306:2191c

Guillou L, Chretiennot-Dinet MJ, Boulben S, Moon-van der Staay SY, Vaulot D (1999) *Symbiomonas scintillans* gen. et sp nov and *Picophagus flagellatus* gen. et sp nov (Heterokonta): two new heterotrophic flagellates of picoplanktonic size. *Protist* 150:383–398

Hackett JD, Yoon HS, Soares MB, Bonaldo MF, Casavant TL, Scheetz TE, Nosenko T, Bhattacharya D (2004) Migration of the plastid genome to the nucleus in a peridinin dinoflagellate. *Curr Biol* 14:213–218

Hackett JD, Yoon HS, Li S, Reyes-Prieto A, Rummele SE, Bhattacharya D (2007) Phylogenomic analysis supports the monophyly of cryptophytes and haptophytes and the association of Rhizaria with Chromalveolates. *Mol Biol Evol* 24:1702–1713

Hagopian JC, Reis M, Kitajima JP, Bhattacharya D, de Oliveira MC (2004) Comparative analysis of the complete plastid genome sequence of the red alga *Gracilaria tenuistipitata* var. *liui* provides insights into the evolution of rhodoplasts and their relationship to other plastids. *J Mol Evol* 59:464–477

Hansen G, Botes L, De Salas M (2007) Ultrastructure and large subunit rDNA sequences of *Lepidodinium viride* reveal a close relationship to *Lepidodinium chlorophorum* comb. nov (=*Gymnodinium chlorophorum*). *Phycol Res* 55:25–41

Harper JT, Keeling PJ (2003) Nucleus-encoded, plastid-targeted glyceraldehyde-3-phosphate dehydrogenase (GAPDH) indicates a single origin for chromalveolate plastids. *Mol Biol Evol* 20:1730

Harris JR, Adrian M, Petry F (2004) Amylopectin: a major component of the residual body in *Cryptosporidium parvum* oocysts. *Parasitol* 128:269–282

Haugen P, Bhattacharya D, Palmer JD, Turner S, Lewis LA, Pryer KM (2007) Cyanobacterial ribosomal RNA genes with multiple, endonuclease-encoding group I introns. *BMC Evol Biol* 7:159

Haugen P, Simon DM, Bhattacharya D (2005) The natural history of group I introns. *Trends Genet* 21:111–119

Hausner G, Olson R, Simon D, Johnson I, Sanders ER, Karol KG, McCourt RM, Zimmerly S (2006) Origin and evolution of the chloroplast *trnK* (*matK*) intron: a model for evolution of group II intron RNA structures. *Mol Biol Evol* 23:380–391

Hess WR, Steglich C, Lichtle C, Partensky F (1999) Phycoerythrins of the oxyphotobacterium *Prochlorococcus marinus* are associated to the thylakoid membrane and are encoded by a single large gene cluster. *Plant Mol Biol* 40:507–521

Hess WR, Rocap G, Ting CS, Larimer F, Stilwagen S, Lamerdin J, Chisholm SW (2001) The photosynthetic apparatus of *Prochlorococcus*: insights through comparative genomics. *Photosynth Res* 70:53–71

Hibberd DJ, Norris RE (1984) Cytology and ultrastructure of *Chlorarachnion reptans* (Chlorarachniophyta divisio nova, Chlorarachniophyceae classis nova). *J Phycol* 20:310–330

Hirokawa Y, Fujiwara S, Suzuki M, Akiyama T, Sakamoto M, Kobayashi S, Tsuzuki M (2007) Structural and physiological studies on the storage β-polyglucan of haptophyte *Pleurochrysis haptonemofera*. *Planta* 227:589–599

Hoef-Emden K, Melkonian M (2003) Revision of the genus *Cryptomonas* (Cryptophyceae): a combination of molecular phylogeny and morphology provides insights into a long-hidden dimorphism. *Protist* 154:371–409

Hoef-Emden K, Marin B, Melkonian M (2002) Nuclear and nucleomorph SSU rDNA phylogeny in the cryptophyta and the evolution of cryptophyte diversity. *J Mol Evol* 55:161–179

Hopkins J, Fowler R, Krishna S, Wilson I, Mitchell G, Bannister L (1999) The plastid in *Plasmodium falciparum* asexual blood stages: a three-dimensional ultrastructural analysis. *Protist* 150:283–295

Horiguchi T, Takano Y (2006) Serial replacement of a diatom endosymbiont in the marine dinoflagellate *Peridinium quinquecorne* (Peridiniales, Dinophyceae). *Phycol Res* 54:193–200

Imanian B, Keeling PJ (2007) The dinoflagellates *Durinskia baltica* and *Kryptoperidinium foliaceum* retain functionally overlapping mitochondria from two evolutionarily distinct lineages. *BMC Evol Biol* 7:172

Inagaki Y, Simpson AGB, Dacks JB, Roger AJ (2004) Phylogenetic artifacts can be caused by leucine, serine, and arginine codon usage heterogeneity: dinoflagellate plastid origins as a case study. *Syst Biol* 53:582–593

Ishida K, Green BR (2002) Second- and third-hand chloroplasts in dinoflagellates: Phylogeny of oxygen-evolving enhancer 1 (PsbO) protein reveals replacement of a nuclear-encoded plastid gene by that of a haptophyte tertiary endosymbiont. *Proc Natl Acad Sci USA* 99:9294–9299

Janse I, Van Rijssel M, Van Hall PJ, Gerwig GJ, Gottschal JC, Prins RA (1996) The storage glucan of *Phaeocystis globosa* (Prymnesiophyceae) cells. *J Phycol* 32:382–387

Jarvis P, Soll M (2001) Toc, Tic, and chloroplast protein import. *Biochim Biophys Acta* 1541:64–79

Joyce PBM, Gray MW (1989) Chloroplast-like transfer RNA genes expressed in wheat mitochondria. *Nucleic Acids Res* 17:5461–5476

Karpov SA, Sogin ML, Silberman JD (2001) Rootlet homology, taxonomy, and phylogeny of bicosoecids based on 18S rRNA gene sequences. *Protistology* 2:34–47

Katz ME, Finkel ZV, Grzebyk D, Knoll AH, Falkowski PG (2004) Evolutionary trajectories and biogeochemical impacts of marine eukaryotic phytoplankton. *Annu Rev Ecol Evol Syst* 35:523–556

Keeling PJ, Archibald JM, Fast NM, Palmer JD (2004) Comment on "The evolution of modern eukaryotic phytoplankton". *Science* 306:2191b

Keeling PJ, Burger G, Durnford DG, Lang BF, Lee RW, Pearlman RE, Roger AJ, Gray MW (2005) The tree of eukaryotes. *Trends Ecol Evol* 20:670–676

Khan H, Archibald JM (2008) Lateral transfer of introns in the cryptophyte plastid genome. *Nucleic Acids Res* 36:3043–3053

Khan H, Parks N, Kozera C, Curtis BA, Parsons BJ, Bowman S, Archibald JM (2007) Plastid genome sequence of the cryptophyte alga *Rhodomonas salina* CCMP1319: lateral transfer of putative DNA replication machinery and a test of chromist plastid phylogeny. *Mol Biol Evol* 24:1832–1842

Kim E, Simpson AGB, Graham LE (2006) Evolutionary relationships of apusomonads inferred from taxon-rich analyses of 6 nuclear encoded genes. *Mol Biol Evol* 23:2455–2466

Kiss JZ, Triemer RE (1988) A comparative study of the storage carbohydrate granules from *Euglena* (Euglenida) and *Pavlova* (Prymnesiida). *J Protozool* 35:237–241

Kiss JZ, Vasconcelos AC, Triemer RE (1986) Paramylon synthesis and chloroplast structure associated with nutrient levels in *Euglena* (Euglenophyceae). *J Phycol* 22:327–333

Kiss JZ, Vasconcelos AC, Triemer RE (1987) Structure of the euglenoid storage carbohydrate, paramylon. *Am J Bot* 74:877–882

Kiss JZ, Roberts EM, Brown RM, Triemer RE (1988) X-ray and dissolution studies of paramylon storage granules from *Euglena*. *Protoplasma* 146:150–156

Kohler S (2005) Multi-membrane-bound structures of Apicomplexa: I. The architecture of the *Toxoplasma gondii* apicoplast. *Parasitol Res* 96:258–272

Kohler S, Delwiche CF, Denny PW, Tilney LG, Webster P, Wilson RJM, Palmer JD, Roos DS (1997) A plastid of probable green algal origin in apicomplexan parasites. *Science* 275:1485–1489

Koike K, Sekiguchi H, Kobiyama A, Takishita K, Kawachi M, Koike K, Ogata T (2005) A novel type of kleptoplastidy in *Dinophysis* (Dinophyceae): presence of haptophyte-type plastid in *Dinophysis mitra*. *Protist* 156:225–237

Kroth PG, Schroers Y, Kilian O (2005) The peculiar distribution of class I and class II aldolases in diatoms and in red algae. *Curr Genet* 48:389–400

Kuhn S, Medlin L, Eller G (2004) Phylogenetic position of the parasitoid nanoflagellate pirsonia inferred from nuclear-encoded small subunit ribosomal DNA and a description of *Pseudopirsonia* n. gen. and *Pseudopirsonia mucosa* (Drebes) comb. nov. *Protist* 155:143–156

La Roche J, van der Staay GWM, Partensky F, Ducret A, Aebersold R, Li R, Golden SS, Hiller RG, Wrench PM, Larkum AWD, Green BR (1996) Independent evolution of the prochlorophyte and green plant chlorophyll *a*/*b* light-harvesting proteins. *Proc Natl Acad Sci USA* 93:15244–15248

Lambowitz AM, Zimmerly S (2004) Mobile group II introns. *Annu Rev Genet* 38:1–35

Larkum AWD, Scaramuzzi C, Cox GC, Hiller RG, Turner AG (1994) Light-harvesting chlorophyll *c*-like pigment in *Prochloron*. *Proc Natl Acad Sci USA* 91:679–683

Larkum AWD, Lockhart PJ, Howe CJ (2007) Shopping for plastids. *Trends Plant Sci* 12:189–195

Leander BS (2004) Did trypanosomatid parasites have photosynthetic ancestors? *Trends Microbiol* 12:251–258

Leander BS, Keeling PJ (2003) Morphostasis in alveolate evolution. *Trends Ecol Evol* 18:395–402

Leander BS, Kuvardina ON, Aleshin VV, Mylnikov AP, Keeling PJ (2003) Molecular phylogeny and surface morphology of *Colpodella edax* (Alveolata): insights into the phagotrophic ancestry of apicomplexans. *J Eukaryot Microbiol* 50:334–340

Lee JJ, Leedale GF, Bradbury P (2000) The illustrated guide to the protozoa. Society of Protozoologists, Lawrence

Lee RE, Kugrens P (1991) *Katablepharis ovalis*, a colorless flagellate with interesting cytological characteristics. *J Phycol* 27:505–513

Lohan AJ, Wolfe KH (1998) A subset of conserved tRNA genes in plastid DNA of nongreen plants. *Genet* 150:425–433

Lukavsky J, Cepak V (1992) DAPI fluorescent staining of DNA material in cyanelles of the rhizopod *Paulinella chromatophora* Lauterb. *Arch Protistenkd* 142:207–212

Maier UG, Fraunholz M, Zauner S, Penny S, Douglas S (2000) A nucleomorph-encoded CbbX and the phylogeny of RuBisCo regulators. *Mol Biol Evol* 17:576–583

Marchessault RH, Deslandes Y (1979) Fine structure of (1–3) β-D-glucans: Curdlan and paramylon. *Carbohydr Res* 75:231–242

Marin B, Nowack ECM, Melkonian M (2005) A plastid in the making: primary endosymbiosis. *Protist* 156:425–432

Marin B, Nowack ECM, Glockner G, Melkonian M (2007) The ancestor of the *Paulinella* chromatophore obtained a carboxysomal operon by horizontal gene transfer from a *Nitrococcus*-like gamma-proteobacterium. *BMC Evol Biol* 7:85

Martin W, Rujan T, Richly E, Hansen A, Cornelsen S, Lins T, Leister D, Stoebe B, Hasegawa M, Penny D (2002) Evolutionary analysis of *Arabidopsis*, cyanobacterial, and chloroplast genomes reveals plastid phylogeny and thousands of cyanobacterial genes in the nucleus. *Proc Natl Acad Sci USA* 99:12246–12251

Maul JE, Lilly JW, Cui LY, dePamphilis CW, Miller W, Harris EH, Stern DB (2002) The *Chlamydomonas reinhardtii* plastid chromosome: islands of genes in a sea of repeats. *Plant Cell* 14:2659–2679

Mazumdar J, Wilson EH, Masek K, Hunter CA, Striepen B (2006) Apicoplast fatty acid synthesis is essential for organelle biogenesis and parasite survival in *Toxoplasma gondii*. *Proc Natl Acad Sci USA* 103:13192–13197

McFadden GI (2001) Primary and secondary endosymbiosis and the origin of plastids. *J Phycol* 37:951–959

McFadden GI, Roos DS (1999) Apicomplexan plastids as drug targets. *Trends Microbiol* 7:328–333

McFadden GI, van Dooren GG (2004) Evolution: red algal genome affirms a common origin of all plastids. *Curr Biol* 14:R514–R516

McFadden GI, Gilson PR, Douglas SE (1994) The photosynthetic endosymbiont in cryptomonad cells produces both chloroplast and cytoplasmic-type ribosomes. *J Cell Sci* 107:649–657

McFadden GI, Gilson PR, Sims IM (1997) Preliminary characterization of carbohydrate stores from chlorarachniophytes (division: Chlorarachniophyta). *Phycol Res* 45:145–151

Mereschkowsky C (1905) Über natur und ursprung der chromatophoren im pflanzenreiche. *Biol Centralbl* 25:593–604

Minnhagen S, Janson S (2006) Genetic analyses of *Dinophysis* spp. support kleptoplastidy. *FEMS Microbiol Ecol* 57:47–54

Miyashita H, Ikemoto H, Kurano N, Miyachi S, Chihara M (2003) *Acaryochloris marina* gen. et sp nov (Cyanobacteria), an oxygenic photosynthetic prokaryote containing Chl *d* as a major pigment. *J Phycol* 39:1247–1253

Moestrup O, Sengco M (2001) Ultrastructural studies on *Bigelowiella natans* gen. et sp. nov., a chlorarachniophyte flagellate. *J Phycol* 37:624–646

Moore RB, Obornik M, Janouskovec J, Chrudimsky T, Vancova M, Green DH, Wright SW, Davies NW, Bolch CJS, Heimann K, Slapeta J, Hoegh-Guldberg O, Logsdon JM, Carter DA (2008) A photosynthetic alveolate closely related to apicomplexan parasites. *Nature* 451:959–963

Morden CW, Golden SS (1989) *psbA* genes indicate common ancestry of prochlorophytes and chloroplasts. *Nature* 337:382–385

Morden CW, Delwiche CF, Kuhsel M, Palmer JD (1992) Gene phylogenies and the endosymbiotic origin of plastids. *Biosystems* 28:75–90

Moriya M, Nakayama T, Inouye I (2000) Ultrastructure and 18S rDNA sequence analysis of *Wobblia lunata* gen. et an. nov., a new heterotrophic flagellate (stramenopiles, Incertae sedis). *Protist* 151:41–55

Moriya M, Nakayama T, Inouye I (2002) A new class of the stramenopiles, Placididea classis nova: description of *Placidia cafeteriopsis* gen. et sp nov. *Protist* 153:143–156

Morse D, Salois P, Markovic P, Hastings JW (1995) A nuclear-encoded form II RuBisCO in dinoflagellates. *Science* 268:1622–1624

Murakami A, Miyashita H, Iseki M, Adachi K, Mimuro M (2004) Chlorophyll *d* in an epiphytic cyanobacterium of red algae. *Science* 303:1633

Nakamura Y, Takahashi J, Sakurai A, Inaba Y, Suzuki E, Nihei S, Fujiwara S, Tsuzuki M, Miyashita H, Ikemoto H, Kawachi M, Sekiguchi H, Kurano N (2005) Some cyanobacteria synthesize semi-amylopectin type α-polyglucans instead of glycogen. *Plant Cell Physiol* 46:539–545

Nelson MJ, Dang YK, Filek E, Zhang ZD, Yu VWC, Ishida K, Green BR (2007) Identification and transcription of transfer RNA genes in dinoflagellate plastid minicircles. *Gene* 392:291–298

Nikolaev SI, Berney C, Fahrni JF, Bolivar I, Polet S, Mylnikov AP, Aleshin VV, Petrov NB, Pawlowski J (2004) The twilight of Heliozoa and rise of Rhizaria, an emerging supergroup of amoeboid eukaryotes. *Proc Natl Acad Sci USA* 101:8066–8071

Not F, Valentin K, Romari K, Lovejoy C, Massana R, Tobe K, Vaulot D, Medlin LK (2007) Picobiliphytes: a marine picoplanktonic algal group with unknown affinities to other eukaryotes. *Science* 315:253–255

Notsu Y, Masood S, Nishikawa T, Kubo N, Akiduki G, Nakazono M, Hirai A, Kadowaki K (2002) The complete sequence of the rice (*Oryza sativa* L.) mitochondrial genome: frequent DNA sequence acquisition and loss during the evolution of flowering plants. *Mol Genet Genomics* 268:434–445

Nowack ECM, Melkonian M, Glöckner G (2008) Chromatophore genome sequence of *Paulinella* sheds light on acquisition of photosynthesis by eukaryotes. *Curr Biol* 18:410–418

Nozaki H, Matsuzaki M, Takahara M, Misumi O, Kuroiwa H, Hasegawa M, Shin-i T, Kohara Y, Ogasawara N, Kuroiwa T (2003) The phylogenetic position of red algae revealed by multiple nuclear genes from mitochondria-containing eukaryotes and an alternative hypothesis on the origin of plastids. *J Mol Evol* 56:485–497

Nudelman MA, Rossi MS, Conforti V, Triemer RE (2003) Phylogeny of Euglenophyceae based on small subunit rDNA sequences: taxonomic implications. *J Phycol* 39:226–235

Ohta N, Matsuzaki M, Misumi O, Miyagishima S, Nozaki H, Tanaka K, Shin-i T, Kohara Y, Kuroiwa T (2003) Complete sequence and analysis of the plastid genome of the unicellular red alga *Cyanidioschyzon merolae*. *DNA Res* 10:67–77

Okamoto N, Inouye I (2005) The katablepharids are a distant sister group of the Cryptophyta: a proposal for Katablepharidophyta divisio nova/Kathablepharida phylum novum based on SSU rDNA and beta-tubulin phylogeny. *Protist* 156:163–179

Oldenburg DJ, Bendich AJ (2004) Most chloroplast DNA of maize seedlings in linear molecules with defined ends and branched forms. *J Mol Biol* 335:953–970

Oudot-Le Secq MP, Grimwood J, Shapiro H, Armbrust EV, Bowler C, Green BR (2007) Chloroplast genomes of the diatoms *Phaeodactylum tricornutum* and *Thalassiosira pseudonana*: comparison with other plastid genomes of the red lineage. *Mol Genet Genomics* 277:427–439

Palmer JD (2003) The symbiotic birth and spread of plastids: How many times and whodunit? *J Phycol* 39:4–11

Palmer JD, Nugent JM, Herbon LA (1987) Unusual structure of *Geranium* chloroplast DNA: a triple-sized inverted repeat, extensive gene duplications, multiple inversions, and 2 repeat families. *Proc Natl Acad Sci USA* 84:769–773

Patron NJ, Keeling PJ (2005) Common evolutionary origin of starch biosynthetic enzymes in green and red algae. *J Phycol* 41:1131–1141

Patron NJ, Inagaki Y, Keeling PJ (2007) Multiple gene phylogenies support the monophyly of cryptomonad and haptophyte host lineages. *Curr Biol* 17:887–891

Petersen J, Teich R, Brinkmann H, Cerff R (2006) A "green" phosphoribulokinase in complex algae with red plastids: Evidence for a single secondary endosymbiosis leading to haptophytes, cryptophytes, heterokonts, and dinoflagellates. *J Mol Evol* 62:143–157

Prechtl J, Kneip C, Lockhart P, Wenderoth K, Maier UG (2004) Intracellular spheroid bodies of *Rhopalodia gibba* have nitrogen-fixing apparatus of cyanobacterial origin. *Mol Biol Evol* 21:1477–1481

Preisig HR, Hibberd DJ (1983) Ultrastructure and taxonomy of *Paraphysomonas* (Chrysophyceae) and related genera, part 3. *Nordic J Bot* 3:695–723

Puerta MVS, Bachvaroff TR, Delwiche CF (2005) The complete plastid genome sequence of the haptophyte *Emiliania huxleyi*: a comparison to other plastid genomes. *DNA Res* 12:151–156

Raven JA (2005) Cellular location of starch synthesis and evolutionary origin of starch genes. *J Phycol* 41:1070–1072

Reid PC, Lancelot C, Gieskes WWC, Hagmeier E, Weichart G (1990) Phytoplankton of the North Sea and its dynamics: a review. *Neth J Sea Res* 26:295–331

Reith M, Munholland J (1993) The ribosomal RNA repeats are nonidentical and directly oriented in the chloroplast genome of the red alga *Porphyra purpurea*. *Curr Genet* 24:443–450

Rice DW, Palmer JD (2006) An exceptional horizontal gene transfer in plastids: gene replacement by a distant bacterial paralog and evidence that haptophyte and cryptophyte plastids are sisters. *BMC Biol* 4:31

Rissler HM, Durnford DG (2005) Isolation of a novel carotenoid-rich protein in *Cyanophora paradoxa* that is immunologically related to the light-harvesting complexes of photosynthetic eukaryotes. *Plant Cell Physiol* 46:416–424

Rodriguez-Ezpeleta N, Brinkmann H, Burey SC, Roure B, Burger G, Loffelhardt W, Bohnert HJ, Philippe H, Lang BF (2005) Monophyly of primary photosynthetic eukaryotes: green plants, red algae, and glaucophytes. *Curr Biol* 15:1325–1330

Rogers MB, Gilson PR, Su V, McFadden GI, Keeling PJ (2007) The complete chloroplast genome of the chlorarachniophyte *Bigelowiella natans*: evidence for independent origins of chlorarachniophyte and euglenid secondary endosymbionts. *Mol Biol Evol* 24:54–62

Rosenblad MA, Samuelsson T (2004) Identification of chloroplast signal recognition particle RNA genes. *Plant Cell Physiol* 45:1633–1639

Rumpho ME, Summer EJ, Manhart JR (2000) Solar-powered sea slugs. Mollusc/algal chloroplast symbiosis. *Plant Physiol* 123:29–38

Saldarriaga JF, McEwan ML, Fast NM, Taylor FJR, Keeling PJ (2003) Multiple protein phylogenles show that *Oxyrrhis marina* and *Perkinsus marinus* are early branches of the dinoflagellate lineage. *Int J Syst Evol Microbiol* 53:355–365

Sanchez-Puerta MV, Bachvaroff TR, Delwiche CF (2007) Sorting wheat from chaff in multi-gene analyses of chlorophyll *c*-containing plastids. *Mol Phylogenet Evol* 44:885–897

Scherer S, Herrmann G, Hirschberg J, Boger P (1991) Evidence for multiple xenogenous origins of plastids: Comparison of *psbA*-genes with a xanthophyte sequence. *Curr Genet* 19:503–507

Schimper AFW (1885) Untersuchungen über die Chlorophyllkö rner und die ihnen homologen Gebilde. *Jahrb Wiss Bot* 16:1–247

Schnepf E, Elbrachter M (1999) Dinophyte chloroplasts and phylogeny: a review. *Grana* 38:81–97

Schweiker M, Elbrachter M (2004) First ultrastructural investigations of the consortium between a phototrophic eukaryotic endocytobiont and *Podolampas bipes* (Dinophyceae). *Phycologia* 43:614–623

Sekiguchi H, Moriya M, Nakayama T, Inouye I (2002) Vestigial chloroplasts in heterotrophic stramenopiles *Pteridomonas danica* and *Ciliophrys infusionum* (Dictyochophyceae). *Protist* 153:157–167

Sheveleva EV, Hallick RB (2004) Recent horizontal intron transfer to a chloroplast genome. *Nucleic Acids Res* 32:803–810

Sheveleva EV, Giordani NV, Hallick RB (2002) Identification and comparative analysis of the chloroplast alpha-subunit gene of DNA-dependent RNA polymerase from seven *Euglena* species. *Nucleic Acids Res* 30:1247–1254

Shimonaga T, Fujiwara S, Kaneko M, Izumo A, Nihei S, Francisco PB, Satoh A, Fujita N, Nakamura Y, Tsuzuki M (2007) Variation in storage alpha-polyglucans of red algae: amylose and semi-amylopectin types in *Porphyridium* and glycogen type in *Cyanidium*. *Mar Biotechnol* 9:192–202

Simon D, Fewer D, Friedl T, Bhattacharya D (2003) Phylogeny and self-splicing ability of the plastid tRNA-Leu group I intron. *J Mol Evol* 57:710–720

Six C, Worden AZ, Rodriguez F, Moreau H, Partensky F (2005) New insights into the nature and phylogeny of prasinophyte antenna proteins: *Ostreococcus tauri*, a case study. *Mol Biol Evol* 22:2217–2230

Stiller JW (2007) Plastid endosymbiosis, genome evolution and the origin of green plants. *Trends Plant Sci* 12:391–396

Stiller JW, Hall BD (1997) The origin of red algae: Implications for plasmid evolution. *Proc Natl Acad Sci USA* 94:4520–4525

Stiller JW, Reel DC, Johnson JC (2003) A single origin of plastids revisited: convergent evolution in organellar genome content. *J Phycol* 39:95–105

Stirewalt VL, Michalowski CB, Loffelhardt W, Bohnert HJ, Bryant DA (1995) Nucleotide sequence of the cyanelle genome from *Cyanophora paradoxa*. *Plant Mol Biol Rep* 13:327–332

Stoebe B, Kowallik KV (1999) Gene-cluster analysis in chloroplast genomics. *Trends Genet* 15:344–347

Stoebe B, Martin W, Kowallik KV (1998) Distribution and nomenclature of protein-coding genes in 12 sequenced chloroplast genomes. *Plant Mol Biol Rep* 16:243–255

Takishita K, Koike K, Maruyama T, Ogata T (2002) Molecular evidence for plastid robbery (Kleptoplastidy) in *Dinophysis*, a dinoflagellate causing diarrhetic shellfish poisoning. *Protist* 153:293–302

Takishita K, Ishida K, Maruyama T (2003) An enigmatic GAPDH gene in the symbiotic dinoflagellate genus *Symbiodinium* and its related species (the order suessiales): possible lateral gene transfer between two eukaryotic algae, dinoflagellate and euglenophyte. *Protist* 154:443–454

Takishita K, Ishida KI, Maruyama T (2004) Phylogeny of nuclear-encoded plastid-targeted GAPDH gene supports separate origins for the peridinin- and the fucoxanthin derivative-containing plastids of dinoflagellates. *Protist* 155:447–458

Takishita K, Ishida KI, Ishikura M, Maruyama T (2005) Phylogeny of the *psbC* gene, coding a photosystem II component CP_{43}, suggests separate origins for the peridinin- and fucoxanthin derivative-containing plastids of dinoflagellates. *Phycologia* 44:26–34

Takahashi F, Okabe Y, Nakada T, Sekimoto H, Ito M, Kataoka H, Nozaki H (2007) Origins of the secondary plastids of Euglenophyta and Chlorarachniophyta as revealed by an analysis of the plastid-targeting, nuclear-encoded gene *psbO*. *J Phycol* 43:1302–1309

Teich R, Zauner S, Baurain D, Brinkmann H, Petersen J (2007) Origin and distribution of Calvin cycle fructose and sedoheptulose bisphosphatases in plantae and complex algae: a single secondary origin of complex red plastids and subsequent propagation via tertiary endosymbioses. *Protist* 158:263–276

Teles-Grilo ML, Tato-Costa J, Duarte SM, Maia A, Casal G, Azevedo C (2007) Is there a plastid in *Perkinsus atlanticus* (phylum Perkinsozoa)? *Eur J Protistol* 43:163–167

Theissen U, Martin W (2006) The difference between organelles and endosymbionts. *Curr Biol* 16:R1016–R1017

Thompson MD, Copertino DW, Thompson E, Favreau MR, Hallick RB (1995) Evidence for the late origin of introns in chloroplast genes from an evolutionary analysis of the genus *Euglena*. *Nucleic Acids Res* 23:4745–4752

Tomitani A, Okada K, Miyashita H, Matthijs HCP, Ohno T, Tanaka A (1999) Chlorophyll *b* and phycobilins in the common ancestor of cyanobacteria and chloroplasts. *Nature* 400:159–162

Tomova C, Geerts WJC, Muller-Reichert T, Entzeroth R, Humbel BM (2006) New comprehension of the apicoplast of *Sarcocystis* by transmission electron tomography. *Biol Cell* 98:535–545

Toso MA, Omoto CK (2007) *Gregarina niphandrodes* may lack both a plastid genome and organelle. *J Eukaryot Microbiol* 54:66–72

Turmel M, Otis C, Lemieux C (2005) The complete chloroplast DNA sequences of the charophycean green algae *Staurastrum* and *Zygnema* reveal that the chloroplast genome underwent extensive changes during the evolution of the Zygnematales. *BMC Biol* 3:22

van Dooren GG, Schwartzbach SD, Osafune T, McFadden GI (2001) Translocation of proteins across the multiple membranes of complex plastids. *Biochim Biophys Acta* 1541:34–53

Viola R, Nyvall P, Pedersen M (2001) The unique features of starch metabolism in red algae. *Proc R Soc Lond Ser B Biol Sci* 268:1417–1422

Vogel K, Meeuse BJD (1968) Characterization of the reserve granules from the dinoflagellate *Thecadinium inclinatum* Balech. *J Phycol* 4:317–318

Von der Heyden S, Chao EE, Cavalier-Smith T (2004) Genetic diversity of goniomonads: an ancient divergence between marine and freshwater species. *Eur J Phycol* 39:343–350

Wakasugi T, Nagai T, Kapoor M, Sugita M, Ito M, Ito S, Tsudzuki J, Nakashima K, Tsudzuki T, Suzuki Y, Hamada A, Ohta T, Inamura A, Yoshinaga K, Sugiura M (1997) Complete nucleotide sequence of the chloroplast genome from the green alga *Chlorella vulgaris*: the existence of genes possibly involved in chloroplast division. *Proc Natl Acad Sci USA* 94:5967–5972

Waller RF, Keeling PJ, van Dooren GG, McFadden GI (2003) Comment on "A green algal apicoplast ancestor". *Science* 301:49a

Walsby AE (1986) Prochlorophytes: Origins of chloroplasts. *Nature* 320:212

Wang D, Wu YW, Shih ACC, Wu CS, Wang YN, Chaw SM (2007) Transfer of chloroplast genomic DNA to mitochondrial genome occurred at least 300 MYA. *Mol Biol Evol* 24:2040–2048

Wilcox LW, Wedemayer GJ (1984) *Gymnodinium acidotum* Nygaard (Pyrrophyta), a dinoflagellate with an endosymbiotic cryptomonad. *J Phycol* 20:236–242

Wilhelm C (1987) Purification and identification of chlorophyll-c_1 from the green alga *Mantoniella squamata*. *Biochim Biophys Acta* 892:23–29

Wolfe KH, Morden CW, Palmer JD (1992) Function and evolution of a minimal plastid genome from a nonphotosynthetic parasitic plant. *Proc Natl Acad Sci USA* 89:10648–10652

Yoon HS, Hackett JD, Bhattacharya D (2002a) A single origin of the peridinin- and fucoxanthin-containing plastids in dinoflagellates through tertiary endosymbiosis. *Proc Natl Acad Sci USA* 99:11724–11729

Yoon HS, Hackett JD, Pinto G, Bhattacharya D (2002b) The single, ancient origin of chromist plastids. *Proc Natl Acad Sci USA* 99:15507–15512

Yoon HS, Reyes-Prieto A, Melkonian M, Bhattacharya D (2006) Minimal plastid genome evolution in the *Paulinella* endosymbiont. *Curr Biol* 16:R670–R672

Yu SK, Blennow A, Bojko M, Madsen F, Olsen CE, Engelsen SB (2002) Physico-chemical characterization of floridean starch of red algae. *Starch* 54:66–74

Zapata M, Garrido JL (1997) Occurrence of phytylated chlorophyll *c* in *Isochrysis galbana* and *Isochrysis* sp. (Clone T-ISO) (Prymnesiophyceae). *J Phycol* 33:209–214

Zeidner G, Preston CM, Delong EF, Massana R, Post AF, Scanlan DJ, Beja O (2003) Molecular diversity among marine picophytoplankton as revealed by *psbA* analyses. *Environ Microbiol* 5:212–216

Zhang ZD, Cavalier-Smith T, Green BR (2002) Evolution of dinoflagellate unigenic minicircles and the partially concerted divergence of their putative replicon origins. *Mol Biol Evol* 19:489–500

Zimmerly S, Hausner G, Wu XC (2001) Phylogenetic relationships among group II intron ORFs. *Nucleic Acids Res* 29:1238–1250

The Chloroplast Envelope Proteome and Lipidome

N. Rolland, M. Ferro, D. Seigneurin-Berny, J. Garin, M. Block, and J. Joyard(✉)

Abstract The lipid and protein components of the two envelope membranes, which delimit the chloroplast from the surrounding cytosol, have been extensively analyzed. Envelope membranes contain a wide diversity of glycolipids, pigments, and prenylquinones and play a key role in their synthesis, and also in the formation of various lipid-derived signaling molecules (chlorophyll precursors, abscisic acid, and jasmonate precursors, for instance). Many of the enzymes involved were identified by proteomics. Here, we present a curated protein list established from chloroplast envelope proteomes analyzed by different groups. The envelope proteome contains key proteins involved in the regulation of metabolic pathways, in cell signaling (and especially in plastid-to-nucleus signaling), in stress responses, etc. A series of transport systems for proteins, metabolites, and ions have also been identified by proteomics. Chloroplasts have had a long and complex evolutionary past and integration of the envelope membranes in cellular functions is the result of this evolution. The lipid and protein equipment of this plastid-specific membrane system reflect both its prokaryotic and eukaryotic origin.

1 Introduction

Chloroplasts present three major structural regions: (a) a highly organized internal membrane network formed of flat compressed vesicles, the thylakoids, (b) an amorphous background rich in soluble proteins and ribosomes, the stroma and (c) a pair of outer membranes, the chloroplast envelope. The two limiting envelope membranes are actually the only permanent membrane structure of the different types of plastids (proplastids, chloroplasts, chromoplasts, etioplasts…); they are present in every plant cell, with very few exceptions (such as the highly specialized male sexual cells). Plastids are

---

J. Joyard
Laboratoire de Physiologie Cellulaire Végétale, iRTSV
CEA; CNRS; INRA; Université Joseph Fourier,
CEA-Grenoble, 38054 Grenoble-cedex 9, France
e-mail: Jacques.joyard@cea.fr

semi-autonomous organelles, with a wide structural and functional diversity, and unique biochemical pathways. They are able to transcribe and translate the information present in their own genome but are strongly dependent on imported proteins that are encoded in the nuclear genome and translated in the cytoplasm. The chloroplast genome encodes about 80–100 proteins, while between 2,500 and 3,500 nuclear-encoded proteins are predicted to be targeted to the chloroplast. The envelope membranes are essential to this process since (a) the tight co-ordination between the expression of plastidial and nuclear genomes requires plastid-to-nucleus signaling and (b) protein transport across the envelope involves a complex transmembrane machinery.

Chloroplasts are crucial for plant cell metabolism. Performing photosynthesis, they are the site of carbon dioxide reduction and its assimilation into carbohydrates, amino acids, fatty acids, and terpenoid compounds. They are also the site of nitrite and sulfate reduction and their assimilation into amino acids. The envelope membranes, as the interface between plastids and the surrounding cytosol, control the uptake of raw material for all synthesis occurring in the plastids and regulate the export to the cytosol of the newly synthesized molecules. Envelope membranes are therefore essential for the integration of plastid metabolism within the cell.

The participation of chloroplast envelope membranes in most of these particular aspects of chloroplast integration within the plant cell has been analyzed in detail in various reviews to which the reader is referred for more specific information (see Douce 1974; Douce and Joyard 1979, 1990; Bölter and Soll 2001; Jarvis and Soll 2002; Hiltbrunner et al. 2001; Aldridge et al. 2005; Weber et al. 2005; Nott et al. 2006; Maple et al. 2005; Maple and Møller 2006; Dörmann 2007; Jouhet et al. 2007; Block et al. 2007; see also other chapters in this volume and references therein).

Our aim is to focus on our understanding of the lipid and protein composition of chloroplast envelope membranes that were analyzed in detail owing to their purification from spinach or *Arabidopsis* chloroplasts. In particular, we want to link the large body of information recently obtained by a series of proteomic studies to our current knowledge of chloroplast envelope functions, and therefore to provide a clearer picture of the role of this membrane system within the plant cell.

2 Chloroplast Envelope Lipidome

Compared to thylakoids, mitochondrial membranes or endomembranes, envelope membranes are very lipid-rich and thus have a low density. In an extensive envelope lipid analysis (performed from 122 mg of dried lipids extracted from purified spinach chloroplast envelope membranes), Siebertz et al. (1979) determined that lipids accounted for 69% of the membrane dry weight. Further analysis of both envelope membranes demonstrated that the outer envelope membranes were mostly responsible for such a high lipid to protein ratio: they contain about 2.5–3-mg lipids/mg proteins, resulting in a very low density (1.08 g/cm^3, Block et al. 1983). The lipid to protein ratio of the inner membrane is around 1-mg lipids/mg proteins, corresponding to a density of 1.13 g/cm^3, Block et al. 1983).

2.1 Glycerolipids

2.1.1 Galactolipids as Major Envelope Lipid Constituents

Like other plastid membranes, envelope membranes are made of polar neutral lipids containing galactose and called galactolipids (Benson 1964). The galactose residue in monogalactosyldiacylglycerol (MGDG) is bound to the *sn*-3 position of the glycerol backbone with a β-anomeric linkage (βGalD), whereas the head group of digalactosyldiacylglycerol (DGDG) is characterized by a terminal α-galactose moiety (1 → 6) linked to the inner β-galactose residue (αGal(1 → 6) βGalD). An-other glycolipid, the sulfolipid SQDG (1,2-di-O-acyl-3-O-(6'-deoxy-6'-sulfo-α-D-glucopyranosyl)-*sn*-glycerol; sulfoquinovosyldiacylglycerol) is also characteristic of plastid membranes. In addition to the main MGDG (1,2-di-*O*-acyl-3-*O*-β–D-galactopyranosyl-*sn*-glycerol) and DGDG (1,2-di-*O*-acyl- 3-*O*-(6'-*O*-α-galacto-pyranosyl-β-D-galactopyranosyl)-*sn*-glycerol), a series of higher homologues formed by sequential addition of galactose residues to C6 of the terminal galactose of the preceding homologue, have been identified in chloroplast subfractions (see below and Table 1). It is now clear that the higher homologues

Table 1 Fatty acids present in a mixture of total envelope lipids and individual components (Siebertz et al. 1979)

	16:0	16:1	3t-16:1	16:2	16:3	18:0	18:1	18:2	18:3	20:0	
	Mol/100 mol										% of total
Total	14.0	0.4	3.6	Trace	9.3	0.6	3.9	8.0	59.9	20.0	–
MGDG	5.2	0.7	0	Trace	17.3	1.2	2.3	3.2	68.9		13.2
DGDG	13.1	0.8	Trace	0.3	5.1	1.7	5.1	5.9	68.0		21.6
SQDG	44.0	Trace	0	0	1.8	1.2	1.9	5.0	46.1		7.5
PG	17.9	0.7	33.6	0	0.5	1.0	2.0	4.4	39.3	0.9	7.3
PC	20.5	1.3	Trace	Trace	0.5	1.5	13.2	25.9	36.8	0.3	16.5
Unknown	31.6	6.7	0	0	1.2	4.5	10.7	16.5	28.7		1.3
PE	30.7	7.2	0	0	0	4.5	13.3	18.4	25.9		Trace
TAG	5.5	0.6	0.5	Trace	13.7	4.7	9.0	12.6	54.5		Trace
DAG	5.7	0.8	1.0	Trace	15.2	0.9	2.9	3.7	69.9		16.8
TGDG	4.8	2.2	0	0	18.8	0.7	2.3	1.7	69.6		5.2
TeGDG	8.8	2.0	0	0	23.1	3.9	4.4	1.9	55.9		2.1
Acylated DGDG	9.7	0	0	0	17.5	1.3	3.4	2.5	64.3		1.8

Lipids were extracted from purified spinach chloroplast envelope membranes containing 122 mg of dried lipids. Lipid proportions are given as mol/100 mol in the last column. Fatty acids are characterized by a number of carbon atoms and double bonds. 16:1 includes *cis*-7 and *cis*-9 isomers. The unknown phospholipid has chromatographic properties similar to PI. For detailed structures of all plant galactolipids and glycolipids molecular species the reader is referred to Heinz (1996) where they have been described and discussed in detail. TGDG, TeGDG, and acyl DGDG are "artificial lipids" formed during the course of envelope preparation (see Table 2). Abbreviations: TAG, triacylglycerol; DAG, diacylglycerol; MGDG, monogalactosyldiacylglycerol; DGDG, digalactosyldiacylglycerol; TGDG, trigalactosyldiacylglycerol; TeGDG, tetragalactosyldiacylglycerol; SQDG, sulfoquinovosyldiacylglycerol; PG, phosphatidylglycerol; PC, phosphatidylcholine; PE, phosphatidylethanolamine trigalactosyldiacylglycerol (TGDG) and tetragalactosyldiacylglycerol (TeGDG) are due to the functioning of a galactolipid:galactolipid galactosyltransferase (GGGT) during the course of envelope purification. The same is true for acylated DGDG which represents a significant proportion of glycerolipids in envelope preparations (Table 1): it is formed at low pH in leaf homogenates after destruction of cellular compartmentation (Siebertz et al. 1979).

Roughly 20–25% of envelope acyl lipids are phospholipids, mostly phosphatidylcholine (PC) and phosphatidylglycerol (PG) representing 5–15% of the total glycerolipids in these membranes (Tables 2, 3). Again, these phospholipids are not distributed evenly in chloroplast membranes. The outer envelope membrane contains significantly more phospholipids than the other chloroplast membranes, mostly PC, which represents 30–35% of the outer envelope membrane lipids. Using intact chloroplasts, the outer surface of the outer envelope membrane can be probed directly with antibodies, proteases, or lipases. For instance, Billecocq et al. (1972) and Billecocq (1975) have shown, by means of specific antibodies, that galactolipids and sulfolipid are present in the cytosolic leaflet of the outer envelope membrane.

Table 2 Lipid composition of envelope membranes from nontreated, thermolysin-treated, and phospholipase C-treated intact spinach chloroplasts

Lipids	Thermolysin-treated	Nontreated	Phospholipase C-treated
MGDG	38	13.5	16
DGDG	30	33.5	33
TGDG	0	3	4
TeGDG	0	1.5	2
SQDG	7	7.5	7
PG	9	9	7
PC	13.5	15	Trace
PI	2	3	3
PE	0	0	
DAG	>0.1	13.5	26

Data from thermolysin-treated chloroplasts and from phospholipase C-treated chloroplasts are from Dorne et al. (1982) and Dorne et al. (1990), respectively. It was shown that thermolysin as well as phospholipase C do not penetrate the outer envelope membranes (in the experimental conditions used), and could be used to probe protein and lipid components of the outer leaflet of the outer envelope membranes. Note that envelope membranes from thermolysin-treated intact chloroplasts are devoid of TGDG, TeGDG, and DAG, and contain more MGDG than DGDG, the phospholipid content is close to that of the control. In phospholipase C-treated intact chloroplasts, there is no PC but large amounts of DAG, in contrast, the galactolipid content is almost identical to that of the control. The thermolysin experiment indicates that a mild proteolytic digestion of the outer surface of the chloroplast envelope hydrolyzes a galactolipid-metabolizing outer envelope protein: this enzyme (GGGT) converts MGDG into DGDG, TGDG, TeGDG, and DAG. This suggests that the genuine envelope lipid composition is close to that from thermolysin-treated chloroplasts. Phospholipase C-treatment (which hydrolyzes PC and forms DAG) indicates that PC is accessible from the cytosolic surface of the envelope, and therefore concentrated in this leaflet of the outer envelope membrane. Numbers are % of total lipids

Table 3 Distribution of lipid compounds in chloroplasts

	Total envelope membranes	Outer envelope membrane	Inner envelope membrane	Thylakoids
Total polar lipids[a] (mg/mg protein)	1.2–1.5	2.5–3	1	0.6–0.8
Polar lipids (% of total)				
MGDG	32	17	55	57
DGDG	30	29	29	27
SQDG	6	6	5	7
PC	20	32	0	0
PG	9	10	9	7
PI	4	5	1	1
PE	0	0	0	0
Total chlorophylls[b] (mg/mg protein)	0.1–0.3	nd	nd	160
Chlorophylls (% of total in the fraction)				
Chlorophyll a	86	nd	nd	72
Chlorophyll b	14	nd	nd	28
Chlorophyll precursors[b] (Protochlorophyllide + Chlorophyllide; mg/mg protein)	0.41	nd	nd	0–0.35
Total carotenoids[c] (mg/mg protein)	6–12	2.9	7.2	20
Carotenoids (% of total)				
β-Carotene	11	9	12	25
Violaxanthin	48	49	47	22
Lutein + Zeaxanthin	21	16	23	37
Antheraxanthin	6	–	5	–
Neoxanthin	13	26	13	16
Total Prenylquinones[d] (mg/mg protein)	4–11	4–12	4–11	4–7
Prenylquinones (% of total)				
α-Tocopherol + α-Tocoquinone	69	81	67	24
Plastoquinone-9 + Plastoquinol	28	18	32	70
Phylloquinone K1	3	1	1	6

Data are average values from spinach and corrected (for polar lipids) from thermolysin-experiments (see Table 2). Adapted from Block et al. (2007). nd, not determined

By using phospholipase C treatment of isolated intact chloroplasts, Dorne et al. (1985, 1990) have demonstrated that the envelope PC is concentrated in the outer leaflet of the outer envelope membrane and absent from the inner envelope membrane and the thylakoids (see also Table 2). In some preparations, very low amounts of PE (phosphatidylethanolamine) can be found. This likely reflects a contamination by extraplastidial membranes such as plant mitochondria or peroxisomes, which contain only phospholipids: mostly PC and PE.

In plants grown under normal conditions, galactolipids are indeed restricted to plastid membranes. Determination of the glycerolipid composition of isolated plant membranes is therefore a good way to probe their purity but this should be used cautiously. In Pi-deprived plants, DGDG strongly and specifically increases, and is found outside plastids (see below under Sect. 3).

2.1.2 Fatty Acids in Envelope Glycerolipids

In contrast to their conserved head group structures, the fatty acids of glycerolipids exhibit a high variability in chain lengths, degree of unsaturation, and distribution to the *sn*-1 and *sn*-2 positions of the glycerol backbone resulting in a high number of different molecular species originating from complex biosynthetic pathways. This diversity depends on the organism, its taxonomic position, and developmental stage. For instance, there are two main classes of glycerolipids (Heinz 1977; Heinz and Roughan 1983) issued from two specific sources of DAG (diacylglycerol) and notably represented at the level of MGDG. The prokaryotic-type class of glycerolipids contains 16-carbon fatty acids at the *sn*-2 position of glycerol whereas it is only 18-carbon fatty acids in eukaryotic-type glycerolipids (Siebertz et al. 1979; Frentzen et al. 1983). Some plants such as *Arabidopsis* and spinach have both prokaryotic-type and eukaryotic-type MGDG ($MGDG_P$ and $MGDP_E$, respectively), and are called 16:3 plants, whereas other plants such as pea or cucumber have only $MGDG_E$, they are called 18:3 plants.

Actually, the fatty acid distribution established for lipids of whole leaves was also found at the level of envelope membranes, as shown by Siebertz et al. (1979) who analyzed the fatty acid composition, positional distribution and pairing of envelope lipids from spinach (Table 1). In spinach, envelope lipids have a slightly increased level of 16:0 fatty acids as compared to thylakoids. The envelope MGDG contains high proportions of 16:3 and 18:3 fatty acids, thus two main species are found in MGDG: 18:3/16:3 (*sn*-1/*sn*-2) and 18:3/18:3, i.e. $MGDG_P$ and $MGDG_E$. This is also the case for envelope DAG, TGDG, and TeGDG, thus providing further evidence for these lipids to derive from MGDG during the course of envelope purification. In contrast, the envelope DGDG has the lowest level of 16:3 fatty acids, but the highest proportion of 16:0 fatty acids. Therefore, the main species in DGDG are 18:3/18:3, 18:3/16:0, and 16:0/18:3 with small amounts of 18:3/16:3 (that could derive from the action of GGGT on MGDG) and of 16:0/16:0 (E. Heinz, personal communication). The percentage of $DGDG_E$ in leaves is highly plant-specific (25% in spinach but almost 100% in *Vicia*, Rullkötter et al. 1975) and is also increased under phosphate deprivation (see below). SQDG has roughly equal proportions of 16:0 and 18:3 fatty acids. Therefore, the major SQDG species identified in spinach envelopes are 16:0/16:0 and both 18:3/16:0 and 16:0/18:3. However, spinach contains a higher proportion of SQDG with a prokaryotic structure, whereas in 18:3 plants, SQDG is exclusively eukaryotic (Bishop et al. 1985). PG, in envelope membranes as well as in thylakoids, is unique since it contains a $16:1_{trans}$ fatty acid at the *sn*-2 position (Fritz et al. 2007). Thus, the two main PG species identified in spinach envelopes contained 18:3/16:1$_{trans}$ or 18:3/16:0, i.e. prokaryotic-type DAG. This is true in 18:3 as well as in 16:3 plants. PC is actually a typical eukaryotic lipid with 16:0 fatty acids (when present) nearly exclusively at the *sn*-1 position of the glycerol backbone. Indeed, the DAG backbone of envelope PC displays many species of similar proportions (mostly 16:0/18:3 and 18:3/18:3 combinations).

In summary of these analyses of spinach envelope lipids, we may say that specificities of DAG backbone known from lipid mixtures extracted from whole leaves are also found in envelope membranes. The occurrence of two main classes of DAG, of the so-called eukaryotic and prokaryotic types, actually reflects the metabolic integration of the prokaryotic ancestor within the eukaryotic ancestor of the plant cell (see reviews by Douce and Joyard 1979, 1990). But the actual pattern is more complex: the structural diversity of DAG backbones (with 18:3/16:3 combination in MGDG, 18:3/16:0 in DGDG, and 18:3/16:1$_{trans}$ in PG for instance) suggests differences in specificities in handling of acyl groups subsequent to the primary incorporation into the DAG backbone of glycerolipids.

2.2 Other Lipid-Soluble Envelope Constituents

Plant membranes, and especially plastid membranes, contain a wide diversity of compounds deriving from the isoprenoid biosynthetic pathway (Lange and Ghassemian 2003). Carotenoids (C40) and chlorophylls (which contain a C20 isoprenoid side-chain) are pigments essential for photosynthesis, whereas plastoquinone, phylloquinone, and ubiquinone (all of which contain long isoprenoid side-chains) participate in electron transport chains. Many of them have been identified as constituents of chloroplast envelope membranes, which play a key role in their synthesis (Douce and Joyard 1990). Furthermore, chloroplasts contain biosynthetic pathways for phytohormones derived from isoprenoid intermediates, such as gibberellins (C20) and abscisic acid (C15), but despite some evidences (Helliwell et al. 2001), we are still missing a global view of the participation of envelope membranes in the production of signaling terpenoid derivatives.

2.2.1 Pigments

In contrast to thylakoids, envelope membranes from chloroplasts and nongreen plastids are yellow, due to the presence of carotenoids and the absence of chlorophyll (Table 2). Carotenoids represent about 0.2% of the total envelope lipid weight (about 10 µg/mg protein) (Siebertz et al. 1979). Violaxanthin is the major carotenoid whereas thylakoids are richer in β-carotene (Jeffrey et al. 1974). Interestingly, whereas most carotenoids are in their stable *trans* configuration, we observed that 9 -*cis*-neoxanthin preferentially accumulates (compared to *trans*-neoxanthin) in chloroplast envelope membranes. In thylakoids, a transmembrane violaxanthin cycle is organized with de-epoxidation taking place on the lumen side and epoxidation on the stromal side of the membrane (Yamamoto et al. 1999). In the envelope, violaxanthin undergoes a light-induced decrease without a corresponding increase in zeaxanthin: the envelope lacks a violaxanthin cycle and the decrease of violaxanthin parallels the decrease in thylakoids (Siefermann-Harms et al. 1978). Therefore, envelope membranes prepared from leaves kept in the dark have up to 3.5 more violaxanthin than lutein + zeaxanthin, whereas in envelope membranes prepared from illuminated leaves, this ratio is much lower (0.75). An exchange of violaxanthin between the thylakoids and envelope but not of zeaxanthin was thus concluded to occur. Recently, Yamamoto (2006) observed that the relative solubility of violaxanthin and zeaxanthin in MGDG, DGDG, and phospholipids could explain the differential partitioning of violaxanthin between the envelope and the thylakoids and hypothesized that the violaxanthin cycle thus links the thylakoids and the envelope for signal transduction of light stress.

Although devoid of chlorophyll, envelope membranes contain low amounts of chlorophyllide and protochlorophyllide (Pineau et al. 1986, 1993), thus suggesting that part of the chlorophyll biosynthetic pathway is present in envelope membranes. Since the development of photosynthetic membranes is dependent upon the synthesis of chlorophylls and their specific integration into photosynthetic complexes in thylakoids, one can question why the envelope membranes contain chlorophyll precursors. Reinbothe et al. (1995, 2000) suggested that protochlorophyllide could regulate plastid import of pPORA, which would couple pPORA import to synthesis of its substrate. Furthermore, there is some evidence that the synthesis of chlorophyll precursors in envelope membranes is involved in intracellular signaling for the control of chloroplast development. The importance of chloroplast envelope membranes in these processes was confirmed by proteomics (see below).

2.2.2 Quinones

Like thylakoids, chloroplast envelope membranes contain several prenylquinones (Lichtenthaler et al. 1981; Soll et al. 1985): plastoquinone-9, phylloquinone K1, α-tocoquinone and the chromanol, α-tocopherol. However, the relative quinone composition of the envelope differs distinctively from that of thylakoid membranes. The outer envelope membrane contains more α-tocopherol than the inner one, although this prenylquinone is the major one in both membranes. On the contrary, plastoquinone-9, the major thylakoid prenylquinone, is present in higher amounts in the inner envelope membrane than in the outer one. Mutant characterization revealed that tocopherol protects plant lipids against oxidative stress (Havaux et al. 2005). The roles of tocopherol in plants are more complex than previously anticipated: further aspects such as interference with signaling pathways, subcellular/subplastidial localization and interactions with the chlorophyll degradation pathway have to be taken into consideration (reviewed by Dörmann 2007).

2.2.3 Sterols

Plastid membranes contain very few sterols (7 µg/mg protein) compared to extraplastidial membranes. Hartmann-Bouillon and Benveniste (1987) found that the major sterol in envelope membranes was stigmat-7-enol, whereas in the microsomes from the same tissue, it is α-spinasterol, thus suggesting that the presence of sterols in envelope membranes is not caused by contamination by sterol-rich membranes (endoplasmic reticulum or plasma membrane). As mentioned above, Siebertz et al. (1979) identified only traces of sterylglycosides in more than 100-mg lipids from envelope membranes.

3 Functions of Chloroplast Envelope Lipids

Glycerolipids synthesized in the envelope are necessary for thylakoid biogenesis (Kobayashi et al. 2007). Understanding the function of these lipids in plant cells is a key question that can be addressed by following the phenotypes of mutants impaired in genes encoding envelope lipid biosynthetic enzymes and/or by analyzing how plants adapt to fluctuating environmental conditions (see below).

MGD1 is the predominant MGD synthase in leaves (Jarvis et al. 2000; Awai et al. 2001; Kobayashi et al. 2007). In the deletion mutant of *mgd1*, invagination of the inner envelope was visible, suggesting a blockage in membrane trafficking from inner envelope to nascent thylakoids in the absence of MGDG synthesis (Kobayashi et al. 2007). Moreover, the deletion mutants of *mgd1* showed a complete impairment of photosynthetic growth with an arrest at embryo development (Kobayashi et al. 2007). DGDG synthases, DGD1 and DGD2, catalyze transfer of galactose from UDP-galactose to MGDG and are present in the outer envelope (Kelly and Dörmann 2002; Kelly et al. 2003). The *dgd1* mutant is strongly deficient in photosynthesis (Dörmann et al. 1995; Härtel et al. 1997) and protein import (Chen and Li 1998). Two proteins involved in PG synthesis, the PG-phosphate synthases PGP1 and PGP2 (Frentzen 2004; Muller and Frentzen 2001), were found in proteomic analyses of *Arabidopsis* chloroplast envelope membranes (Ferro et al. 2003; see below). The PGP1 protein is essential for chloroplast differentiation and for the biosynthesis of thylakoids PG (Babiychuk et al. 2003). SQDG was demonstrated to interact with an annexin in a Ca^{2+}-dependent manner on the outer surface of chloroplast, suggesting a role of SQDG in the binding of this protein (Seigneurin-Berny et al. 2000). Along with PG, SQDG contributes to maintaining a negatively charged lipid–water interface, which is presumably required for proper function of photosynthetic membranes (Frentzen 2004).

Galactolipids synthesized in the envelope are also important for overall membrane lipid homeostasis of the plant cell. Under Pi starvation, a form of DGDG with a specific fatty acid signature, for example with 16:0 at the *sn*-1 position of glycerol and 18:2 at the *sn*-2 position, corresponding to a eukaryotic-type of DGDG, is particularly abundant in the cell (Härtel et al. 1998, 2000; Klaus et al. 2002).

DGDG can thus form up to 20–25% of the lipid content of mitochondria membranes (Jouhet et al. 2004), and plasma membrane (Andersson et al. 2003) and is also present in the tonoplast (Andersson et al. 2005). It was shown that mitochondrial DGDG is formed in the plastid envelope and then transported to mitochondria (Jouhet et al. 2004).

The particular concentration of PC in the outer leaflet of the outer envelope membrane suggests a role of PC in a connection with extra-plastidic compartments. Studies with labeled lipid precursors indicated that PC synthesized in ER provides its DAG-backbone to chloroplast eukaryotic glycerolipids (Heinz and Harwood 1977; Slack et al. 1977). Although PC phospholipases are not present in chloroplasts, envelope PC can represent an intermediate step in the lipid transfer process.

The formation and trafficking of lipid synthesis intermediates in the chloroplast envelope may coordinate chloroplast development with overall cell development. DAG and PA are both key intermediates of the glycerolipid metabolism and both, and especially PA, are now recognized as major signaling lipids in plants (Testerink and Munnik 2005). Eukaryotic PA is not detected in the envelope under standard conditions. Only a very small pool of prokaryotic PA can be detected in chloroplasts (Fritz et al. 2007). Prokaryotic PA is the last common precursor shared by prokaryotic galactolipids and plastidial PG, which is exclusively prokaryotic. However, a recent report indicated that artificial formation of eukaryotic PA in the envelope can induce conversion of eukaryotic PA into eukaryotic PG which severely reduced plant growth (Fritz et al. 2007). Several envelope proteins recently characterized, TGD1, TGD2, and TGD3 contribute to the transport of PA in the envelope membranes (Xu et al. 2005; Awai et al. 2006; Lu et al. 2007). Lipid metabolism is strongly affected in the three *tgd* mutants, with the largest effects being the reduction of the contents of eukaryotic MGDG and DGDG. It was proposed that TGD proteins can control the flux of eukaryotic PA between the two envelope membranes and, altogether, establish a link between lipid metabolism in ER and thylakoid development.

4 Chloroplast Envelope Proteome

4.1 From Protein Lists to a Snapshot of the Chloroplast Envelope Proteome

Proteomics, by combining the interest of targeted approaches (made possible by the preparation of envelope membranes from highly purified chloroplasts, especially from *Arabidopsis*) together with the availability of an increasing number of genome sequences (see for instance *The Arabidopsis Genome Initiative* 2000), proved to be a formidable tool to identify new proteins and therefore new functions residing in chloroplast envelope membranes. Before, we only had a very limited knowledge of the envelope protein equipment, despite many biochemical and physiological studies that characterized enzymatic and transport activities in purified envelope membranes and chloroplasts (Douce and Joyard 1979, 1990). Table 4 shows our present knowledge of the chloroplast envelope proteome obtained by combining data from Ferro et al. (2002, 2003) and Froehlich et al. (2003). This "curated" list (see legend to Table 4) contains 226 proteins that we expect to be envelope proteins. We suspect that proteins possibly residing in another cell compartment (stroma, thylakoids, mitochondria, nuclei, etc.) represent about 20% of the proteins identified by Ferro et al. (2002, 2003) and about 45% of those identified by Froehlich et al. (2003). This difference is mostly due to differences in sample preparation of envelope membrane fractions prior to proteomic analyses (see below). Key questions are raised by the comparison between the original (published) protein lists and the list in Table 4.

One generally considers that all membrane proteins should be hydrophobic, but the majority of the proteins identified by proteomics in membranes are not. Thus, where do we put the limit between a "true" envelope protein, a protein of the intermembrane space, an envelope-bound protein and a stromal protein, for instance? This is an open question and the answer is not simple. Another question is that of dual localization, although this is likely to concern only a minor proportion of the proteins.

Furthermore, it remains rather difficult to distinguish just by sequence analyses between envelope proteins that are likely residing within the membranes or are just envelope-bound. Proteins expected to be inner envelope membrane intrinsic proteins are metabolite and ionic transporters, components of the protein import machinery, and enzymes. Proteins of the last two categories as well as channel-forming proteins are expected to be located in the outer envelope membrane, most of them being devoid of any predictable transit sequence. Envelope preparations are expected to contain many "envelope-bound proteins": this is probably the case for some proteases and chaperones, for several putative RNA-binding proteins as well as enzymes like carbonic anhydrases and lipoxygenase. Indeed, many enzymes of carbon metabolism that are clearly identified as stroma enzymes are active in the vicinity of the inner face of the inner envelope membrane. It would not be too surprising that this would be the case for several proteins among the hundred identified by Froehlich et al. (2003) that we suspect to be stroma proteins.

4.2 Strategies for Membrane Proteomics Reflect Physico-Chemical Properties of Envelope Proteins

The main differences between published envelope proteomes reflect the strategies used (Fig. 1). Froehlich et al. (2003) used *Arabidopsis* envelope preparation directly, without any fractionation, together with off-line multidimensional protein identification technology (off-line MUDPIT) to analyze the envelope proteome. In contrast, Ferro et al. (2002, 2003) analyzed envelope proteins from spinach and *Arabidopsis* chloroplasts by using a set of complementary methods (Seigneurin-Berny et al. 1999; Ferro et al. 2002, 2003; Ephritikhine et al. 2004) involving

Table 4 The chloroplast envelope proteome

Gene	Description	Ferro et al. (2002, 2003)	Froehlich et al. (2003)	TargetP	cTP	TM	Calc. GRAVY	Calc. M_w (kD)	Calc. PI
Metabolism (lipid)									
At1g01090	PDH (E1) ALPHA E1 alpha subunit, pyruvate DH complex, chloroplast precursor	−	+	C	Y	ND	−0.269	47.17	7.16
At1g30120	PDH (E1) BETA (HP44b) E1 beta subunit, pyruvate DH complex, chloroplast precursor (EC 1.2.4.1)	+	−	C	Y	ND	−0.093	44.24	5.91
At3g25860	LAT2 (E2) Dihydrolipoamide acetyltransferase, pyruvate DH complex, chloroplast precursor	−	+	C	Y	ND	0.045	50.08	8.33
At1g34430	LTA2 (E2, EMB3003) Dihydrolipoamide acetyltransferase, pyruvate DH complex, chloroplast precursor	−	+	C	Y	ND	0.13	48.3	8.8
At4g16155	PTLPD2 (E3) Dihydrolipoamide dehydrogenase 2, pyruvate DH complex, chloroplast precursor	−	+	C	Y	ND	0.026	60.14	7.29
At2g38040	ACCDa (HP88b, CAC3) Acetyl-CoA carboxylase alpha chain, chloroplast precursor	+	+	C	Y	ND	−0.524	85.3	5.61
AtCg00500	ACCDb Acetyl-CoA carboxylase beta chain, chloroplast encoded	+	−	C	N	ND	−0.402	55.61	5.89
At5g35360	CAC2 Acetyl-CoA carboxylase, biotin carboxylase subunit	−	+	C	Y	ND	−0.135	58.38	6.84
At4g25050	ACP4 ACP-like acyl carrier-like protein, chloroplast precursor	+	+	C	Y	ND	−0.013	14.54	4.73
At1g77590	LACS9 (HP76) Long-chain-fatty-acid-CoA ligase family protein, chloroplast precursor	−	+	C	N	ND	−0.064	76.17	6.53
At3g23790	Long-Chain Acyl-CoA Synthetase, chloroplast precursor	−	+	C	Y	ND	−0.229	81.14	8.73
At4g14070	HP81 (AAE15) Long-Chain Acyl-CoA synthetase, chloroplast precursor	+	+	C	Y	ND	−0.219	81.46	8.89
At3g11170	FAD3C (FAD7, FADD) omega-3 fatty acid desaturase, chloroplast precursor	+	+	C	Y	4	−0.299	51.17	8.14
At4g30950	FAD6C (FADC) Omega-6 fatty acid desaturase, chloroplast precursor	+	+	C	Y	3	−0.07	51.22	9.01

AGI	Name								
At4g30580	LPAT1 (LPAAT, ACT2) Lysophosphatidic 1-acylglycerol-Phosphate acyltransferase, chloroplast precursor	+	–	C	N	3	0.029	39.39	9.87
At4g31780	MGD1 (MGDA) monogalactosyldiacylglycerol synthase, chloroplast precursor	+	+	C	Y	ND	-0.157	58.53	9.29
At5g01220	SQD2 sulfolipid synthase/UDP-sulfoquinovose:DAG sulfoquinovosyltransferase, chloroplast precursor	–	+	C	Y	1	-0.108	56.63	8.61
At3g60620	CDP-Diacylglycerol synthetase, chloroplast precursor	–	+	C	N	7	0.336	43.25	9.3
At2g39290	PGPS1 (HP32c, PGP1) phosphatidylglycerolphosphate synthase, dual targeted	+	–	C	Y	2	0.224	32.15	10.3
At3g55030	PGPS2 (HP25b, PGP) phosphatidylglycerolphosphate synthase	+	–	–	Y	4	0.458	25.22	9.61
At3g10840	Lysophospholipase (EC 3.1.1.5), hydrolase, alpha/beta fold family protein	–	+	C	N	ND	-0.053	50.88	8.59
At1g52570	PLDA2 (PLD2) phospholipase D alpha 2 (EC 3.1.4.4)	–	+	–	N	ND	-0.408	91.59	5.82
At1g33810	GDSL-lipase 1, GDSL-motif lipase, hydrolase family protein	–	+	–	N	ND	-0.781	15.69	8.82
At3g14210	GDSL-like Lipase, Acylhydrolase like protein (ESM1)	–	+	S	N	ND	-0.145	44.06	7.58
At1g07420	SMO2 sterol 4-alpha-methyl-oxidase 2	–	+	S	N	4	0.098	30.92	8.63
At3g25760	AOC1 (ERD12-1) Allene oxide cyclase 1, chloroplast precursor	–	+	C	Y	ND	-0.265	27.8	9.11
At3g25770	AOC2 (ERD12-2) Allene oxide cyclase 2, chloroplast precursor	–	+	C	Y	ND	-0.231	27.63	6.9
At5g42650	CP74A (AOS) Allene oxide synthase, chloroplast precursor	+	+	C	Y	ND	-0.231	58.19	8.75
At4g15440	HPL-like hydroperoxide lyase-like protein	–	+	–	N	ND	-0.077	42.66	5.64
At5g16010	3-Oxo-5-alpha-steroid 4-dehydrogenase (steroid 5-alpha-reductase) family protein	–	+	–	N	6	0.446	30.12	9.37
At3g45140	LOX2 Lipoxygenase, chloroplast precursor (1.13.11.12)	–	+	C	Y	ND	-0.465	102.04	5.42
Metabolism (vitamins and pigments)									
At4g14210	CRTI (PDS3) Phytoene dehydrogenase, chloroplast precursor (EC 1.14.99.-)	–	+	C	Y	ND	-0.135	62.96	6.07
At3g53130	LUT1 Carotenoid epsilon-ring hydroxylase, chloroplast precursor, Cytochrome P450-like	–	+	C	Y	ND	-0.181	60.55	6.04

(continued)

Table 4 (continued)

Gene	Description	Ferro et al. (2002, 2003)	Froehlich et al. (2003)	TargetP	cTP	TM	Calc. GRAVY	Calc. M_w (kD)	Calc. PI
At3g04870	ZDS1 (ZCD) Zeta-carotene desaturase, chloroplast precursor (EC 1.14.99.30)	–	+	C	Y	ND	–0.109	61.63	7.03
At5g67030	ZEP (LOS6, ABA1) Zeaxanthin epoxidase, chloroplast precursor	–	+	C	Y	ND	–0.295	73.84	6.6
At3g09580	HP52b (PDS-like) weak similarity to phytoene dehydrogenase	+	–	C	Y	ND	0.005	52.25	7.02
At1g67080	ABA4 (HP23 conserved protein), Neoxanthin synthase, similar to Sll0354 protein (*Synechocystis* sp.)	+	+	C	N	4	0.316	24.62	9.44
At3g14110	FLU protein, tetratricopeptide repeat (TPR)-containing protein	–	+	C	N	1	–0.311	34.58	9.09
At5g18660	DVR [3,8-DV]-Chlide a 8-vinyl reductase	–	+	C	Y	ND	–0.101	45.89	7.47
At3g56940	CHL27 (CRD1, AT103) magnesium-protoporphyrin IX monomethylester [oxidative] cyclase	–	+	C	Y	ND	–0.348	47.63	8.55
At5g14220	PPOII (PPOX, HEMG2/MEE61) protoporphyrinogen IX oxidase	–	+	–	N	ND	–0.303	55.63	8.35
At4g01690	PPOC (PPOX) protoporphyrinogen oxidase, chloroplast precursor (EC 1.3.3.4)	–	+	M	Y	1	–0.131	57.69	9.13
At4g25080	CHLM Magnesium-protoporphyrin IX methyltransferase, chloroplast precursor (EC 2.1.1.1)	–	+	C	Y	ND	–0.055	33.79	7.68
At4g27440	PORB Protochlorophyllide reductase B, chloroplast precursor	+	+	C	Y	ND	–0.198	43.36	9.22
At1g03630	PORC Protochlorophyllide reductase C, chloroplast precursor	–	+	C	Y	ND	–0.351	43.88	9.18
At3g11950	AtHST (HP43) homogentisate solanesyl-geranylgeranyl-farnesyl-transferase, chloroplast precursor	+	–	–	Y	7	–0.155	106.65	5.15
At3g63410	VTE3 (APG1, IEP37) SAM-dependent methyl transferase involved in ubiquinone, menaquinone biosynthesis	+	+	C	Y	1	–0.268	37.92	9.18

AGI	Name								
Metabolism (others)									
At3g60750	TKL-1 (TKTC) transketolase-1, chloroplast precursor	+	–	C	Y	2	–0.238	79.96	5.93
At3g01500	CAHC (CA1) beta carbonic anhydrase 1, chloroplast precursor (EC 4.2.1.1)	+	+	–	N	ND	–0.025	29.5	5.53
At5g14740	CA2 (CAH2) beta carbonic anhydrase 2, chloroplast precursor (EC 4.2.1.1)	–	+	–	Y	ND	–0.073	28.34	5.36
At1g74640	Conserved plant and cyanobacterial, hydrolase, alpha/beta fold family protein, similar to Slr1235 protein (*Synechocystis* sp.)	–	+	C	Y	ND	–0.221	41.05	8.3
At5g38520	Hydrolase, alpha/beta fold family protein, low similarity to hydrolase	–	+	C	Y	ND	0	39.57	6.54
At3g06510	HP70 (SFR2) glycosyl hydrolase family 1 protein, similar to beta-galactosidase	+	+	S	N	1	–0.258	70.77	8.79
At5g35170	AK adenylate kinase, chloroplast precursor (EC 2.7.4.3)	–	+	C	Y	ND	–0.307	65.73	8.84
At1g10510	HP73 (EMB2004) similar to ribonuclease inhibitor, contains Leucine Rich Repeat domains	+	+	C	Y	1	–0.134	64.72	7.18
At3g63170	Weak similarity to Chalcone isomerase	–	+	C	Y	ND	–0.118	30.39	6.97
At5g39410	SCPDH Probable mitochondrial saccharopine dehydrogenase family (EC 1.5.1.9)	–	+	–	Y	ND	–0.178	49.68	8.53
Plastid division and positioning									
At5g42480	ARC6 (DNAJ, FTN2) plastid division protein, chloroplast precursor	–	+	C	Y	2	–0.157	88.26	4.76
At2g21280	GC1 plastid division protein (Giant Chloroplast 1), chloroplast precursor	–	+	C	Y	ND	0.022	37.74	9.31
At5g10470	TH65 kinesin motor protein-related TH65 protein	–	+	C	Y	ND	–0.403	141.03	5.86
At1g07120	CHUP1-like Chloroplast Unusual Positioning, essential for Proper Chloroplast Positioning	–	+	–	N	ND	–0.891	44.93	9.28
At1g03780	Xklp2-like, similar to microtubule-associated protein	–	+	–	N	ND	–0.846	78.65	9.22
Protein degradation and regeneration (protease and chaperone)									
At5g02500	HSP70-1 (HSC71) Heat shock cognate 70-kDa protein 1	–	+	–	N	ND	–0.436	71.35	5.02
At1g08640	DNAJ domain containing protein	–	+	C	Y	3	–0.072	32.9	9.82
At2g30950	FtsH2 (VAR2) protease, chloroplast precursor	–	+	C	Y	1	–0.141	74.15	5.99

(continued)

Table 4 (continued)

Gene	Description	Ferro et al. (2002, 2003)	Froehlich et al. (2003)	TargetP	cTP	TM	Calc. GRAVY	Calc. M_w (kD)	Calc. PI
At3g16290	FtsH protease family protein	–	+	C	N	2	-0.484	99.86	9.27
At3g47060	FtsH7 protease, chloroplast precursor	+	–	C	Y	2	-0.257	87.8	8.3
At5g42270	FtsH5 (FtsH2, VAR1) cell division protease, chloroplast precursor (EC 3.4.24.-)	–	+	C	Y	ND	-0.116	75.23	5.36
At5g64580	AAA-type ATPase family protein, similar to zinc dependent protease	–	+	C	Y	1	-0.359	96.85	5.66
At3g04340	FtsH (EMB2458) protease family protein	–	+	–	–	ND	-0.325	111.01	8.01
At5g35210	SREBP-like peptidase M50 family protein	–	+	–	N	4	-0.289	174.28	6.48
At4g25370	ClpS1 Clp amino terminal domain-containing protein, chloroplast precursor	–	+	C	Y	ND	-0.273	26.05	9.24
At4g24280	cpHSP70-1 (DnaK homologue) heat shock protein 70-1, chloroplast precursor	–	+	C	Y	ND	-0.32	76.5	5.06
At5g49910	cpHSP70-2 (Dnak homologue) heat shock protein 70-7, chloroplast precursor	–	+	C	Y	ND	-0.358	76.99	5.17
At5g50920	ClpC1 (Hsp100) ATP-dependent Clp protease ATP-binding subunit, chloroplast precursor	+	+	C	Y	ND	-0.394	103.45	6.36
At3g48870	ClpC2 (Hsp93-III) ATP-dependent Clp protease ATP-binding subunit	–	+	C	N	ND	-0.313	105.77	6.06
At1g66670	ClpP3 (nClpP3) ATP-dependent Clp protease proteolytic subunit, chloroplast precursor	–	+	C	Y	ND	-0.313	33.92	7.59
At5g45390	ClpP4 (nClpP4) ATP-dependent Clp protease proteolytic subunit, chloroplast precursor	–	+	C	Y	ND	-0.043	31.49	5.37
At1g02560	ClpP5 (nClpP1) ATP-dependent Clp protease proteolytic subunit, chloroplast precursor	+	+	C	Y	ND	-0.197	32.35	8.34
At1g11750	ClpP6 (nClpP6) ATP-dependent Clp protease proteolytic subunit, chloroplast precursor	–	+	C	Y	ND	-0.128	29.38	9.37
At1g12410	ClpR2 (nClpP2) ATP-dependent Clp protease proteolytic subunit, chloroplast precursor	+	+	C	Y	ND	-0.428	31.2	9.19

AGI	Description								
At1g09130	ClpR3 (nClpP8) ATP-dependent Clp protease proteolytic subunit, chloroplast precursor	–	+	C	Y	2	−0.145	36.3	8.63
At4g17040	ClpR4 (nClpP9) ATP-dependent Clp protease proteolytic subunit, chloroplast precursor	–	+	C	Y	ND	−0.207	33.44	9.33
At4g12060	ClpS2 Clp amino terminal domain-containing protein	–	+	C	Y	ND	−0.166	26.56	8.85
At3g13470	Cpn60-beta-1 chaperonin, similar to RuBisCO subunit binding-protein beta subunit, chloroplast precursor	–	+	C	Y	ND	−0.056	63.34	5.59
At5g20720	CH10C (Cpn21, Cpn20, CH1C) chloroplast Cpn21 chaperonin	+	+	C	Y	ND	−0.082	26.8	8.85
At3g62030	CP20C (ROC4) peptidyl-prolyl cis-trans isomerase, chloroplast precursor	–	+	C	Y	ND	−0.169	28.2	8.83
Protein modification										
At1g02980	CUL2 cullin family protein, similar to cullin 1 from Homo sapiens	–	+	–	N	ND	−0.402	85.96	7.26
At1g76370	APK1A-like putative protein kinase, similar to protein kinase APK1A	–	+	–	N	ND	−0.284	42.31	9.25
Protein targeting										
At5g28750	Tha4 mttA/Hcf106 family, similar to thylakoid assembly 4 protein, chloroplast precursor	–	+	C	Y	1	−0.26	15.71	9.22
At4g03320	Tic20-IV subunit of the translocon at the inner envelope membrane, chloroplast precursor	–	+	C	Y	4	0.155	32.52	9.75
At4g33350	Tic22 (HP27) subunit of the translocon at the inner envelope membrane, chloroplast precursor	+	+	C	Y	ND	−0.36	30.1	9.31
At5g16620	Tic40 (PDE120) translocon Tic40-like protein, chloroplast precursor	+	+	C	Y	1	−0.55	48.9	5.37
At2g24820	Tic55 subunit of the translocon at the inner envelope membrane, Rieske (2Fe-2S) domain-containing protein, chloroplast precursor	+	+	C	Y	2	−0.281	60.6	8.94
At1g06950	Tic110 (HP112) subunit of the translocon at the inner envelope membrane, chloroplast precursor	+	+	C	Y	2	−0.3	112.12	5.72
At2g47840	Tic20-like (IEP16) low similarity to Tic20, chloroplast precursor	+	+	C	Y	4	0.461	22.91	10.28

(continued)

Table 4 (continued)

Gene	Description	Ferro et al. (2002, 2003)	Froehlich et al. (2003)	TargetP	cTP	TM	Calc. GRAVY	Calc. M_W (kD)	Calc. PI
At4g25650	Tic55-like (HP62, ACD1-Like) Rieske (2Fe-2S) protein domain-containing protein, chloroplast precursor	+	–	C	Y	2	–0.34	61.26	8.9
At2g34460	Ycf39-like (HP26c) (Tic62-like) NAD-dependent epimerase, dehydratase family protein	+	+	C	N	ND	–0.053	30.47	9.03
At3g46780	Ycf39-like (Tic62-like) NAD-dependent epimerase, dehydratase family protein	–	+	C	Y	ND	–0.349	54.35	8.9
At2g15290	Tic21 (Cia5) or PIC1 translocon at the inner envelope membrane or Iron transporter, chloroplast precursor	–	+	C	Y	4	0.315	31.27	10.23
At2g28900	OEP16-1 (HP15) mitochondrial import inner membrane translocase subunit Tim17/Tim22/Tim23 family protein	+	+	–	N	ND	0.12	15.48	9.16
At5g55510	OEP16-like (HP22) mitochondrial import inner membrane translocase subunit Tim17/Tim22/Tim23 family protein	+	–	–	N	ND	0.04	22.52	9.08
At3g49560	OEP16-like (HP30) mitochondrial import inner membrane translocase subunit Tim17/Tim22/Tim23 family protein, dual targeted ?	+	+	–	N	3	–0.124	27.98	9.49
At5g24650	OEP16-like (HP30-2) mitochondrial import inner membrane translocase subunit Tim17/Tim22/Tim23 family protein, dual targeted	+	+	–	N	3	–0.125	27.77	9.6
At4g26670	OEP16-like mitochondrial import inner membrane translocase subunit Tim17/Tim22/Tim23 family protein	+	–	–	N	2	0.136	21.81	7.62
At1g02280	Toc33 (HP32b) GTP-binding protein, chloroplast outer envelope translocon subunit (EC 3.6.5.-)	+	+	–	N	ND	–0.139	32.92	9.1
At5g05000	Toc34 (OEP34) GTP-binding protein, chloroplast outer envelope translocon subunit (EC 3.6.5.-)	+	+	–	N	ND	–0.151	34.7	9.42
At3g17970	Toc64-III (HP64b) chloroplast outer envelope translocon subunit	+	+	S	N	ND	–0.137	64.02	8.61

AGI	Description								
At3g46740	Toc75-III (OEP75-3) chloroplast outer envelope import-associated channel, chloroplast precursor	+	+	C	Y	ND	-0.365	89.19	8.93
At5g19620	Toc75-V (OEP85, OEP80, IAP75) chloroplast outer envelope import-associated channel, chloroplast precursor	–	+	C	Y	ND	-0.365	79.93	8.41
At4g02510	Toc159 (OEP86) chloroplast outer envelope protein import component	+	+	–	N	ND	-0.479	160.82	4.42
Redox									
At3g26060	PrxQ peroxiredoxin Q, chloroplast precursor	–	+	C	Y	ND	-0.402	23.67	9.53
At3g59840	P1-like NADP-dependent oxidoreductase P1-like protein (EC 1.3.1.74)	–	+	C	N	ND	-0.31	10.5	4.64
At4g13010	ceQORH (IE41) quinone oxidoreductase, uncleaved chloroplast precursor	+	–	–	N	ND	0.034	34.43	9.04
At1g06690	HP52 aldo/keto reductase family protein	+	+	C	Y	ND	-0.275	41.49	8.85
At2g27680	Conserved protein of the aldo/keto reductase family	–	+	C	Y	ND	-0.292	43.15	8.2
At1g71500	Rieske (2Fe-2S) domain-containing protein	–	+	C	Y	ND	-0.226	31.72	8.88
At3g11630	PrxB (BAS1) 2-cys peroxiredoxin, chloroplast precursor	–	+	C	Y	ND	-0.112	29.09	6.91
At4g35460	TRXB1 (NTR1, NRTB) NADPH-dependent thioredoxin reductase 1 (EC 1.8.1.9)	–	+	C	N	ND	-0.038	39.62	6.95
RNA binding									
At2g43630	Conserved glycine-rich plant protein	–	+	C	Y	1	-0.597	30.7	8.29
At5g22640	MORN repeat-containing, glutamic acid-rich protein (EMB1211) restricted to Arabidopsis, no similarity to other protein (even plant proteins)	–	+	–	N	ND	-0.882	99.95	4.36
At4g13130	CHP-rich zinc finger protein, DC1 domain-containing protein	–	+	–	N	ND	-0.341	89.62	6.02
At1g09340	Rap38 (CSP41B) putative RNA-binding protein	–	+	–	N	ND	-0.364	42.62	8.18
At3g63140	Rap41 (CSP41A) mRNA binding protein, chloroplast precursor	–	+	C	Y	ND	-0.17	43.93	8.54
At2g37220	ROC1 (CP29B) putative RNA-binding protein, chloroplast precursor	–	+	C	Y	ND	-0.323	30.71	5.05
At1g55480	TPR repeat containing protein, similarity to tyrosine phosphatase	–	+	C	Y	ND	-0.709	37.41	8.18

(continued)

Table 4 (continued)

Gene	Description	Ferro et al. (2002, 2003)	Froehlich et al. (2003)	TargetP	cTP	TM	Calc. GRAVY	Calc. M_W (kD)	Calc. PI
Signaling (other than lipid signaling)									
At3g51140	HP28 (AtCDF1-like) cell growth defect factor, domain HSP DnaJ, similar to Slr1918 protein (*Synechocystis* sp.)	+	+	C	Y	4	−0.016	31.53	10.66
At5g23040	HP28-like (AtCDF1) cell growth defect factor, domain HSP DnaJ, similar to Slr1918 protein (*Synechocystis* sp.)	−	+	C	Y	3	0.163	28.81	9.68
At5g64430	Octicosapeptide/Phox/Bem1p (PB1) domain-containing protein	−	+	−	N	ND	−0.732	56.44	5.6
Stress (biotic and abiotic)									
At3g18890	UOS1-like similar to UV-B and ozone regulated protein 1	−	+	C	Y	ND	−0.37	68.34	8.27
At3g12570	SLT1-like protein, sodium–lithium tolerant protein 1, contains domain HSP20-like chaperone	−	+	−	N	ND	−0.537	55.27	6.04
At1g30360	ERD4 early-responsive to dehydration stress protein 4	−	+	S	N	10	0.298	81.93	9.28
At2g43940	TMT2-like similar to thiol methyltransferase 2	−	+	−	N	ND	−0.25	25.04	5.66
Stress (oxidative)									
At2g25080	GPX2 (GPX1, PHGPx) phospholipid hydroperoxide glutathione peroxidase 1, chloroplast precursor	+	+	C	Y	ND	−0.173	26.01	9.41
At1g11840	GLX1 glyoxalase I-2, lactoylglutathione lyase-like, putative	−	+	−	N	ND	−0.336	31.92	5.19
At4g35000	APX3 L-ascorbate peroxidase 3, dual targeting to chloroplast and mitochondria	+	+	−	N	1	−0.365	31.57	6.46
Translation (cytosolic)									
At1g06380	RSL1D1-like Ribosomal L1 domain-containing protein	−	+	C	−	ND	−0.255	28.71	9.48
Transporters									
At2g42770	HP25 PMP22-like peroxisomal membrane 22-kDa family protein	+	+	−	Y	2	−0.144	25.91	9.81

AGI	Description							
At5g19750	HP25-like (HP30c) PMP22-like peroxisomal membrane 22-kDa family protein	+	–	C	Y	3	0.202	30.36	10.23
At5g62720	IEP18 integral membrane HPP family protein, weak similarity to various transporter families	+	+	C	Y	5	0.52	25.82	10.14
At5g52540	HP47 conserved membrane protein family, weak similarity to various transporter families	+	–	C	Y	9	0.757	47.64	9.96
At1g32080	HP45 LrgB-like family protein, conserved membrane protein	+	+	C	Y	12	0.701	54.01	9.63
At3g60590	HP36b weak similarity to branched-chain amino acid transport system II carrier protein	+	–	M	Y	5	0.654	26.27	9.6
At5g58270	ABC (STA1, STARIK 1) mitochondrial half-ABC transporter	–	+	M	Y	6	–0.017	80.42	9.27
At5g03910	ABC (ATATH12) ABC-2 homologe 12, similar to sll1276 protein (Synechocystis sp.)	–	+	C	Y	6	0.23	69.19	9.05
At4g25450	NAP8 (HP77) Nonintrinsic ABC protein 8, chloroplast precursor	+	+	M	Y	5	0.205	77.92	9.12
At4g33460	NAP13 (EMB2751) Nonintrinsic ABC protein 13, chloroplast precursor	–	+	C	Y	ND	–0.046	29.62	8.68
At1g80300	TLC1 (AATP1, AtNTT1) chloroplast ADP, ATP carrier protein 1, chloroplast precursor	–	+	C	Y	11	0.368	68.13	9.39
At1g15500	TLC2 (AATP2, AtNTT2) chloroplast ADP, ATP carrier protein 2, chloroplast precursor	+	–	M	Y	11	0.412	67.53	9.51
At4g37270	HMA1 (AHM1) metal-transporting P-type ATPase, chloroplast precursor	+	–	M	Y	6	0.119	88.19	8.05
At5g59250	HP59 sugar transporter family	+	+	C	Y	11	0.549	59.83	9.02
At5g16150	IEP62 (GLT1/PGLCT) putative glucose transporter, chloroplast precursor	+	+	C	Y	12	0.592	56.97	9.06
At4g00290	MscS (MsL) mechanosensitive ion channel domain-containing protein	–	+	M	Y	5	0.092	53.88	9.06
At1g48460	Weak similarity to MscS Mechanosensitive ion channel	–	+	C	Y	5	0.313	38.07	9.68
At5g43745	HP88-like ion channel DMI1-like, castor/pollux family	–	+	C	Y	2	–0.086	92.21	9.14
At5g02940	HP88 ion channel DMI1-like, castor/pollux family	+	+	M	N	3	–0.172	92.13	7.53
At1g01790	KEA1 (HP64) K+ efflux antiporter, putative	+	+	–	N	13	0.665	64.98	6.1

(continued)

Table 4 (continued)

Gene	Description	Ferro et al. (2002, 2003)	Froehlich et al. (2003)	TargetP	cTP	TM	Calc. GRAVY	Calc. M_w (kD)	Calc. PI
At4g00630	KEA2 (HP64-like) K+ efflux antiporter, putative	–	+	M	N	12	0.628	66.41	7.07
At3g20320	TGD2 (HP41b) phosphatidic acid-binding protein, chloroplast precursor	+	+	C	Y	1	–0.04	41.63	8.9
At5g17520	MEX1 (RCP1) maltose transporter (root cap 1), chloroplast precursor	+	–	C	Y	9	0.409	45.28	9.37
At5g01500	MCF (HP45b) mitochondrial substrate carrier family protein	+	–	C	Y	3	–0.079	45.09	9.81
At2g35800	MCF (HP90) mitochondrial substrate carrier family protein	+	–	–	N	2	–0.125	90.62	8.95
At3g51870	MCF (HP42) mitochondrial substrate carrier family protein, similar to peroxisomal Ca-dependent solute carrier	+	–	C	Y	2	0.002	41.81	9.75
At5g22830	AtMRS2-11 (GMN10) magnesium transporter 10, CorA-like family protein	–	+	C	Y	2	–0.182	51.09	5.23
At2g26900	IEP36 bile acid:sodium symporter family protein	+	–	C	N	9	0.641	43.6	8.95
At3g56160	IEP36-like weak similarity to bile acid:sodium symporter family	–	+	C	Y	7	0.57	46.53	9.91
At5g64290	DiT2-1 glutamate/malate translocator, chloroplast precursor	+	+	C	Y	10	0.607	59.99	9.33
At5g12860	DiT1 (IEP45) 2-oxoglutarate/malate translocator, chloroplast precursor	+	+	C	Y	14	0.775	59.21	9.75
At3g26570	PHt2-1 (IEP60) phosphate transporter, chloroplast precursor	+	+	–	N	12	0.47	64.78	9.26
At5g33320	PPT (IEP33, CUE1) phosphate/phosphoenolpyruvate translocator, chloroplast precursor	+	–	C	Y	6	0.434	44.22	10.16
At5g46110	TPT (IEP30, APE2) phosphate/triose-phosphate translocator, chloroplast precursor	+	+	M	Y	8	0.549	44.63	9.75

Gene ID	Description								
At4g39460	MCF SAMC1/SAMT1 (HP35) S-Adenosylmethionine transporter, dual targeted (chloroplast, mitochondria)	+	–	C	Y	5	0.1	34.86	9.68
At1g76405	OEP21-1 ATP-regulated anion-selective solute channel	+	+	–	N	ND	−0.769	19.45	9.49
At3g52230	OEP24 (OMP24) chloroplast outer envelope high-conductance solute channel	+	+	–	N	ND	−0.99	16.12	4.51
At5g42960	OEP24-II chloroplast outer envelope high-conductance solute channel	–	+	–	N	ND	−0.271	23.41	9.26
At2g43950	OEP37 (HP44) chloroplast outer envelope ion channel	+	+	C	Y	ND	−0.538	38.83	9.16
Vesicle formation and trafficking									
At2g20890	THF1 (TF1, PSB29) thylakoid formation1 protein, chloroplast precursor	–	+	C	Y	ND	−0.387	33.79	9.2
At1g65260	Vipp1 (IM30, PspA, HCF155, PTAC4) vesicle-inducing protein in plastids 1	+	+	C	Y	ND	−0.561	36.39	9.17
At5g17670	PGAP1-like conserved plant and cyanobacteria protein, lipase protein family	–	+	M	Y	ND	−0.131	33.4	5.57
At2g40060	Weak similarity to Clathrin subfamily proteins	–	+	–	N	ND	−0.916	28.83	4.9
At3g22520	CAP1-like protein, CDPK adapter-like protein	–	+	–	N	ND	−0.811	67.57	5.29
Unknown									
At5g03900	Conserved plant and cyanobacterial protein, similar to Slr1603 protein (*Synechocystis* sp.), contains HesB/YadR/YfhF (Iron sulfur assembly protein IscA) domain	–	+	C	Y	ND	−0.34	59.5	9.19
At4g31530	Conserved plant and cyanobacterial protein, similar to Sll0096 protein (*Synechocystis* sp.), contains NAD-dependent epimerase/dehydratase domain	–	+	C	Y	ND	−0.173	35.22	7.67
At3g11560	Conserved plant protein, contains InterPro and PFAM domain LETM1-like protein	–	+	C	Y	ND	−0.373	97.79	6.46
At2g36835	Conserved plant protein, no similarity to characterized protein	–	+	C	N	ND	−0.053	13.38	7.84
At1g42960	HP17 conserved protein, similar to Ssl2009 protein (*Synechocystis* sp.), no similarity to characterized protein	+	+	C	Y	1	−0.188	17.82	8.54

(continued)

Table 4 (continued)

Gene	Description	Ferro et al. (2002, 2003)	Froehlich et al. (2003)	TargetP	cTP	TM	Calc. GRAVY	Calc. M_w (kD)	Calc. PI
At5g16660	HP17-like conserved protein, similar to Ssl2009 protein (*Synechocystis* sp.), no similarity to characterized protein	–	+	C	Y	1	–0.479	18.17	8.51
At4g27990	HP24 Ycf19-like YGGT family protein, conserved plant and cyanobacterial protein (chloroplast encoded in algae), similarity to ssr2142 protein (*Synechocystis* sp.)	+	–	C	Y	3	0.393	23.7	11.28
At3g07430	HP24-like Ycf19-like YGGT family protein, conserved plant and cyanobacterial protein, similar to Ycf19 protein (*Synechocystis* sp.)	–	+	C	Y	ND	0.362	24.92	10.66
At3g57280	HP26b conserved 14c-like transmembrane protein	+	+	C	Y	4	0.108	24.34	9.17
At2g38550	HP26b-like (HP36c) conserved 14c-like transmembrane protein	+	+	C	Y	4	–0.155	36.7	6.99
At3g43520	HP26b-like conserved 14c-like transmembrane protein	–	+	C	Y	4	0.093	24.76	9.13
At3g32930	HP27b conserved plant protein, no similarity to characterized protein	+	–	C	Y	ND	–0.361	27.43	9.6
At4g13590	HP28b conserved protein and cyanobacteria protein, similar to sll0615 protein (*Synechocystis* sp.), no similarity to characterized protein	+	–	C	Y	6	0.439	37.93	8.82
At3g61870	HP29b conserved plant and cyanobacteria protein, similar to Slr1676 protein (*Synechocystis* sp.), no similarity to characterized protein	+	+	C	Y	4	0.093	29.58	9.34
At5g13720	HP29c conserved plant protein, contains InterPro domain IPR000218 Ribosomal protein L14b/L23e	+	–	C	Y	3	0.324	28.91	9.02
At1g78620	HP34 conserved plant and cyanobacteria membrane protein, similar to protein sll0875 (*Synechocystis* sp.)	+	–	C	Y	6	0.414	34.87	9.83
At3g08640	HP35b (LCD1-like) conserved plant protein, no similarity to characterized protein	+	–	C	Y	3	0.099	35.25	9

AGI	Description								
At5g12470	HP35b-like (HP40, LCD1-like) conserved plant protein, no similarity to characterized protein	+	+	C	Y	4	0.103	41.32	7.01
At5g22790	HP35b-like (HP46, LCD1-like) conserved plant protein, no similarity to characterized protein	+	–	C	Y	3	–0.127	46.83	5.14
At5g24690	HP35b-like (HP56b, LCD1-like) conserved plant protein, no similarity to characterized protein	+	+	C	Y	3	–0.133	56.81	9.14
At1g20830	HP40b conserved plant protein, no similarity to characterized protein	+	–	C	Y	1	–0.622	39.41	9.09
At2g44640	HP50 (PDE320-like) conserved plant protein, no similarity to characterized protein	+	+	C	N	ND	–0.236	49.83	8.8
At5g08540	HP53 conserved plant protein, no similarity to characterized protein	+	+	C	N	1	–0.404	38.65	5.67
At5g01590	HP65 expressed protein, no similarity to characterized protein	+	+	C	Y	ND	–0.731	48.87	6.38
At5g23890	HP103 conserved protein, S-layer domain-containing protein, similar to slr2000 protein (*Synechocystis* sp.)	+	+	C	Y	1	–0.518	103.92	4.61
At4g13670	PTAC5 DnaJ domain-containing protein, similar to Alr2745 protein (*Anabaena* sp.)	–	+	C	Y	ND	–0.595	44.02	4.84
At1g16790	Conserved plant protein, no similarity to characterized protein	–	+	–	N	ND	–0.41	15.45	9.55
At3g18420	HP35c tetratricopeptide repeat (TPR) containing protein	+	–	C	Y	ND	–0.325	35.63	5.24
At2g37400	TPR repeat containing protein, weak similarity with the Slr1644 protein (*Synechocystis* sp.)	–	+	C	Y	ND	–0.463	38.15	6.7
At2g01220	Conserved protein, no similarity to characterized protein	–	+	–	Y	ND	–0.037	42.14	6.27
At3g63160	OEP6 chloroplast outer envelope membrane protein, no similarity to characterized protein	+	+	–	N	1	–0.157	7.25	9.04
At2g34585	OM14 similar to Outer envelope membrane protein (*Pisum sativum*)	+	–	S	–	ND	–0.13	8.57	4.03
At2g11910	Aspolin1-like protein, aspartic acid-rich protein,	–	+	–	N	1	–1.407	18.49	3.59
At1g19100	Conserved eukaryote protein, ATPase-like domain	–	+	–	N	ND	–0.426	74.17	7.7
At3g22620	Conserved plant protein, weak similarity with nonspecific lipid-transfer protein (LTP) family protein	+	–	S	Y	2	0.151	20.77	6.09

(continued)

Table 4 (continued)

Gene	Description	Ferro et al. (2002, 2003)	Froehlich et al. (2003)	TargetP	cTP	TM	Calc. GRAVY	Calc. M_w (kD)	Calc. PI
At1g67700	Conserved plant protein, no similarity to characterized protein, oligopeptidase/protease domain	–	+	M	N	ND	–0.475	26.01	9.57
At2g31190	HP48 (EMB1879-like) conserved plant protein, no similarity to characterized protein	+	–	–	N	2	–0.081	48.25	7.7
At1g13930	Conserved plant protein, no similarity to characterized protein	–	+	–	N	ND	–0.883	16.16	4.82
At4g00640	HP57 conserved plant protein, weak similarity with protein sll1021 (*Synechocystis* sp.), no similarity to characterized protein	+	–	–	N	1	–0.523	51.41	4.82
At3g28220	Conserved plant protein with PPR signature	–	+	–	–	1	–0.425	42.88	8.59
At2g37930	Expressed protein, no similarity to characterized protein, Transcriptional factor tubby domain	–	+	–	N	ND	–0.665	51.96	8.83
At2g47750	GH39 auxin-responsive GH3 family protein	–	+	–	N	ND	–0.11	66.15	6.08
At2g40550	Conserved protein, restricted to eukaryotes	–	+	–	N	ND	–0.267	65.75	5.18
At5g51320	Protein restricted to Arabidopsis, no similarity to any other protein (even plant proteins)	–	+	–	N	ND	0.101	13.1	10.12

This protein list is derived from data published by Ferro et al. (2002, 2003) and Froehlich et al. (2003). It is a "curated" list: we have removed from published envelope proteomes those proteins we suspect to be contaminants (see Fig. 2). This was performed manually for each protein identified in envelope fractions by searching in databases (see for instance Schwacke et al. 2003; Friso et al. 2003), in published protein lists from various origins and in proteomic analyses we (and others) performed on different chloroplast subfractions (especially stroma) prepared from *Arabidopsis* chloroplasts (to be published). Target P was used to determine where the protein was targeted (C: chloroplast; M: mitochondria). Various information (i.e. the calculated GRAVY index, *M*w and P*i*) are from PPDB; a Plant Proteome DataBase for *Arabidopsis thaliana* and *Zea mays*; Friso et al. (2004). The presence of a chloroplast transit peptide cTP was determined by ChloroP (Emanuelsson et al. 1999). The number of transmembrane domains (TM) was determined using Aramemnon (Schwacke et al. 2003). This protein list is only a snapshot at the time of publication: more genuine envelope proteins are expected to be identified with the increasing number of proteome analyses performed by different groups combined with recent and fast technological developments in mass spectrometry different extraction procedures and analytical techniques, i.e. solubilization in chloroform/methanol, and alkaline and saline treatments, prior to mass spectrometry analyses. Therefore, Ferro et al. (2002, 2003) chose to introduce a bias in their studies, since no "crude" envelope preparations were analyzed. The rationale behind this choice was to focus on the core of envelope membrane proteins, the more peripheral proteins being excluded since it was difficult to choose – as mentioned above – between stromal (or to a lesser extent cytosolic) contamination and functional association to the envelope membrane. The envelope subfractions obtained were separated by 1D SDS-PAGE, followed by in-gel trypsin digestion and the tryptic fragments were analyzed by LC-MS/MS. For more details concerning this strategy, the reader is referred to several recent methodological reviews (Marmagne et al. 2006; Salvi et al. 2008a,b; Seigneurin-Berny et al. 2008).

4.2.1 Hydrophobic and Soluble Proteins in Envelope Membranes

The main interest of using different procedures to extract envelope proteins is to cover a wide range of protein dynamics and hydrophobicity. Ferro et al. (2003) identified more than 100 proteins in envelope membranes from *Arabidopsis* chloroplasts: about two thirds (69/112) were found from only one extraction method and

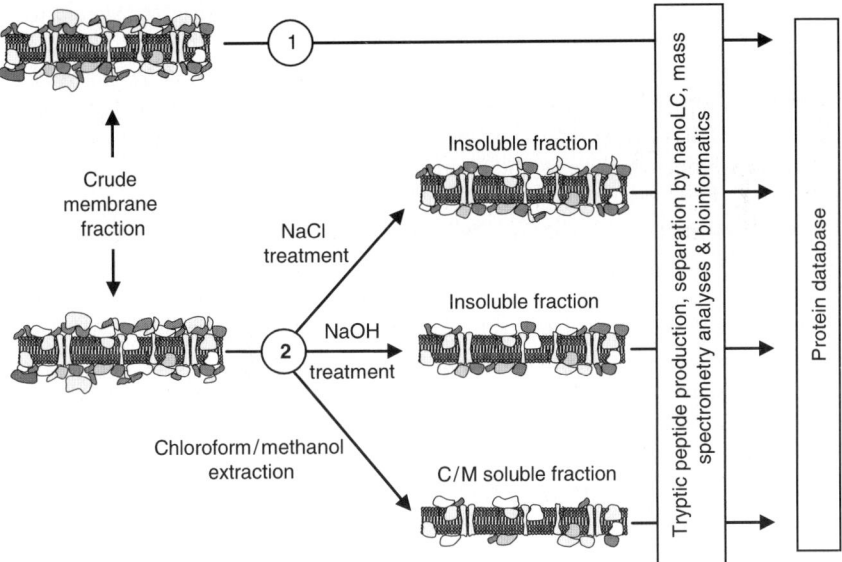

Fig. 1 Strategies for envelope proteomics. Scheme 1: Froehlich et al. (2003) directly used envelope preparation without any fractionation together with off-line multidimensional protein identification technology (off-line MUDPIT) to analyze the envelope proteome. Scheme 2: Ferro et al. (2002, 2003) analyzed envelope proteins from spinach and *Arabidopsis* chloroplasts by using a set of complementary methods, i.e. solubilization in chloroform/methanol, and alkaline and saline treatments, prior to mass spectrometry analyses

not from any of the two other ones. Indeed chloroform/methanol extraction, alkaline and saline washings of *Arabidopsis* envelope membranes allowed the identification of 37, 51, and 74 proteins, respectively, by mass spectrometry (Ferro et al. 2003).

Chloroform/methanol is the most stringent method and alkaline treatment is more stringent than saline treatment with regard to the recovery of hydrophobic proteins. Organic solvent extraction appears to select for low Mr and high hydrophobicity with a low contamination with hydrophilic proteins (Ferro et al. 2002, 2003). This extraction method allowed the identification of proteins that are likely to be low abundance transporters. For instance, among the chloroform/methanol-soluble proteins, Ferro et al. (2003) identified a new phosphate transporter representing only a minute fraction (about 0.1%) of the total envelope proteins. Although being less stringent with respect to hydrophobicity, alkaline and saline treatments allowed the identification of a few highly hydrophobic proteins altogether with more hydrophilic ones. Most of the proteins identified from NaCl treatment are predicted to contain no or only one transmembrane domain (Ephritikhine et al. 2004). NaOH treatment appears to be a good compromise to retrieve a wide range of proteins with different physico-chemical properties: this procedure clearly selects for more intrinsic proteins when compared to saline treatment since less hydrophilic contaminants are recovered (Ephritikhine et al. 2004). In addition, chloroform/methanol solubilize the most hydrophobic envelope proteins: almost all *Arabidopsis* envelope proteins identified in the chloroform/methanol extract have a Res/TM (Res stands for number of amino acid residues; TM for predicted transmembrane domains) ratio below 200 (31 among 37 identified proteins). As expected, more "soluble" proteins (Res/TM above 600) were identified in NaCl-washed envelope membranes, even when compared to NaOH-washed membranes (Ferro et al. 2003). In contrast, the envelope proteome identified by using unfractionated envelope preparations contains a large number of proteins (about 350) among which a very high proportion (70%) are devoid of any transmembrane domain (Froehlich et al. 2003). This value would even be much higher if the expected contaminant proteins (see Table 4) were taken into account. As expected, the number of transmembrane domains is related to protein functions (Fig. 2). Putative transporters represent the majority of envelope proteins with several α-helices constituting transmembrane domains: most of the 5% of envelope proteins having more than ten transmembrane domains are putative transporters. Within this functional class, only a small proportion is devoid of transmembrane domains: these proteins have, in general, a β-barrel structure and were shown to reside in the outer envelope. In contrast, only a few of the numerous envelope proteins of the "Metabolism" class have transmembrane domains. This is also true for proteins of the "Chaperone & protease" class. However, one should keep in mind that not all "true" membrane proteins have transmembrane domains: such domains are lacking in porin-type proteins, monotopic proteins, or proteins post-translationally modified by lipidic anchors. For instance, AtMGD1, the enzyme involved in the synthesis of MGDG is a monotopic inner envelope membrane protein without any transmembrane α-helice (Miège et al. 1999).

4.2.2 Chloroplast Envelope Membranes are Rich in Basic Proteins

Table 4 demonstrates that most of the proteins identified in envelope fractions were shown to be basic, whatever the extraction procedure (Ferro et al. 2003). A similar conclusion was made from in silico studies by Sun et al. (2004). Almost half of the envelope proteins likely to reside at the inner membrane have an isoelectric point higher than 9.0. When only putative transporters are considered, this value goes up to 70%. Thus, selection of basic proteins by chloroform/methanol extraction in chloroplast envelope membranes (Ferro et al. 2002, 2003) is due to the actual nature of the envelope proteins rather than to a specific bias of the extraction procedure.

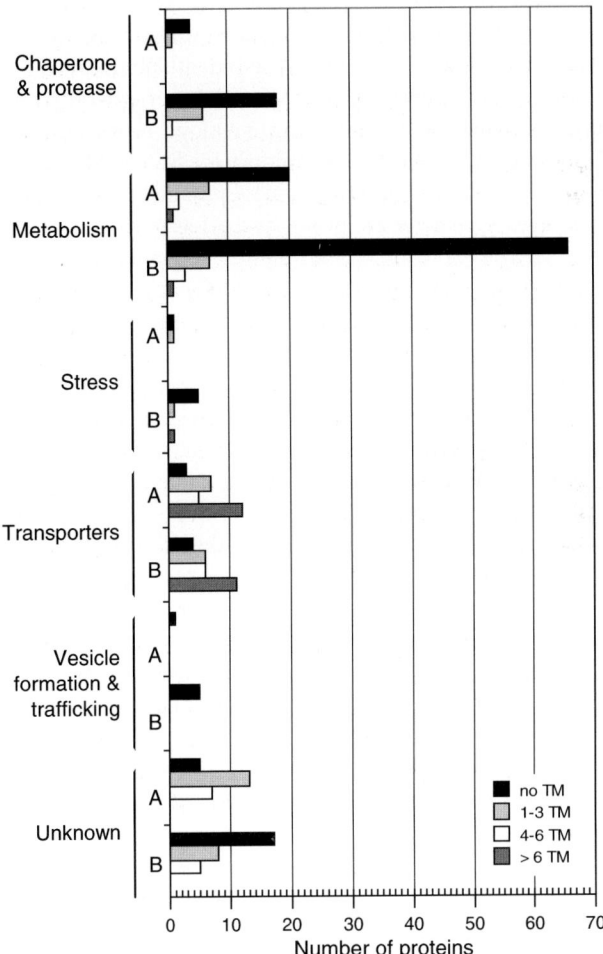

Fig. 2 Functional classification of envelope proteins identified by proteomics. The number of putative transmembrane domains is indicated for each protein class. (**a**) Data from Froehlich et al. (2003), (**b**) data from Ferro et al. (2002, 2003)

This was confirmed in a comparative survey of different plant membrane proteins extracted by chloroform/methanol (Ephritikhine et al. 2004): from all plant cell membranes analyzed, the envelope membranes contain the highest proportion of basic proteins.

4.2.3 In silico Analyses

Bioinformatics was also used to identify envelope membrane proteins (Koo and Ohlrogge 2002; Ferro et al. 2002, 2003; Rolland et al. 2003; Sun et al. 2004). For instance, Koo and Ohlrogge (2002) made attempts to predict plastid envelope proteins from the *Arabidopsis* nuclear genome by using computational methods and criteria such as the presence of *N*-ter plastid-targeting peptide (cTP) and of membrane-spanning domains (known thylakoid membrane proteins being subtracted). Using a combination of predictors and experimentally derived parameters, four plastid subproteomes, including envelope proteomes, were predicted from the fully annotated *Arabidopsis* genome by Sun et al. (2004). They were evaluated for distribution of physical-chemical parameters: they differ strongly in protein size and average isoelectric point, as well as transmembrane distribution. Removal of the cleavable, *N*-ter transit peptide sequences greatly affected isoelectric point and size distribution. Sun et al. (2004) observed that the Cys content was much lower for thylakoid proteomes than for the inner envelope and related this observation to the role of the thylakoid membrane in light-driven electron transport.

Ferro et al. (2002) observed that chloroplast envelope transporters are synthesized as precursor proteins with a predicted cTP; they have a Res/TM ratio below 100; they contain more than four transmembrane domains; and have an isoelectric point above 8.8. These parameters were used to mine the AMPL database: Ferro et al. (2002) found that only 136 proteins (of the 25,498 predicted *Arabidopsis* proteins) meet all these requirements. About 35% of these proteins correspond to proteins belonging to plant transporter families, a few percent of these proteins are proteins involved in lipid or pigment metabolisms, whereas the remaining 50% are hypothetical proteins whose function could not be predicted on the basis of their primary structure. About 15% of these 136 proteins correspond to proteins identified by means of the proteomic approach (Ferro et al. 2002). This is very close to the 137 membrane proteins probably targeted to plastids that were classified by Weber et al. (2005) as members of the transporter family.

5 Functions of Chloroplast Envelope Proteins

Envelope proteins in Table 4 are classified among functional classes: almost 25% are involved in lipid metabolism (fatty acids, glycerolipids, pigments, and terpenoids); another 25% are likely involved in metabolite and ion transport across the envelope, whereas more than 10% of the envelope proteome are either known components of the import machineries or homologous to them. In the following paragraphs, it is not our goal to address in detail all the envelope functions. For this, the readers are referred to all other chapters in this book that specifically address the main aspects of the role of envelope membranes in plastid division, chloroplast nucleus communication, lipid trafficking and signaling, metabolite transport, etc. (see also Block et al. 2007). Here, we will try to link the actual functions of envelope membranes with the proteins identified by Ferro et al. (2002, 2003) and Froehlich et al. (2003). Classification of proteins in given functional classes is also arbitrary. For instance, some proteins involved in fatty acid or pigment metabolism are also involved in signaling, mechanosensitive (MscS) ion channels also play a role in controlling chloroplast size and shape and proteins in redox metabolism are often related to stress responses. One also should keep in mind that only putative functions of the proteins are indicated in Table 4. As these are based on available genome annotations and databases analysis, the actual function of a given protein can therefore be strikingly different. However, it is the link between the proteomic data and the actual envelope functions that provides insight into the roles of envelope membranes for the chloroplast and for the plant cell. This is of special interest for controversial aspects of envelope metabolism that can only be addressed by protein identification.

5.1 *Metabolism*

5.1.1 Fatty Acid and Glycerolipid Metabolisms

The question of a possible role of envelope membranes in the first steps of fatty acid biosynthesis was unexpected but was raised by the presence in envelope preparations of components of the plastid pyruvate dehydrogenase complex (ptPDC) and acetyl-CoA carboxylase (ACCase) both considered as stromal enzymes. Table 4 shows that α (At1g01090) and β (At1g30120) subunits of E1, dihydrolipoamide dehydrogenase, two E2 isoforms (At3g25860, At1g34430) and E3 (At4g16155) from ptPDC are present in envelope preparations. This is somewhat puzzling. The plastid form of PDC provides acetyl-CoA and NADH for fatty acid biosynthesis (Mooney et al. 2002) and is sensitive to light/dark changes in the redox state of the chloroplast stroma (Tovar-Mendez et al. 2003). The presence of ptPDC components in envelope membranes (Table 4) raises the question of whether acetyl CoA formation (mostly for fatty acid synthesis) actually occurs in the vicinity of the envelope or whether this just reflects contamination by stroma. Ferro et al. (2003) identified an E1 β subunit of ptPDC in NaCl-washed envelope fractions, thus suggesting that it could be at least associated with the inner envelope membranes. Furthermore, the α (At2g38040) and β (AtCg00500) subunits and the biotin carboxylase subunit (CAC2, At5g35360) of the heteromeric chloroplastic acetyl-coenzyme A carboxylase (ACCase) (which converts acetyl-CoA into malonyl-CoA) are also found in envelope preparation (Table 4). This is in agreement with a series of observations (Thelen and Ohlrogge 2002) suggesting that ACCase is anchored to the chloroplast envelope through nonionic interactions with the carboxyltransferase subunits. Therefore, the presence of both ptPDC and ACCase in (or close to) inner envelope membrane could favor channeling of acetyl-CoA for its conversion into malonyl-CoA, an essential step in fatty acid biosynthesis.

Providing molecular evidence for previous biochemical studies showing that envelope membranes are the site of galactolipid, SQDG, and PG synthesis (for reviews, Douce and Joyard 1979, 1990; Andersson and Dörmann 2008), envelope proteomics demonstrated the occurrence of all key enzymes for the biosynthesis of chloroplast glycerolipids (Table 4): an enzyme of the Kornberg–Pricer pathway, 1-acylglycerol-3-phosphate *O*-acyltransferase (At4g30580), that catalyzes phosphatidic acid synthesis; of MGDG synthase (AtMGD1, At4g31780) and SQDG synthase (SQD2, At5g01220); and of at least three enzymes involved in PG synthesis, i.e. a CDP-diacylglycerol synthetase (At3g60620), and the two PG synthases, PGP1 (At2g39290) and PGP2 (At3g55030). In *Arabidopsis thaliana*, PG synthases are present in the three different compartments but are encoded by two related genes, *PGP1* and *PGP2* (Müller and Frentzen 2001). Although the presence of PGP1 in envelope membranes was expected, that of PGP2 in the envelope proteome is somewhat surprising since *PGP2* encodes the microsomal isozyme, whereas *PGP1* encodes a preprotein that is targeted to both plastids and mitochondria (Müller and Frentzen 2001; Babiychuk et al. 2003). PGP1 is essential for plastidial PG biosynthesis and for photoautotrophic growth, whereas it is redundant for the biosynthesis of PG in mitochondria (Frentzen 2004).

Proteomics also revealed a wide diversity of lipid metabolism enzymes: lipases (phospholipase, lysophospholipase, acyl hydrolases), a long-chain acyl-CoA synthetase (LACS9, see also Schnurr et al. 2002), two desaturases (omega-3 and omega-6 fatty acid desaturases), and enzymes involved in fatty acid hydroperoxide metabolism (lipoxygenase – LOX2, allene oxide synthase, allene oxide cyclase, and hydroperoxide lyase-like enzyme). Allene oxide synthase and hydroperoxide lyase are known envelope enzymes (Blée and Joyard 1996), they constitute a branch point leading specifically from 13(*S*)-hydroperoxy-9(*Z*),11(*E*),15(*Z*)-octadecatrienoic acid to 12-oxo-phytodienoic acid, the precursor of jasmonic acid.

Altogether, these results provide further evidence for chloroplast envelope membranes being a key actor in several aspects of lipid metabolism (biosynthesis, transfer, desaturation, oxidation...) and in the production of lipid-derived plant growth regulators and defense compounds in response to extracellular stimuli.

5.1.2 Carotenoid, Chlorophyll, and Prenylquinone Metabolism

The importance of chloroplast envelope membranes in the formation of isoprenoid chloroplast lipid constituents such as carotenoids, chlorophylls, and quinones, previously raised by biochemical studies (for a review see Douce and Joyard 1990), and often considered as controversial in the literature, was confirmed unambiguously by proteomics (see below).

Firstly, envelope proteomics detected several enzymes of carotenoid metabolism, such as phytoene dehydrogenase (At4g14210), ζ-carotene desaturase (At3g04870), carotene ε-ring hydroxylase (At3g53130), zeaxanthin epoxidase (At5g67030), and neoxanthin synthase (At1g67080) (Table 4). The two desaturases are on the common pathway leading to synthesis of lycopene, the precursor for xanthophylls (the main carotenoid group of chloroplasts). The carotene ε-ring hydroxylase is on the α-carotene branch of the pathway leading to the formation of lutein. The two other enzymes, zeaxanthin epoxidase and neoxanthin synthase, are on the branch leading to the formation of precursors of abscisic acid. Altogether, these results provide further support to a key role of envelope membranes in carotenoid biosynthesis, as proposed by Costes et al. (1979).

Secondly, several enzymes of chlorophyll biosynthesis are present in the envelope membranes (Table 4): protoporphyrinogen oxidase (At5g14220, At4g01690), Mg-protoporphyrin IX (Mg-Proto IX) monomethylester cyclase (At3g56940), CHLM (Mg-Proto IX) methyltransferase (At4g25080), [3,8-DV]-Chlide a 8-vinyl reductase (At5g18660), and protochlorophyllide reductases (At4g27440, At1g03630). The dual localization of this biosynthetic pathway in the envelope and in thylakoids raised the question of the physiological significance of the envelope pathway. In fact, some of the intermediates formed in the envelope may play a role in signaling between chloroplast and nucleus in order to coordinate chloroplast development and nuclear gene expression (see for instance Block et al. 2007). Indeed, envelope membranes contain key enzymes for the formation of one of the plastid signals, the chlorophyll intermediate Mg-Proto IX, which acts as a negative regulator of photosynthetic gene expression in the nuclei (Strand et al. 2003). This function of Mg-Proto IX was confirmed by phenotypic analyses of a knock-out mutant and the possible implication of Mg-Proto IX methyl ester, the product of CHLM, in chloroplast-to-nucleus signaling was also suggested (Pontier et al. 2007). Furthermore, the FLU (for *flu*orescent) protein (At3g14110) was also identified in envelope membranes. This protein appears to regulate specifically the Mg^{2+} branch of the tetrapyrrole pathway and operates independently of heme (Meskauskiene et al. 2001). In *Chlamydomonas*, Flu-like proteins were shown to act as regulators of chlorophyll synthesis, and their expression is controlled by light and plastid signals (Falciatore et al. 2005). Therefore, owing to the presence of the chlorophyll biosynthetic pathway that release specific chlorophyll-derived signals, chloroplast envelope membranes play a key role in mediating the plastid-to-nucleus regulatory pathway (for reviews, Nott et al. 2006; Block et al. 2007; Dietzel et al. 2008).

Finally, envelope membranes also contain enzymes involved in the biosynthesis of prenylquinones. Although Soll et al. (1985) demonstrated that all enzymes involved in the last steps of α-tocopherol and plastoquinone-9 biosynthesis are localized on the inner envelope membrane; this has been a matter of debate (review by Block et al. 2007). Proteomic analyses (Table 4) demonstrate that, with the exception of 4-hydroxyphenylpyruvate dioxygenase (HPPD), a cytosolic enzyme (Garcia et al. 1997), and tocopherol cyclase (VTE1) associated to plastoglobules (Vidi et al. 2006; Ytterberg et al. 2006), the other enzymes of prenylquinone biosynthesis were found in envelope membranes. The first committed reaction in prenylquinone biosynthesis is the condensation of homogentisic acid with prenyl (solanesyl, geranylgeranyl or farnesyl) diphosphate. AtHST (At3g1195) represents homogentisate solanesyltransferases involved in plastoquinone-9 biosynthesis (Sadre et al. 2006) and was localized in envelope membranes by proteomics (Table 4) and more recently by confocal microscopy (Tian et al. 2007). Overexpression of the enzyme in *A. thaliana* can affect tocopherol biosynthesis as well (Sadre et al. 2006). IEP37 (At3g63410, VTE3) is a major inner envelope membrane protein with a SAM-dependent methyltransferase activity (Teyssier et al. 1996), committed to the biosynthesis of plastid prenylquinones (Motohasi et al. 2003).

5.2 Transporters

Since early work on intact chloroplasts, the inner envelope membrane is known to represent the actual permeability barrier between plastids and the surrounding cytosol whereas the outer envelope membrane is expected to be freely permeable to small molecules owing to the presence of porins (see Weber et al. 2005). This is probably not as simple since substrate-specific gated pore-forming proteins were characterized in the outer envelope membrane (see Bölter and Soll 2001). Altogether, the combined proteomic and *in silico* approaches suggest that a series of known or putative transport systems are likely to be localized in the chloroplast envelope (Seigneurin-Berny et al. 1999; Ferro et al. 2002, 2003; Weber et al. 2005).

The major inner envelope proteins are triose-P/Pi translocators (reviewed by Weber et al. 2005), identified in all proteomic studies (Table 4). Such transporters are essential for controlling the Pi level in the stroma and the homeostasis required to initiate the Benson and Calvin cycle, especially during the dark/light and light/dark transitions. Interestingly, whereas some transporters, like the members of the triose-P/Pi, PEP/Pi, or Glucose-6P/Pi translocators catalyze an equimolar exchange of Pi, others like the putative H^+/Pi transporter could catalyze a net import of Pi to the chloroplast. Identification of these new phosphate transport systems in chloroplasts is expected to lead to a better understanding of their role in cell metabolism.

Envelope membranes contain two members of the 2-oxoglutarate/malate translocator family: DiT2-1 (At5g64290) and DiT1 (At5g12860). These proteins could differ in substrate specificity as DiT2-1was demonstrated to catalyze the transport of glutamate/malate (Renne et al. 2003). The identification of several other proteins is consistent with transport activities already associated with the chloroplast envelope. For example, several classes (belonging to at least two superfamilies) of amino acid transporters were identified in plants (Ortiz-Lopez et al. 2001). Identification of some members of the amino acid transporter families in envelope membranes (Table 4) provides some clues for distinguishing these transporters with overlapping substrate specificity. Several ABC proteins and NAP (nonintrinsic ABC protein) were also identified (Table 4). They are expected to play a role in the transport of a wide variety of substrates (metabolites, lipids, etc.) across the envelope. However, one of them, named STA1 (At5g58270), was expected to reside within mitochondria and was shown to be essential for transport of iron/sulfur clusters and iron homeostasis (Kushnir et al. 2001).

Several members of the Mitochondrial Carrier Family (MCF) were identified in envelope membranes (Table 4), many of them being only hypothetical proteins. In fact, not all MCF members are located in mitochondria: some are also present in peroxisomes, glyoxysomes, and plastids (review by Haferkamp 2007) and their characteristic features are mostly structural (see Millar and Heazlewood 2003). Therefore, the MCF denomination is misleading. In plants, MCF proteins are involved in the transport of solutes like nucleotides, phosphate, di- and tricarboxylates and therefore exhibit physiological functions similar to known isoforms from animal or yeast mitochondria. Interestingly, several MCF proteins mediate the transport of different substrates like folates, S-adenosylmethionine, ADPglucose or ATP, ADP and AMP in plastids (see Weber and Fischer 2008). S-adenosylmethionine is formed exclusively in the cytosol but is imported to the plastid where it plays a major role (Ravanel et al. 2004). An S-adenosylmethionine transporter was characterized by Bouvier et al. (2006) and is located in the chloroplast envelope, as previously suggested by bioinformatics (Koo and Ohlrogge 2002) and demonstrated by proteomics (Ferro et al. 2002, 2003). This transporter is probably dual targeted since it was also recently demonstrated to reside within mitochondria (Palmieri et al. 2006). Although the mitochondria were demonstrated to be the sole site of dihydrofolate synthesis in the plant cell, folate-mediated reactions were identified in the cytosol, the mitochondria, and the plastids (Ravanel et al. 2001), suggesting that folate must be imported into chloroplasts. Indeed, the AtFOLT1 protein (At5g66380), identified by bioinformatics as a candidate for folate transport across the chloroplast envelope (Ferro et al. 2002), was validated by Bedhomme et al. (2005). However, it has not yet been identified by proteomics and the exact function of the majority of the MCF proteins in *Arabidopsis* is still unknown.

Several ion channels were identified by envelope proteomics (Table 4). MSL proteins probably control plastid size and shape during plant development by altering ion flux across the envelope membranes (Haswell and Meyerowitz 2006; see above). KEA1 and KEA2 (At 1g01790, At4g 00630, respectively) are two of the putative K^+ efflux antiporters in *Arabidopsis* (Mäser et al. 2001). Two DMI1-like proteins are present in envelope membranes (At5g43745 and At5g02940) and they belong to the novel class of plastid-localized "Castor and Pollux" family that are absolutely required for early signal transduction events leading to endosymbioses (Imaizumi-Anraku et al. 2005). The *Arabidopsis* protein encoded by gene At5g49960, before the two DMI1-like proteins, is the most similar protein to *Medicago truncatula* proteins Castor and Pollux and DMI1. Therefore, one can question whether the two DMI1-like envelope proteins could play a role similar to that of DMI1. Imaizumi-Anraku et al. (2005) also raised the question of how are these proteins implicated in a process that starts with the perception of microbial signals at the plasma membrane and leads to changes in gene expression in the nucleus, among other responses.

Metal ions are essential for chloroplast development and function. Unfortunately little is known about their transport across the chloroplast envelope. Envelope proteome shows several metal transporters (Table 4). For instance, AtMRS2-11

(At5g22830) is a Mg transporter of the corA group of the bacterial Mg transporter family. It has been shown to complement a yeast double mutant and restores Al tolerance (Li et al. 2001). Several chloroplast P_{IB}-type ATPases are involved in metal ion transport across the envelope: in addition to PAA1 (Shikanai et al. 2003), HMA1 – that was identified by envelope proteomics (Ferro et al. 2003) – is also involved in Cu import into the chloroplast under high light conditions, thus allowing plants to respond to oxidative stress generated in these conditions (Seigneurin-Berny et al. 2006; see below).

Outer envelope membrane also contains substrate-specific gated pore-forming proteins (see Bölter and Soll 2001) and several of these high-conductance solute channels were identified by proteomics (Table 4): OEP21-1 (At1g76405), OEP24 (At3g52230), OEP24-II (At5g42960) and OEP37 (At2g43950). All of them have a β-barrel structure. The OEP24 channels were the first outer envelope channel to be functionally characterized. They allow the flux of triose phosphate, dicarboxylic acids, positively or negatively charged amino acids, sugars, ATP, and Pi (Pohlmeyer et al. 1998). Hemmler et al. (2006) analyzed the Oep21 channel properties by biochemical and electrophysiological methods and proposed that it is involved in the regulation of the flux of phosphorylated carbohydrates, like 3-phosphoglycerate and glyceraldehyde 3-phosphate, across the outer membrane. OEP37 was reconstituted in vitro and was demonstrated to form a rectifying high conductance cation-selective channel (Goetze et al. 2006). OEP37 may constitute a novel peptide-sensitive ion channel in the outer envelope of plastids with function during embryogenesis and germination (Goetze et al. 2006).

5.3 Protein Targeting: the Protein Import Machinery of the Chloroplast Envelope

One of the key functions in chloroplast envelope membranes is the transport of nuclear-encoded proteins residing within the chloroplast. Numerous components of the chloroplast import machinery have been characterized, i.e. the translocon complexes at the outer (TOC complex) and inner (TIC complex) envelope membranes (for reviews see Hofmann and Theg 2005; Schleiff and Soll 2005; Kessler and Schnell 2006; Aronsson and Jarvis 2008). Table 4 demonstrates that envelope preparations contain more than 20 proteins related to components of the protein import machinery. They can be classified into (a) known components of the TOC complex (Toc 33, Toc34, Toc 64, Toc75, Toc159); (b) known components of the TIC complex (Tic 20, Tic 21, Tic40, Tic55, Tic 110); (c) Tic20-like (IEP16), Tic55-like, and Tic-62-like proteins; and (d) a series of OEP16-like proteins (HP15, HP22, HP30, HP30-2) with some homology to components of the mitochondrial inner membrane import machinery (Tim17/Tim22/Tim 23). To this list one should also add chaperones that are likely to be involved in protein import (see below).

Other envelope proteins identified through proteomic analyses are homologous to known components of the chloroplast or of the mitochondrial import machinery. The actual participation of these new proteins to protein import mechanisms into chloroplasts remains to be demonstrated. Are they true components of the TIC/TOC complexes or of a distinct import system that could be present in chloroplast envelope membranes? Proteomic analyses led to the characterization of inner envelope proteins that do not use the classical TOC machinery for transport across the outer envelope membrane: ceQORH (At4g13010) that was devoid of cleavable N-terminal transit sequences and contained internal targeting information (Miras et al. 2002, 2007) and Tic32, which contained an N-terminal but uncleavable pre-sequence (Nada and Soll 2004). Finally, Villarejo et al. (2005) recently showed that a chloroplast-located protein takes an alternative route through the secretory pathway, and becomes *N*-glycosylated before entering the chloroplast. Therefore, the actual components of these new import machineries need to be identified.

5.4 Envelope Proteins and Lipid Trafficking

Lipid trafficking between the chloroplast and the endomembrane system and within plastids are key processes in plant development (reviews by Benning et al. 2006; Jouhet et al. 2007). Proteomic data only provide few elements in understanding the processes (Table 4; Andersson and Dörmann (2008)). Firstly, the long-chain acyl-CoA synthetase (At1g77590) could be responsible for the export of chloroplast-synthesized fatty acids to the endomembrane system (two other proteins encoded by *At3g23790* and *At4g14070* genes are also annotated as acyl-CoA synthetases). Secondly, glycerolipid synthesis requires that the eukaryotic type of DAG backbone, necessary for the formation of eukaryotic chloroplast glycerolipids, comes from endomembrane glycerolipids. The TGD protein (At3g20320) in the chloroplast envelope proteome promises new insights into lipid-trafficking between the envelope and the endomembrane system (review by Benning et al. 2006) Thirdly, thylakoid formation requires a massive transfer of the different lipids synthesized in the envelope to the thylakoids and selective lipid transport from the inner envelope membrane to the thylakoids has been demonstrated (Rawyler et al. 1995). This process might involve VIPP1 (Vesicle Inducing Protein in Plastids 1, encoded by At1g65260) and/or THF1 (THYLAKOID FORMATION 1 protein, encoded by At2g20890), proteins involved in thylakoid formation through vesicular trafficking. The intriguing point is that THF1 is localized to both the outer envelope membrane and the stroma and is supposed to interact with GPA1, a Plasma Membrane G-Protein, in a putative sugar-signaling mechanism that would be essential for thylakoid formation (Huang et al. 2006 and references therein).

5.5 Protein Degradation and Regeneration (Chaperone and Protease)

In *Arabidopsis*, plastids contain more than 11 types of proteases encoded by more than 50 genes (Sakamoto 2006). Froehlich et al. (2003) and, to a lesser extent, Ferro et al. (2003) identified a series of HSP70, ATP-dependent ClpP and ClpR as well as proteases of the GTP-dependent FtsH family in envelope preparations (Table 4). Of the 15 plastid-localized Clp proteolytic proteins (Adam et al. 2006), about ten were found in envelope preparations (ClpS1, ClpC1-6, and ClpR2-4). In addition, FtsH2, 5, 7 were identified in envelope proteomes (Table 4). Such proteases are involved in numerous aspects of the biogenesis and maintenance of chloroplasts, such as the removal and degradation of signal sequences, the degradation of partially assembled complexes or damaged proteins, adaptation to changes in environmental conditions, and the breakdown of chloroplasts in senescing leaves (Adam 2005). It is generally assumed that most of these proteases reside within the stroma; however, some of them are associated with envelope-specific processes. For instance, ClpC1 (HSP 100), the first protease demonstrated as being associated with the chloroplast protein translocation machinery in the inner envelope (see for instance Nielsen et al. 1997), was indeed identified by proteomics (Table 4). It is possible that the diverse set of envelope-bound Clp found in proteomic studies is required for the translocation of many proteins across the chloroplast envelope membranes, presumably by providing energy in the form of ATP hydrolysis for protein precursor translocation across the envelope membranes.

5.6 Plastid Division and Positioning

Proteins involved in chloroplast division were identified by searching possible homologues of components of known prokaryotic division machinery (e.g. FtsZ, MinD, MinE, and GC1/SulA), by detailed analyses of the *arc* (accumulation and replication of chloroplasts) mutants, and of components that affect chloroplast division but were not directly involved in the division machinery (MscS-like or MSL proteins) (for reviews see Aldrige et al. 2005; Maple and Møller 2006; Maple et al. 2008). Proteomic studies identified some of these proteins in envelope preparations (Froehlich et al. 2003; Table 4). For instance, ARC6 (homologous to a bacterial cell division protein), which was identified from *arc6* mutants that contain only one giant chloroplast per cell (Vitha et al. 2003) and GC1 (for Giant Chloroplast1), which also has a prokaryotic origin (Maple et al. 2004). Two MSL proteins in *Arabidopsis thaliana* envelope membranes, MSL2 and MSL3 (Haswell and Meyerowitz 2006), are related to MSL1 (At4g00290), which was identified by proteomics (Table 4). In the model proposed by Haswell and Meyerowitz (2006), MSL2 and MSL3 control plastid size, shape, and perhaps division by altering ion flux in response to changes in membrane tension.

Table 4 also shows the presence of a CHUP1-like protein (At1g07120). CHUP (chloroplast unusual positioning) proteins are required for organellar positioning and movement in plant cells and were identified from the screening of *Arabidopsis* mutants defective in chloroplast photorelocation movement (Oikawa et al. 2003; Suetsugu and Wada 2008). Proteomics also revealed TH65 kinesin related protein (At5g10470) and a kinesin-like protein (Xklp2-like protein similar to microtubule associated protein, At1g03780) in envelope preparations. Functional studies are necessary to determine whether such proteins are actually motor proteins that could transport chloroplasts along microtubule tracks. Like CHUP1, they exhibit coiled-coil domains that are expected to play a role in subcellular infrastructure maintenance and trafficking control (Rose et al. 2005).

5.7 Signaling

A key role for envelope membranes is the production of lipid- and pigment-derived signaling molecules (see above). For instance, envelope membranes contain the whole set of enzymes that convert polyunsaturated fatty acids into jasmonate precursors; the cleavage of xanthophylls formed in envelope membranes is responsible for the formation of abscisic acid; envelope-synthesized chlorophyll precursors mediate plastid-to-nucleus signaling.

Several envelope proteins are probable candidates as components of cellular signaling pathways. Two homologous hypothetical proteins (AtCDF1-like proteins, encoded by *At3g51140* and *At5g23040* genes) were identified in envelope membranes (Table 4). Interestingly, Kawai-Yamada et al. (2005) demonstrated that overexpression of *Cdf1* gene (*At5g23040*) of *Arabidopsis* in yeast could induce apoptosis-like cell death in yeast but here, CDF1 fused to GFP was expressed in mitochondria.

Furthermore, several envelope proteins contain putative domains that are present in components of signaling pathways, such as the protein encoded by gene *At5g64430* with an octicosapeptide/Phox/Bem1p (PB1) domain (Moscat et al. 2006). In addition, the identification in envelope membranes of two DMI1-like proteins (Table 4), putative members of the "Castor and Pollux" family of proteins that is required for early signal transduction events leading to endosymbioses, clearly opens a new field of research (see above).

5.8 Stress and Redox Metabolism

Plants are submitted to a wide variety of environmental stresses, like high light, drought, nutrient deficiency, and temperature that can induce oxidative stress. In addition, intermediates in electron transfer reactions may react with other cell constituents and produce radicals and reactive oxygen species. Reactive oxygen species are responsible for major damages to membrane lipids and proteins. Consequently, an efficient antioxidant network is essential for protecting cell functions, particularly in chloroplasts. A whole set of metabolites such as glutathione and ascorbate, carotenoids, tocopherols, and enzymes such as superoxide dismutase (SOD), ascorbate peroxidase (APX), and glutathione peroxidase are known elements of antioxidant metabolism (for review, see Mullineaux and Karpinski 2002). The presence of α-tocopherol and xanthophylls in envelope membranes (see above) is a first element for this system to play a role in oxidative stress responses. Furthermore, several proteins that could be involved in oxidative stress responses were identified in *Arabidospis* envelope membranes (Table 4): namely a phospholipid hydroperoxide glutathione peroxidase (GPX2, At2g25080) and an ascorbate peroxydase (APX3, At4g35000). Two members of the peroxiredoxins (Prx, for review see Dietz 2007) class of enzymatic antioxidants, PrxA (At3g11630) and PrxQ (At3g26060), were also identified (Table 4). However, PrxQ was also assigned to be bound to thylakoids (Lamkemeyer et al. 2006). In fact, most of these proteins do not have any transmembrane domains and are probably bound to, and active at, the stromal surface of the envelope membranes.

The stromal Cu/Zn SOD contains Cu, which is toxic at high concentration, due to generation of oxygen and hydroxyl free radicals and the oxidation of dithiols to disulfide in proteins (see for instance Shingles et al. 2004). Indeed, Seigneurin-Berny et al. (2006) demonstrated that *Arabidopsis* plants with a mutation in the gene encoding the envelope protein HMA1, a member of the metal-transporting P1B-type ATPases family (see above), have a photosensitivity phenotype under high light. This demonstrates that Cu homeostasis in chloroplast, and therefore Cu transport across the envelope, is essential for an efficient photoprotection, especially under high light.

One original way to prevent photodamage in chloroplasts when plants are exposed to high light levels could be chloroplast avoidance movement, in which chloroplasts move from the cell surface to the periphery of cells under high light conditions. Kasahara et al. (2002) have shown that *Arabidopsis chup1* mutants (devoid of CHUP1 envelope protein, see above) are defective in chloroplast avoidance movement and are therefore more susceptible to damage in high light than wild-type plants.

Finally, several envelope (or envelope-bound) proteins are induced upon stress. For example, GLX1 (At1g11840) and ERD4 (At1g30360) are induced upon drought (Seki et al. 2001), and the damages caused to membrane proteins submitted to an oxidative stress require the presence of active repair mechanisms, such as chaperones and proteases (see above).

6 Conclusion

Our understanding of the essential role of chloroplast envelope membranes in the plant cell is strongly improving. As discussed by Sun et al. (2004) "the assembly of rigorously curated subcellular proteomes is in itself also important as a parts list for plant and systems biology." The picture emerging from our present understanding of chloroplast envelope membranes is that of a major node for integration of metabolic and ionic networks in plant cell metabolism and of a key player in chloroplast biogenesis and signaling for the co-ordinated gene expression of chloroplast-specific protein. Furthermore, the lipid and protein constituents of chloroplast envelope membranes reflect both their prokaryotic and eukaryotic origin.

Finally, one should keep in mind that the protein list in Table 4 is only a snapshot at a given time: more genuine envelope proteins are expected to be identified with the increasing number of proteome analyses performed by different groups combined with recent and fast technological developments in mass spectrometry.

Acknowledgments We are grateful to Prof. Roland Douce for his communicative enthusiasm and his continuous mentoring for pursuing the search he has initiated for deciphering new functions in chloroplast envelope membranes. We also would like to thank Prof. Ernst Heinz for his comments and advices on envelope lipids.

References

Adam Z (2005) The chloroplast proteolytic machinery. In: Møller SG (ed) Plastids. Blackwell, Oxford, pp 214–236
Adam Z, Rudella A, van Wijk K (2006) Recent advances in the study of Clp, FtsH and other proteases located in chloroplasts. Curr Opin Plant Biol 9:234–240
Aldridge C, Maple J, Møller SG (2005) The molecular biology of plastid division in higher plants. J Exp Bot 56:1061–1077
Andersson MX, Dörmann P (2008) Chloroplast membrane lipid biosynthesis and transport. Plant Cell Monogr., doi:10.1007/7089_2008_18
Andersson MX, Stridh MH, Larsson KE, Liljenberg C, Sandelius AS (2003) Phosphate-deficient oat replaces a major portion of the plasma membrane phospholipids with the galactolipid digalactosyldiacylglycerol. FEBS Lett 537:128–132
Andersson MX, Larsson KE, Tjellstrom H, Liljenberg C, Sandelius AS (2005) Phosphate-limited oat. The plasma membrane and the tonoplast as major targets for phospholipid-to-glycolipid replacement and stimulation of phospholipases in the plasma membrane. J Biol Chem 280:27578–27586
Aronsson H, Jarvis P (2008) The chloroplast protein import apparatus, its components, and their roles. Plant Cell Monogr., doi:10.1007/7089_2008_40
Awai K, Maréchal E, Block MA, Brun D, Masuda T, Shimada H, Takamiya K, Ohta H, Joyard J (2001) Two types of MGDG synthase genes, found widely in both 16:3 and 18:3 plants, differentially mediate galactolipid syntheses in photosynthetic and nonphotosynthetic tissues in *Arabidopsis thaliana*. Proc Natl Acad Sci USA 98: 10960–10965
Awai K, Xu C, Tamot B, Benning C (2006) A phosphatidic acid-binding protein of the chloroplast inner envelope membrane involved in lipid trafficking. Proc Natl Acad Sci USA 103:10817–10822
Babiychuk E, Muller F, Eubel H, Braun HP, Frentzen M, Kushnir S (2003) *Arabidopsis* phosphatidylglycerophosphate synthase 1 is essential for chloroplast differentiation, but is dispensable for mitochondrial function. Plant J 33:899–909
Bedhomme M, Hoffmann M, McCarthy EA, Gambonnet B, Moran RG, Rebeille F, Ravanel S (2005) Folate metabolism in plants: an *Arabidopsis* homolog of the mammalian mitochondrial folate transporter mediates folate import into chloroplasts. J Biol Chem 280:34823–34831
Benning C, Xu C, Awai K, (2006) Non-vesicular and vesicular lipid trafficking involving plastids. Curr Opin Plant Biol 9:241–247
Benson AA, (1964) Plant membrane lipids. Annu Rev Plant Physiol 15:1–16
Billecocq A, (1975) Structure des membranes biologiques: localisation du sulfoquinovosyldiglycéride dans les diverses membranes des chloroplastes au moyen des anticorps spécifiques. Ann Imunol (Institut Pasteur) 126C:337–352
Billecocq A, Douce R, Faure M, (1972) Structure des membranes biologiques: localisation des galactosyldiglycérides dans les chloroplastes au moyen des anticorps spécifiques. CR Acad Sci Paris 275:1135–1137

Bishop DG, Sparace SA, Mudd JB, (1985) Biosynthesis of sulfoquinovosyldiacylglycerol in higher plants: the origin of the diacylglycerol moiety. Arch Biochem Biophys 240:851–858

Blée E, Joyard J (1996) Envelope membranes from spinach chloroplasts are a site of metabolism of fatty acid hydroperoxides. Plant Physiol 110:445–454

Block MA, Dorne AJ, Joyard J, Douce R (1983) Preparation and characterization of membrane fractions enriched in outer and inner envelope membranes from spinach chloroplasts. I-Electrophoresis and immunochemical analysesJ Biol Chem 258:13273–13280

Block MA, Douce R, Joyard J, Rolland N (2007) Chloroplast envelope membranes: a dynamic interface between plastids and the cytosol. Photosynth Res 92:225–244

Bölter B, Soll J (2001) Ion channels in the outer membranes of chloroplasts and mitochondria: open doors or regulated gates? EMBO J 20:935–940

Bouvier F, Linka N, Isner JC, Mutterer J, Weber AP, Camara B (2006) *Arabidopsis* SAMT1 defines a plastid transporter regulating plastid biogenesis and plant development. Plant Cell 18:3088–3105

Chen LJ, Li HM (1998) A mutant deficient in the plastid lipid DGD is defective in protein import into chloroplasts. Plant J 16, 33–39

Costes C, Burghoffer C, Joyard J, Block MA, Douce R (1979) Occurrence and biosynthesis of violaxanthin in isolated spinach chloroplast envelope. FEBS Lett 103:17–21

Dietz KJ (2007) The dual function of plant peroxiredoxins in antioxidant defence and redox signaling. Subcell Biochem 44:267–294

Dietzel L, Steiner S, Schröter Y, Pfannschmidt T (2008) Retrograde signalling. Plant Cell Monogr., doi:10.1007/7089_2008_41

Dörmann P (2007) Functional diversity of tocochromanols in plants. Planta 225:269–276

Dörmann P, Hoffmann-Benning S, Balbo I, Benning C (1995) Isolation and characterization of an *Arabidopsis* mutant deficient in the thylakoid lipid digalactosyl diacylglycerol. Plant Cell 7:1801–1810

Dorne AJ, Block MA, Joyard J, Douce R (1982) The galactolipid:galactolipid galactosyltransferase is located on the outer surface of the outer chloroplast envelope. FEBS Lett 145:30–34

Dorne AJ, Joyard J, Block MA, Douce R (1985) Localization of phosphatidylcholine in outer envelope membrane of spinach chloroplasts. J Cell Biol 100:1690–1697

Dorne AJ, Joyard J, Douce R (1990) Do thylakoids really contain phosphatidylcholine? Proc Natl Acad Sci USA 87:71–74

Douce R (1974) Site of galactolipid synthesis in spinach chloroplasts. Science 183:852–853

Douce R, Joyard J (1979) Structure and function of the plastid envelope. Adv Bot Res 7:1–116

Douce R, Joyard J (1990) Biochemistry and function of the plastid envelope. Annu Rev Cell Biol 6:173–216

Emanuelsson O, Nielsen H, von Heijne G (1999) ChloroP, a neural network-based method for predicting chloroplast transit peptides and their cleavage sites. Protein Science 8:978–984

Ephritikhine G, Ferro M, Rolland N (2004) Plant membrane proteomics. Plant Physiol Biochem 42:943–962

Falciatore A, Merendino L, Barneche F, Ceol M, Meskauskiene R, Apel K, Rochaix JD (2005) The FLP proteins act as regulators of chlorophyll synthesis in response to light and plastid signals in *Chlamydomonas*. Genes Dev 19:176–187

Ferro M, Salvi D, Rivière-Rolland H, Vermat T, Seigneurin-Berny D, Grunwald D, Garin J, Joyard J, Rolland N (2002) Integral membrane proteins of the chloroplast envelope: identification and subcellular localization of new transporters. Proc Natl Acad Sci USA 99:11487–11492

Ferro M, Salvi D, Brugière S, Miras S, Kowalski S, Louwagie M, Garin J, Joyard J, Rolland N (2003) Proteomics of the chloroplast envelope membranes from *Arabidopsis thaliana*. Mol Cell Proteomics 2:325–345

Frentzen M (2004) Phosphatidylglycerol and sulfoquinovosyldiacylglycerol: anionic membrane lipids and phosphate regulation. Curr Opin Plant Biol 7:270–276

Frentzen M, Heinz E, McKeon TA, Stumpf PK (1983) Specificities and selectivities of glycerol-3-phosphate acyltransferase and monoacylglycerol-3-phosphate acyltransferase from pea and spinach chloroplasts. Eur J Biochem 129:629–636

Friso G, Ytterberg AJ, Giacomelli L, Peltier JB, Rudella A, Sun Q, van Wijk KJ (2004) In-depth analysis of the thylakoid membrane proteome of *Arabidopsis thaliana* chloroplasts; new proteins, functions and a plastid proteome database. Plant Cell 16:478–499

Fritz M, Lokstein H, Hackenberg D, Welti R, Roth M, Zahringer U, Fulda M, Hellmeyer W, Ott C, Wolter FP, Heinz E (2007) Channeling of eukaryotic diacylglycerol into the biosynthesis of plastidial phosphatidylglycerol. J Biol Chem 282:4613–4625

Froehlich JE, Wilkerson CG, Ray WK, McAndrew RS, Osteryoung KW, Gage DA, Phinney BS (2003) Proteomic study of the *Arabidopsis thaliana* chloroplastic envelope membrane utilizing alternatives to traditional two-dimensional electrophoresis. J Proteome Res 2:413–425

Garcia I, Rodgers M, Lenne C, Rolland A, Sailland A, Matringe M (1997) Subcellular localization and purification of a *p*-hydroxyphenylpyruvate dioxygenase from cultured carrot cells and characterization of the corresponding cDNA. Biochem J 325:761–769

Goetze TA, Philippar K, Ilkavets I, Soll J, Wagner R (2006) OEP37 is a new member of the chloroplast outer membrane ion channels. J Biol Chem 281:17989–17998

Haferkamp I (2007) The diverse members of the mitochondrial carrier family in plants. FEBS Lett 581:2375–2379

Härtel H, Lokstein H, Dörmann P, Grimm B, Benning C (1997) Changes in the composition of the photosynthetic apparatus in the galactolipid-deficient dgd1 mutant of *Arabidopsis thaliana*. Plant Physiol 115:1175–1184

Härtel H, Essigmann B, Lokstein H, Hoffmann-Benning S, Peters-Kottig M, Benning C (1998) The phospholipid-deficient pho1 mutant of *Arabidopsis thaliana* is affected in the organization, but not in the light acclimation, of the thylakoid membrane. Biochim Biophys Acta 1415:205–218

Härtel H, Dörmann P, Benning C (2000) DGD1-independent biosynthesis of extraplastidic galactolipids after phosphate deprivation in *Arabidopsis*. Proc Natl Acad Sci USA 97:10649–10654

Hartmann-Bouillon MA, Benveniste P (1987) Plant membrane sterols: isolation, identification, and biosynthesis. Methods Enzymol 148:632–650

Haswell ES, Meyerowitz EM (2006) MscS-like proteins control plastid size and shape in *Arabidopsis thaliana*. Curr Biol 16:1–11

Havaux M, Eymery F, Porfirova S, Rey P, Dörmann P (2005) The protective functions of vitamin E against photooxidative stress in *Arabidopsis thaliana*. Plant Cell 17:3451–3469

Heinz E (1977) Enzymatic reactions in galactolipid biosynthesis. In: Tevini A, Lichtenthaler HK (eds) Lipids and lipid polymers in higher plants. Springer, Berlin, pp 102–120

Heinz E (1996) Plant Glycolipids. In: Christie WW (ed) Advances in lipid methodology – three, chap 6. The Oily, Dundee, pp 211–332

Heinz E, Harwood JL (1977) Incorporation of carbon dioxide, acetate and sulphate into the glycerolipids of *Vicia faba* leaves. Hoppe Seylers Z Physiol Chem 358:897–908

Heinz E, Roughan PG (1983) Similarities and differences in lipid metabolism of chloroplasts isolated from 18:3 and 16:3 plants. Plant Physiol 72:273–279

Helliwell CA, Sullivan JA, Mould RM, Gray JC, Peacock WJ, Dennis ES (2001) A plastid envelope location of *Arabidopsis ent*-kaurene oxidase links the plastid and endoplasmic reticulum steps of the gibberellin biosynthesis pathway. Plant J 28:201–208

Hemmler R, Becker T, Schleiff E, Bölter B, Stahl T, Soll J, Götze TA, Braams S, Wagner R (2006) Molecular properties of Oep21, an ATP-regulated anion-selective solute channel from the outer chloroplast membrane. J Biol Chem 281:12020–12029

Hiltbrunner A, Bauer J, Alvarez-Huerta M, Kessler F (2001) Protein translocon at the *Arabidopsis* outer chloroplast membrane. Biochem Cell Biol 79:629–635

Hofmann NR, Theg SM (2005) Chloroplast outer membrane protein targeting and insertion. Trends Plant Sci. 10:450–457

Huang J, Taylor JP, Chen JG, Uhrig JF, Schnell DJ, Nakagawa T, Korth KL, Jones AM (2006) The plastid protein THYLAKOID FORMATION1 and the plasma membrane G-protein GPA1 interact in a novel sugar-signaling mechanism in *Arabidopsis*. Plant Cell 18:1226–1238

Imaizumi-Anraku H, Takeda N, Charpentier M, Perry J, Miwa H, Umehara Y, Kouchi H, Murakami Y, Mulder L, Vickers K, Pike J, Downie JA, Wang T, Sato S, Asamizu E, Tabata S, Yoshikawa M, Murooka Y, Wu GJ, Kawaguchi M, Kawasaki S, Parniske M, Hayashi M (2005) Plastid proteins crucial for symbiotic fungal and bacterial entry into plant roots. Nature 433:527–531

Jarvis P, Soll J (2002) Toc, tic, and chloroplast protein import. Biochim Biophys Acta 1590:177–189

Jarvis P, Dörmann P, Peto CA, Lutes J, Benning C, Chory J (2000) Galactolipid deficiency and abnormal chloroplast development in the *Arabidopsis* MGD synthase 1 mutant. Proc Natl Acad Sci USA 97:8175–8179

Jeffrey SW, Douce R, Benson AA (1974) Carotenoid transformations in the chloroplast envelope. Proc Natl Acad Sci USA 71:807–810

Jouhet J, Maréchal E, Baldan B, Bligny R, Joyard J, Block MA (2004) Phosphate deprivation induces transfer of DGDG galactolipid from chloroplast to mitochondria. J Cell Biol 167:863–874

Jouhet J, Maréchal E, Block MA (2007) Glycerolipid transfer for the building of membranes in plant cells. Prog Lipid Res 46:37–55

Kasahara M, Kagawa T, Oikawa K, Suetsugu N, Miyao M, Wada M (2002) Chloroplast avoidance movement reduces photodamage in plants. Nature 420:829–832

Kawai-Yamada M, Saito Y, Jin L, Ogawa T, Kim K-M, Yu L-H, Tone Y, Hirata A, Umeda M, Uchimiya H (2005) A novel *Arabidopsis* gene causes Bax-like lethality in *Saccharomyces cerevisiae*. J Biol Chem 280:39468–39473

Kelly AA, Dörmann P (2002) DGD2, an *Arabidopsis* gene encoding a UDP-galactose-dependent digalactosyldiacylglycerol synthase is expressed during growth under phosphate-limiting conditions. J Biol Chem 277:1166–1173

Kelly AA, Froehlich JE, Dörmann P (2003) Disruption of the two digalactosyldiacylglycerol synthase genes *DGD1* and *DGD2* in *Arabidopsis* reveals the existence of an additional enzyme of galactolipid synthesis. Plant Cell 15:2694–2706

Kessler F, Schnell DJ (2006) The function and diversity of plastid protein import pathways: a multilane GTPase highway into plastids. Traffic 7:248–257

Klaus D, Hartel H, Fitzpatrick LM, Froehlich JE, Hubert J, Benning C, Dörmann P (2002) Digalactosyldiacylglycerol synthesis in chloroplasts of the *Arabidopsis* dgd1 mutant. Plant Physiol 128:885–895

Kobayashi K, Kondo M, Fukuda H, Nishimura M, Ohta H (2007) Galactolipid synthesis in chloroplast inner envelope is essential for proper thylakoid biogenesis, photosynthesis, and embryogenesis. Proc Natl Acad Sci USA 104:17216–17221

Koo AJ, Ohlrogge JB (2002) The predicted candidates of *Arabidopsis* plastid inner envelope membrane proteins and their expression profiles. Plant Physiol 130:823–836

Kushnir S, Babiychuk E, Storozhenko S, Davey MW, Papenbrock J, Rycke RD, Engler G, Stephan UW, Lange H, Kispal G, Lill R, Montagu MV (2001) A mutation of the mitochondrial ABC transporter Sta1 leads to dwarfism and chlorosis in the *Arabidopsis* mutant *starik*. Plant Cell 13:89–100

Lamkemeyer P, Laxa M, Collin V, Li W, Finkemeier I, Schöttler MA, Holtkamp V, Tognetti VB, Issakidis-Bourguet E, Kandlbinder A, Weis E, Miginiac-Maslow M, Dietz KJ (2006) Peroxiredoxin Q of *Arabidopsis thaliana* is attached to the thylakoids and functions in context of photosynthesis. Plant J 45:968–981

Lange BM, Ghassemian M (2003) Genome organization in *Arabidopsis thaliana*: a survey for genes involved in isoprenoid and chlorophyll metabolism. Plant Mol Biol 51:925–948

Li L, Tutone AF, Drummond RS, Gardner RC, Luan S (2001) A novel family of magnesium transport genes in *Arabidopsis*. Plant Cell 13:2761–2775

Lichtenthaler HK, Prenzel H, Douce R, Joyard J (1981) Localization of prenylquinones in the envelopes of spinach chloroplasts. Biochim Biophys Acta 641:99–105

Lu B, Xu C, Awai K, Jones AD, Benning C (2007) A small ATPase protein of *Arabidopsis*, TGD3, involved in chloroplast lipid import. J Biol Chem Dec 282:35945–35953

Maple J, Møller SG (2006) Plastid division: evolution, mechanism and complexity. Ann Bot (Lond) 99:565–579
Maple J, Fujiwara MT, Kitahata N, Lawson T, Baker NR, Yoshida S, Møller SG (2004) GIANT CHLOROPLAST 1 is essential for correct plastid division in *Arabidopsis*. Curr Biol 14:776–781
Maple J, Aldridge C, Møller SG (2005) Plastid division is mediated by combinatorial assembly of plastid division proteins. Plant J 43:811–823
Maple J, Mateo A, Møller SG (2008) Plastid division regulation and interactions with the environment. Plant Cell Monogr., doi:10.1007/7089_2008_20
Marmagne A, Salvi D, Rolland N, Ephritikhine G, Joyard J, Barbier-Brygoo H (2006) Proteomics of *Arabidopsis* membrane proteins. In: Salinas J, Sanchez-Serrano JJ (eds) Arabidopsis protocols, 2nd edn. Methods in molecular biology, vol. 323. Humana, Totowa, pp 403–420
Mäser P, Thomine S, Schroeder JI, Ward JM, Hirschi K, Sze H, Talke IN, Amtmann A, Maathuis FJ, Sanders D, Harper JF, Tchieu J, Gribskov M, Persans MW, Salt DE, Kim SA, Guerinot ML (2001) Phylogenetic relationships within cation transporter families of *Arabidopsis*. Plant Physiol 126:1646–1667
Meskauskiene R, Nater M, Goslings D, Kessler F, op den Camp R, Apel K (2001) FLU: a negative regulator of chlorophyll biosynthesis in *Arabidopsis thaliana*. Proc Natl Acad Sci 98:12826–12831
Miège C, Maréchal E, Shimojima M, Awai K, Block MA, Ohta H, Takamiya K, Douce R, Joyard J (1999) Biochemical and topological properties of type A MGDG synthase, a spinach chloroplast envelope enzyme catalyzing the synthesis of both prokaryotic and eukaryotic MGDG. Eur J Biochem 265:990–1001
Millar AH, Heazlewood JL (2003) Genomic and proteomic analysis of mitochondrial carrier proteins in *Arabidopsis*. Plant Physiol 131:443–453
Miras S, Salvi D, Ferro M, Grunwald D, Garin J, Joyard J, Rolland N (2002) Non-canonical transit peptide for import into the chloroplast. J Biol Chem 277:47770–47778
Miras S, Salvi D, Piette L, Seigneurin-Berny D, Grunwald D, Reinbothe C, Joyard J, Reinbothe S, Rolland N (2007) Toc159- and Toc75-independent import of a transit sequence-less precursor into the inner envelope of chloroplasts. J Biol Chem 282:29482–29492
Mooney BP, Miernyk JA, Randall DD (2002) The complex fate of alpha-ketoacids. Annu Rev Plant Biol 53:357–375
Moscat J, Diaz-Meco MT, Albert A,Campuzano1 S (2006) Cell signaling and function organized by PB1 domain interactions. Mol Cell 23:631–640
Motohasi R, Ito T, Kobayashi M, Taji T, Nagata N, Asami T, Yoshida S, Yamaguchi-Shinozaki K, Shinozaki K (2003) Functional analysis of the 37 kDa inner envelope membrane polypeptide in chloroplast biogenesis using Ds-tagged *Arabidopsis* pale-green mutant. Plant J 34:719–731
Muller F, Frentzen M (2001) Phosphatidylglycerophosphate synthases from *Arabidopsis thaliana*. FEBS Lett 509:298–302
Mullineaux P, Karpinski S (2002) Signal transduction in response to excess light: getting out of the chloroplast. Curr Opin Plant Biol 5:43–48
Nada A, Soll J (2004) Inner envelope protein 32 is imported into chloroplasts by a novel pathway. J Cell Sci 117:3975–3982
Nielsen E, Akita M, Davila-Aponte J, Keegstra K (1997) Stable association of chloroplastic precursors with protein translocation complexes that contain proteins from both envelope membranes and a stromal hsp 100 molecular chaperone. EMBO J 16:935–946
Nott A, Jung HS, Koussevitzky S, Chory J (2006) Plastid-to-nucleus retrograde signaling. Annu Rev Plant Biol 57:739–759
Oikawa K, Kasahara M, Kiyosue T, Kagawa T, Suetsugu N, Takahashi F, Kanegae T, Niwa Y, Kadota A, Wada M (2003) Chloroplast unusual positioning1 is essential for proper chloroplast positioning. Plant Cell 15:2805–2815
Ortiz-Lopez A, Chang HC, Bush DR (2001) Amino acid transporters in plants. Biochim Biophys Acta 1465:275–280

Palmieri L, Arrigoni R, Blanco E, Carrari F, Zanor MI, Studart-Guimaraes C, Fernie AR, Palmieri F (2006) Molecular identification of an *Arabidopsis* S-adenosylmethionine transporter. Analysis of organ distribution, bacterial expression, reconstitution into liposomes, and functional characterization. Plant Physiol 142:855–865

Pineau B, Dubertret G, Joyard J, Douce R (1986) Fluorescence properties of the envelope membranes from spinach. J Biol Chem 261:9210–9215

Pineau B, Gérard-Hirne C, Douce R, Joyard J (1993) Identification of the main species of tetrapyrrolic pigments in envelope membranes from spinach chloroplasts. Plant Physiol 102:821–828

Pohlmeyer K, Soll J, Grimm R, Hill K, Wagner R (1998) A high-conductance solute channel in the chloroplastic outer envelope from Pea. Plant Cell 10:1207–1216

Pontier D, Albrieux C, Joyard J, Lagrange T, Block MA (2007) Knock-out of the magnesium protoporphyrin IX methyltransferase gene in *Arabidopsis*. Effects on chloroplast development and on chloroplast-to-nucleus signaling. J Biol Chem 282:2297–2304

Ravanel S, Cherest H, Jabrin S, Grunwald D, Surdin-Kerjan Y, Douce R, Rébeillé F (2001) Tetrahydrofolate biosynthesis in plants: molecular and functional characterization of dihydrofolate synthetase and three isoforms of folylpolyglutamate synthetase in *Arabidopsis thaliana*. Proc Natl Acad Sci USA 98:15360–15365

Ravanel S, Block MA, Rippert P, Jabrin S, Curien G, Rébeillé F, Douce R (2004) Methionine metabolism in plants: chloroplasts are autonomous for de novo methionine synthesis and can import S-adenosylmethionine from the cytosol. J Biol Chem 279:22548–22557

Rawyler A, Meylan-Bettex M, Siegenthaler PA (1995) (Galacto) lipid export from envelope to thylakoid membranes in intact chloroplasts. II. A general process with a key role for the envelope in the establishment of lipid asymmetry in thylakoid membranes. Biochim Biophys Acta 1233:123–133

Reinbothe S, Reinbothe C, Holtorf H, Apel K (1995) Two NADPH:Protochlorophyllide oxidoreductases in barley: evidence for the selective disappearance of PORA during the light-induced greening of etiolated seedlings. Plant Cell 7:1933–1940

Reinbothe S, Mache R, Reinbothe C (2000) A second, substrate-dependent site of protein import into chloroplasts. Proc Natl Acad Sci USA 97:9795–9800

Renne P, Dressen U, Hebbeker U, Hille D, Flügge UI, Westhoff P, Weber AP (2003) The *Arabidopsis* mutant dct is deficient in the plastidic glutamate/malate translocator DiT2. Plant J 35:316–331

Rolland N, Ferro M, Seigneurin-Berny D, Garin J, Douce R, Joyard J (2003) Proteomics of chloroplast envelope membranes. Photosynth Res 78:205–230

Rose AK, Schraegle SJ, Stahlberg EA, Meier I (2005) Coiled-coil protein composition of 22 proteomes – differences and common themes in subcellular infrastructure and traffic control. BMC Evol Biol 5:66

Rullkötter J, Heinz E, Tulloch AP (1975) Combination and positional distribution of fatty acids in plant digalactosyl diglycerides. Z Pflanzenphysiol 76:163–175

Sadre R, Gruber J, Frentzen M (2006) Characterization of homogentisate prenyltransferases involved in plastoquinone-9 and tocochromanol biosynthesis. FEBS Lett 580:5357–5362

Sakamoto W (2006) Protein degradation machineries in plastids. Annu Rev Plant Biology 57:599–621

Salvi D, Rolland N, Joyard J, Ferro M (2008a) Purification and proteomic analysis of chloroplasts and their sub-organellar compartments. In: P ieger D, Rossier J (eds) Organelle proteomics. Methods in molecular biology, vol 432. Humana, Totowa

Salvi D, Rolland N, Joyard J, Ferro M (2008b) Assessment of organelle purity using antibodies and specific assays: the example of the chloroplast envelope. In: P ieger D, Rossier J (eds) Organelle proteomics. Methods in molecular Biology, vol 432. Humana, Totowa

Schleiff E, Soll J (2005) Membrane protein insertion: mixing eukaryotic and prokaryotic concepts. EMBO Rep 6:1023–1027

Schnurr JA, Shockey JM, de Boer GJ, Browse JA (2002) Fatty acid export from the chloroplast. Molecular characterization of a major plastidial acyl-coenzyme A synthetase from *Arabidopsis*. Plant Physiol 129:1700–1709

Schwacke R, Schneider A, Van Der Graaff E, Fischer K, Catoni E, Desimone M, Frommer WB, Flügge UI, Kunze R (2003) ARAMEMNON, a novel database for *Arabidopsis* integral membrane proteins. Plant Physiol 131:16–26

Seigneurin-Berny D, Rolland N, Garin J, Joyard J (1999) Differential extraction of hydrophobic proteins from chloroplast envelope membranes: a subcellular-specific proteomic approach to identify rare intrinsic membrane proteins. Plant J 19:217–228

Seigneurin-Berny D, Rolland N, Dorne AJ, Joyard J (2000) Sulfolipid is a potential candidate for annexin binding to the outer surface of chloroplast. Biochem Biophys Res Commun 272:519–524

Seigneurin-Berny D, Gravot A, Auroy P, Mazard C, Kraut A, Finazzi G, Grunwald D, Rappaport F, Vavasseur A, Joyard J, Richaud P, Rolland N (2006) HMA1, a new Cu-ATPase of the chloroplast envelope, is essential for growth under adverse light conditions. J Biol Chem 281:2882–2892

Seigneurin-Berny D, Salvi D, Joyard J, Rolland N (2008) Purification of chloroplasts from two model plants: Arabidopsis and spinach. Curr Protocols Cell Biol 33 (in press)

Seki M, Narusaka M, Abe H, Kasuga M, Yamaguchi-Shinozaki K, Carninci P, Hayashizaki Y, Shinozaki K (2001) Monitoring the expression pattern of 1,300 Arabidopsis genes under drought and cold stresses by using a full-length cDNA microarray. Plant Cell 13, 61–72

Shikanai T, Muller-Moule P, Munekage Y, Niyogi KK, Pilon M (2003) PAA1, a P-type ATPase of *Arabidopsis*, functions in copper transport in chloroplasts. Plant Cell 15:1333–1346

Shingles R, Wimmers LE, McCarty RE (2004) Copper transport across pea thylakoid membranes. Plant Physiol 135:145–151

Siebertz HP, Heinz E, Linscheid M, Joyard J, Douce R (1979) Characterization of lipids from chloroplast envelopes. Eur J Biochem 101:429–438

Siefermann-Harms D, Joyard J, Douce R (1978) Light-induced changes of the carotenoid levels in chloroplast envelopes. Plant Physiol 61:530–533

Slack CR, Roughan PG, Balasingham N (1977) Labelling studies in vivo on the metabolism of the acyl and glycerol moieties of the glycerolipids in the developing maize leaf. Biochem J 162:289–296

Soll J, Schultz G, Joyard J, Douce R, Block MA (1985) Localization and synthesis of prenylquinones in isolated outer and inner envelope membranes from spinach chloroplasts. Arch Biochem Biophys 238:290–299

Strand A, Asami T, Alonso J, Ecker JR, Chory J (2003) Chloroplast to nucleus communication triggered by accumulation of Mg-protoporphyrinIX. Nature 421:79–83

Suetsugu N, Wada M (2008) Chloroplast photorelocation movement. Plant Cell Monogr., doi:10.1007/7089_2008_34

Sun Q, Emanuelsson O, van Wijk KJ (2004) Analysis of curated and predicted plastid subproteomes of *Arabidopsis*. Subcellular compartmentalization leads to distinctive proteome properties. Plant Physiol 135:723–734

Testerink C, Munnik T (2005) Phosphatidic acid: a multifunctional stress signaling lipid in plants. Trends Plant Sci 10:368–375

Teyssier E, Block MA, Douce R, Joyard J (1996) Is E37, a major polypeptide of the inner membrane from plastid envelope, an *S*-adenosyl methionine-dependent methyltransferase? Plant J 10:903–912

Tian L, DellaPenna D, Dixon RA (2007) The pds2 mutation is a lesion in the *Arabidopsis* homogentisate solanesyltransferase gene involved in plastoquinone biosynthesis. Planta 226:1067–1073

The *Arabidopsis* Genome Initiative (2000) Analysis of the genome sequence of the flowering plant *Arabidopsis thaliana*. Nature 408:796–815

Thelen JJ, Ohlrogge JB (2002) The multisubunit acetyl-CoA carboxylase is strongly associated with the chloroplast envelope through non-ionic interactions to the carboxyltransferase subunits. Arch Biochem Biophys 400:245–257

Tovar-Méndez A, Miernyk JA, Randall DD (2003) Regulation of pyruvate dehydrogenase complex activity in plant cells. Eur J Biochem. 270:1043–1049

Vidi PA, Kanwischer M, Baginsky S, Austin JR, Csucs G, Dörmann P, Kessler F, Bréhélin C (2006) Proteomics identify *Arabidopsis* plastoglobules as a major site in tocopherol synthesis and accumulation. J Biol Chem 281:11225–11234

Villarejo A, Burén S, Larsson S, Déjardin A, Monné M, Rudhe C, Karlsson J, Jansson S, Lerouge P, Rolland N, von Heijne G, Grebe M, Bako L, Samuelsson G (2005) Evidence for a protein transported through the secretory pathway en route to the higher plant chloroplast. Nat Cell Biol. 7:1224–1231

Vitha S, Froehlich JE, Koksharova O, Pyke KA, van Erp H, Osteryoung KW (2003) ARC6 is a J-domain plastid division protein and an evolutionary descendant of the cyanobacterial cell division protein Ftn2. Plant Cell 15:1918–1933

Weber AP, Schwacke R, Flugge UI (2005) Solute transporters of the plastid envelope membrane. Annu Rev Plant Biol 56:133–164

Weber APM, Fischer K (2008) The role of metabolite transporters in integrating chloroplasts with the metabolic network of plant cells. Plant Cell Monogr., doi:10.1007/7089_2008_19

Xu CC, Fan J, Froehlich JE, Awai K, Benning C (2005) Mutation of the TGD1 chloroplast envelope protein affects phosphatidate metabolism in *Arabidopsis*. Plant Cell 17:3094–3110

Yamamoto HY (2006) Functional roles of the major chloroplast lipids in the violaxanthin cycle. Planta 224:719–724

Yamamoto HY, Bugos RC, Hieber AD (1999) Biochemistry and molecular biology of the xanthophyll cycle. In: Frank HA, Young AJ, Britton G, Cogdell RJ (eds) Advances in photosynthesis. The photochemistry of carotenoids, vol 8. Kluwer, Dordrecht, pp 293–303

Ytterberg AJ, Peltier J-B, van Wijk KJ (2006) Protein profiling of plastoglobules in chloroplasts and chromoplasts. A surprising site for differential accumulation of metabolic enzymes. Plant Physiol 140:984–997

The Chloroplast Protein Import Apparatus, Its Components, and Their Roles

H. Aronsson(✉) and P. Jarvis

Abstract According to the endosymbiont theory, an early eukaryotic cell engulfed the ancestor of present-day chloroplasts – a relative of extant cyanobacteria. This was the start of a new era for the chloroplast progenitor, because it was placed under the control of the host cell. Many of the genes found originally in the cyanobacterial genome are today present in the cell nucleus. Thus, over time, the chloroplast has learned to live with less and less "home-made" proteins. Nevertheless, the chloroplast retains many of the functions found in cyanobacteria (e.g., photosynthesis, fatty acid and amino acid production). To maintain these functions, many proteins have to be transported "back" to the chloroplast. An import machinery drives this transport process, and this consists of translocons located in the outer and the inner envelope membranes, called TOC and TIC (*T*ranslocon of the *O*uter/*I*nner envelope membrane of *C*hloroplasts). This chapter focuses on these translocons, and summarizes how they mediate import of nucleus-encoded proteins into the chloroplast from the surrounding cytoplasm.

1 Evolution of Chloroplasts and the Import Machinery

More than a billion years ago, an ancestral free-living cyanobacterium was taken up by a eukaryotic cell (Olson 2006). The resulting symbiosis gave the host cell access to valuable resources produced by photosynthesis (e.g., carbohydrates), and in return the cyanobacterium received a stable environmental milieu. Disadvantages of this arrangement included the tendency for oxidative damage to the organellar DNA and the absence of sexual recombination, rendering the plastome vulnerable to the accumulation of serious mutations (Lynch and Blanchard 1998; Martin and Herrman 1998). To avoid this scenario, many genes were transferred from the

---

H. Aronsson
Department of Plant and Environmental Sciences, University of Gothenburg,
Box 461, 405 30 Gothenburg, Sweden
e-mail: henrik.aronsson@dpes.gu.se

plastome to the nucleus, which also gave overall control to the host cell. Modern chloroplasts retain many metabolic pathways from the original endosymbiont, and so proteins that were previously synthesized inside the chloroplast are now transported "back" to the organelle to maintain its former functionality. The system required for the translocation of proteins into the chloroplast consists of at least 20 protein components in *Arabidopsis thaliana* (Jackson-Constan and Keegstra 2001; Jarvis 2008), called either TOC or TIC (*T*ranslocon of the *O*uter/*I*nner envelope membrane of *C*hloroplasts; Schnell et al. 1997).

In cyanobacteria, homologues for only a few of these proteins exist; e.g., Toc75, Tic20, Tic22 and Tic55 (Reumann et al. 1999; 2005). Interestingly, the related transport system in cyanobacteria (based on a homologue of the Toc75 channel) is proposed to mediate secretion from the cell, which is opposite to the direction of transport during chloroplast import. This directionality change might be related to the relocation of the Toc75 gene to the nucleus (von Heijne 1995). The transit peptide (TP) (which is needed to bring preproteins to the chloroplast) may be derived from a secretory peptide in the endosymbiont, which was recognized and secreted by the ancestral Toc75 (Reumann et al. 1999). In relation to the evolution of the import machinery, there are still many unanswered questions; for example, many components of the translocons are not present in cyanobacteria (e.g., Toc34, Toc159, and Tic110). However, this topic has been nicely reviewed by Reumann et al. (2005) and is not within the scope of this chapter.

2 Overview of Chloroplast Protein Import

Around 3,000 different proteins are predicted to exist within chloroplasts, and ~95% of these are nucleus-encoded and so must be imported post-translationally (Abdallah et al. 2000). Most imported proteins are thought to utilize the TOC and TIC translocons (Soll and Schleiff 2004; Bédard and Jarvis 2005; Reumann et al. 2005; Kessler and Schnell 2006). Interestingly, there appear to be at least two different TOC/TIC import pathways, and it is now clear that TOC/TIC-independent or "non-canonical" protein targeting to chloroplasts also occurs. Thus, the notion of a "general import pathway" for all chloroplast proteins is somewhat outdated.

All proteins that follow the TOC/TIC route have a cleavable, N-terminal TP. This acts as a targeting flag, directing the preprotein exclusively to the chloroplast (Smeekens et al. 1986). It has been suggested that the TP can be divided into three domains: the N-terminus is mainly uncharged and proposed to play a role in recognition; the central part lacks acidic residues and mediates translocation over the envelope; finally, the C-terminus is enriched in arginines and involved in TP cleavage inside the chloroplast (von Heijne et al. 1989; Rensink et al. 1998). Despite the presence of common, general features in TPs (a preponderance of hydroxylated residues and a deficiency in acidic residues, giving an overall positive charge), no consensus sequence or structure exists, making it difficult to predict chloroplast location by sequence analysis (Bruce 2000). However, in the last ten years several algorithms have become available for predicting chloroplast localization with reasonable confidence; e.g., ChloroP (Emanuelsson et al. 1999), PSORT (Nakai and Horton 1999), TargetP (Emanuelsson et al. 2000) and Predotar (Small et al. 2004).

In the TOC/TIC pathway, the binding of the preprotein to the chloroplast outer envelope membrane (OEM) is mediated by the TP. In the absence of an energy source, binding to the import apparatus is reversible and no translocation will occur (Perry and Keegstra 1994). This step may also involve interactions between the TP and the outer envelope lipids (Bruce 2000). In the presence of GTP, and low concentrations of ATP (≤ 100 μM), the binding step is irreversible and an early import intermediate is formed (Olsen and Keegstra 1992; Young et al. 1999). At this stage, the preprotein has penetrated the OEM and is in contact also with the inner envelope membrane (IEM) (Wu et al. 1994; Ma et al. 1996). To achieve complete translocation, high ATP concentrations (>100 μM) are required in the stroma, and this is thought to be consumed by stromal molecular chaperones (Pain and Blobel 1987; Theg et al. 1989). In addition to the essential role of the TP, the mature part of the preprotein has also been reported to influence the interaction between the preprotein and the translocon (Dabney-Smith et al. 1999).

3 Initial Contact Between Preproteins and the Outer Membrane

Several hypotheses exist for the transport of nucleus-encoded proteins from the cytosol, where they are synthesized, to the chloroplast surface. One proposal involves a so-called "guidance complex," which brings the preprotein to the TOC components. A second hypothesis entails contact with OEM lipids, which might induce changes in the bilayer to facilitate contact with a nearby TOC complex (Bruce 2000). Another possibility is direct interaction with the TOC complex, mediated by membrane-integrated receptors (Toc34 or Toc159). A variation on the latter involves a soluble form of Toc159, which first recognizes the preprotein in the cytosol and, like the guidance complex, brings the preprotein to the TOC machinery (Fig. 1; Hiltbrunner et al. 2001b). Finally, a putative third TOC component, Toc64, has been suggested to act as a receptor for a subset of proteins pre-bound by Hsp90; however, the relevance of this idea is debated (Qbadou et al. 2006, 2007; Aronsson et al. 2007; see Sect. 4 for TOC receptors).

3.1 The Guidance Complex

The guidance complex consists of 14-3-3 and Hsp70 proteins; it interacts with the TP and targets it to the chloroplast OEM (May and Soll 2000). Formation of the guidance complex is proposed to be dependent on the phosphorylation of a serine or threonine residue within the TP (Waegemann and Soll 1996; May and Soll 2000). Use of the guidance complex might "fast track" certain preproteins (e.g.,

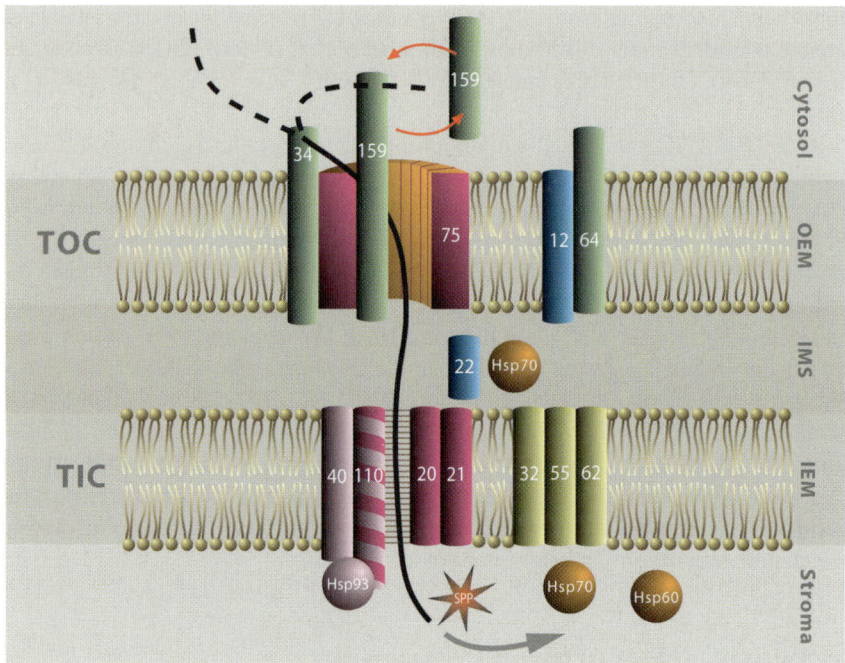

Fig. 1 *Chloroplast protein translocation machinery.* The preprotein (*black line*) approaches the outer envelope membrane (OEM) and is recognized by one of the TOC receptors (*green*): either (A) the membrane bound Toc34 receptor, or (B) a soluble form of the Toc159 receptor (*dashed lines*). Then, the preprotein enters the Toc75 channel (*magenta/orange*) and reaches the intermembrane space (IMS), where the TOC and TIC meet each other. IMS associated proteins (*blue*) mediate movement of the preprotein towards TIC channel components (*magenta*). Redox-related TIC subunits (*yellow*) may assist the transfer of some preproteins through the inner envelope membrane (IEM). Finally, the preprotein is driven into the stroma with help from a putative motor complex (*pink*), and the transit peptide (*grey arrow*) is removed by the stromal processing peptidase (SPP)

small subunit of Rubisco; pSS) that are needed at high levels, since import in the presence of the guidance complex becomes ~3–4-times faster (May and Soll 2000). However, only a few proteins have been experimentally tested for usage of the guidance complex (May and Soll 2000). Moreover, in a study by Nakrieko et al. (2004) using GFP-tagged pSS proteins, no effects of phosphorylation site mutations within the TP could be shown. Thus, the guidance complex is not a prerequisite for import; indeed, preproteins in many in vitro studies lack the guidance complex and still show functional import (Dabney-Smith et al. 1999; Aronsson et al. 2001).

3.2 The Involvement of Galactolipids

The galactolipids, monogalactosyldiacylglycerol (MGDG) and digalactosyldiacylglycerol (DGDG), are the most abundant, non-proteinaceous constituents of chloroplast membranes (Douce and Joyard 1990). The MGDG lipid has been suggested to influence protein import into chloroplasts (for pSS and pFerredoxin) as well as insertion into the chloroplast OEM (for OEP7) (van't Hof et al. 1991, 1993; Chupin et al. 1994; Pilon et al. 1995; Pinnaduwage and Bruce 1996; Bruce 1998; Schleiff et al. 2001; Hofmann and Theg 2005a). However, other reports suggested no specific role for MGDG in import (Inoue et al. 2001; Schleiff et al. 2003b). Chloroplast protein import, or insertion into the OEM, in an *Arabidopsis* MGDG-deficient mutant, *mgd1*, showed equal performance as compared with wild type (Aronsson et al. 2008). Thus, in vivo data suggest that MGDG is not as important for chloroplast protein targeting as was implied by earlier in vitro studies. A major difference between the in vivo analysis and the previous studies was the use of intact chloroplasts instead of artificial lipid layers or vesicles that mimic OEM lipid composition (the latter lack the many proteins present in the OEM). Nevertheless, one must bear in mind that the *mgd1* mutant contains only ~40% less MGDG than wild type (Jarvis et al. 2000). A recently described MGD1 null mutant displays a very severe phenotype, and is so sick that it is unlikely it could be used for import studies (Kobayashi et al. 2007).

In contrast, similar experiments using the DGDG-deficient mutant, *dgd1*, did reveal a significant defect in protein import (Chen and Li 1998; Aronsson and Jarvis 2002). These results can be interpreted in two different ways. Firstly, they may indicate that DGDG is relatively more important for chloroplast import than MGDG. This possibility is supported by the observation that DGDG was the only galactolipid associated to the TOC complex (Schleiff et al. 2003b), and by the fact that in *mgd1* the level of DGDG is not altered (Jarvis et al. 2000). However, such an interpretation would be in disagreement with a host of in vitro studies, which suggested a particularly important role for MGDG (van't Hof et al. 1991, 1993; Chupin et al. 1994; Pilon et al. 1995; Pinnaduwage and Bruce 1996; Bruce 1998; Schleiff et al. 2001). An alternative explanation for the *dgd1* results is that they simply reflect the fact that this mutant has a much more severe lipid defect: the *dgd1* mutant shows a 90% reduction in DGDG levels (Dörmann et al. 1995).

4 Receptors at the TOC Complex

The Toc34 and Toc159 components are related GTPases, and were first identified in pea as being involved in preprotein recognition and binding (Hirsch et al. 1994; Kessler et al. 1994; Seedorf et al. 1995). Initially, Toc159 was seen as a fragment of 86 kDa, and so the name Toc86 was used. However, due to the identification of a larger homologue in *Arabidopsis*, the pea Toc86 protein was further scrutinized and shown to have a native size comparable to the *Arabidopsis* protein; the 86-kDa pea protein was produced by proteolytic degradation during the experiments (Bölter et al. 1998a; Chen et al. 2000). A putative third receptor, Toc64, was also identified in pea (Sohrt and Soll 2000), but its role is less well defined.

4.1 Toc34

So far, only one Toc34 isoform has been identified in pea (*Pisum sativum* Toc34, or psToc34), but two homologues exist in *Arabidopsis thaliana* (Table 1; atToc33 and atToc34); both *Arabidopsis* proteins are very similar to psToc34 (~60% identity) (Jarvis et al. 1998; Voigt et al. 2005). Toc34 consists of a cytosolic GTPase (G)

Table 1 The *Arabidopsis* proteins most closely related to putative or actual components of the pea chloroplast protein import apparatus, their proposed functions, and other homologues

Component (generic name)	Main isoform in *Arabidopsis*	AGI no.	Domains or motifs	Proposed function(s)	Additional homologues in *Arabidopsis*
Toc12	atToc12	At1g80920	DnaJ	Co-chaperone	None detected
Toc34	atToc33	At1g02280	GTPase	Preprotein receptor	atToc34
Toc64	atToc64-III	At3g17970	Amidase; TPR[a]	Receptor; unknown	atToc64-I, -V
Toc75	atToc75-III	At3g46740	POTRA[a]; ß-barrel	Import channel	atToc75-IV, -V
Toc159	atToc159	At4g02510	GTPase; A and M domains	Preprotein receptor; import motor	atToc90, 120, 132
Tic20	atTic20-I	At1g04940	Polytopic, α-helical TMDs[a]	Import channel	atTic20-II, -IV, -V
Tic21	atTic21	At2g15290	Polytopic, α-helical TMDs[a]	Import channel; permease	None detected
Tic22	atTic22-IV	At4g33350	None detected	TOC–TIC interaction	atTic22-III
Tic32	atTic32-IVa	At4g23430	NADPH-dependent short chain dehydrogenase	Redox/calcium sensing	atTic32-IVb[b]
Tic40	atTic40	At5g16620	TPR[a]; Sti1	Co-chaperone	None detected
Tic55	atTic55-II	At2g24820	Rieske iron-sulfur centre; mononuclear iron site	Redox sensing	None detected[b]
Tic62	atTic62	At3g18890	NAD(P)H dehydrogenase; FNR-binding site	Redox sensing	None detected[b]
Tic110	atTic110	At1g06950	TP-binding site; Tic40-binding site	Import channel; Chaperone recruitment	None detected
Hsp93	atHsp93-V	At5g50920	Walker ATPase, ClpC/Hsp100	Import motor	atHsp93-III
SPP	atSPP	At5g42390	Zinc-binding, pitrilysin	TP cleavage	None detected

For a more comprehensive table, see Jarvis (2008)
[a] *TPR*, tetratricopeptide repeat; *POTRA*, polypeptide transport associated domain; *TMDs*, transmembrane domains
[b] Please also refer to Kalanon et al. (2008) and references therein

domain and a short, membrane-spanning helix at the C-terminal end (Fig. 1) (Kessler et al. 1994; Seedorf et al. 1995). Binding of GTP is thought to be necessary for Toc34 to receive incoming preproteins, and for it to act as a receptor (Kouranov and Schnell 1997; Sveshnikova et al. 2000b). According to one model for preprotein recognition, the structure of Toc34 changes upon GTP hydrolysis and the preprotein is released towards Toc159 and the Toc75 channel. In another model, Toc34 functions as a receptor for an incoming preprotein-Toc159 complex (Sect. 4.2).

Crystals of psToc34 in the GDP-bound state showed that the receptor can dimerize. It was suggested that each GTPase within a dimer acts as GTPase-activating protein (GAP) for the opposing monomer (Sun et al. 2002; Bos et al. 2007); however, it was subsequently reported that mutation of the putative arginine finger (R130 in atToc33; arginine fingers are critical mediators of GAP function) did not affect GTP hydrolysis (Weibel et al. 2003). Because Toc34 and Toc159 share significant homology within their G-domains, it has been suggested that both receptors dimerize, and, even more interestingly, that heterodimerization may be an important component of the import mechanism (Kessler and Schnell 2002). Assembly of the TOC translocon, as well as precursor transport, is also suggested to be dependent on heterodimerization (Wallas et al. 2003). However, there is considerable disagreement concerning the consequences of dimerization for receptor activity (Sun et al. 2002; Weibel et al. 2003; Reddick et al. 2007; Yeh et al. 2007). Interestingly, atToc33 was found to have unusual properties, exhibiting affinity for both GTP and XTP in its wild-type state (Aronsson et al. 2003a). Clearly, the structure of Toc34 and its GTPase activity are not yet fully understood.

The binding of GTP to Toc34 has been proposed to be controlled by receptor phosphorylation. When Toc34 is phosphorylated, GTP is not bound and the action of the receptor is inhibited (Sveshnikova et al. 2000b). The phosphorylation site has been identified as serine 113 in psToc34, and as serine 181 in atToc33 (Jelic et al. 2002; 2003). When S181 was substituted with alanine, aspartate or glutamate (predicted to create constitutively active or inactive mutants), the in vivo activity of the protein was not detectably altered (Aronsson et al. 2006). While a more recent study suggested that these mutations may have subtle consequences during the earliest stages of the development (Oreb et al. 2007), the fact remains that the role of the phosphorylation is far from clear. It could be more important at earlier stages of plant development when the need for import is higher (Dahlin and Cline 1991), but it is difficult to understand why the regulation of two orthologues (psToc34 and atToc33) might be mediated by phosphorylation at very different positions (both of which lack conservation in other species), and, presumably, through completely different mechanisms. A kinase was shown to phosphorylate Toc34 in vitro, but its identity is unknown (Fulgosi and Soll 2002).

The function of Toc34 in preprotein recognition is widely accepted, but whether Toc34 or Toc159 is the primary receptor for the preprotein is debated. Two models have been proposed: the "motor model" places Toc34 in this role, while the "targeting model" has Toc159 as the primary receptor (Sect. 4.2). Evidence arguing for Toc34 as the primary receptor includes: preprotein interaction with Toc34 in vitro (Jelic et al. 2002; Becker et al. 2004b; Reddick et al. 2007); greater quantities of Toc34 (versus Toc159) in the envelope membrane (Schleiff et al. 2003b; Kikuchi et al. 2006; Chen and Li 2007); proposed interaction with the guidance complex (Qbadou et al. 2006); and, the lack of direct input from Toc34 during translocation over the membrane (Schleiff et al. 2003a).

Whether or not Toc34 is the primary receptor, it seems that different Toc34 isoforms have specific preferences for certain preproteins. Proteomic studies by Kubis et al (2003) on an atToc33 null mutant, named *ppi1* (*plastid protein import 1*; Jarvis et al. 1998), showed that photosynthetic proteins are specifically deficient in the mutant, whereas non-photosynthetic, housekeeping proteins are rather stable. The same is true for an atToc159 mutant, termed *ppi2* (Bauer et al. 2000), and so import into *Arabidopsis* chloroplasts is proposed to follow two different pathways (Sect. 4.2): atToc33 and atToc159 preferentially import photosynthetic proteins; atToc34 and atToc132/atToc120 import housekeeping proteins. Nevertheless, "cross-talk" between the pathways seems to occur (Hiltbrunner et al. 2001a).

While the atToc33 knockout mutant (*ppi1*) is pale (Jarvis et al. 1998), an atToc34 null mutant (*ppi3*) shows no obvious phenotypes in aerial, photosynthetic tissues, but its roots (which are non-photosynthetic) are shorter than normal (Constan et al. 2004b). In spite of their specialization, double-mutant studies on the homologues revealed a considerable degree of functional overlap: the atToc33/atToc34 double-null genotype is embryo-lethal, while plants homozygous for *ppi1* and heterozygous for *ppi3* are even paler than *ppi1* alone. Multiple Toc34 isoforms also exist in moss (*Physcomitrella patens*), maize and spinach, and in the latter case evidence suggests that the isoforms exhibit functional specialization (Voigt et al. 2005).

4.2 Toc159

The Toc159 family consists of four members in *Arabidopsis* – atToc159, atToc132, atToc120 and atToc90 (Table 1; Jackson-Constan and Keegstra 2001; Hiltbrunner et al. 2001a) – while in pea presently only psToc159 (formerly Toc86) has been identified (Hirsch et al. 1994; Kessler et al. 1994; Perry and Keegstra 1994). The psToc159 protein is most similar to atToc159 (48% identity), and so these two are believed to be functional orthologues (Bauer et al. 2000). Toc159 proteins have three domains: an N-terminal acidic (A) domain which is very sensitive to proteolysis, giving rise to an 86-kDa fragment (Hirsch et al. 1994; Kessler et al. 1994; Schnell et al. 1994); a central GTPase (G) domain related to the Toc34 G-domain; and, a hydrophilic M-domain that anchors atToc159 in the membrane (Bauer et al. 2000; Hiltbrunner et al. 2001a). The M-domain is a 52-kDa protease-resistant region that does not carry typical hydrophobic, transmembrane helices, and so is unusual (Hirsch et al. 1994; Kessler et al. 1994; Bauer et al. 2000).

In the "targeting model" of preprotein recognition, Toc159 acts as the primary receptor (Fig. 1). This is partly based on cross-linking studies using preproteins arrested during the binding step of import: under these conditions, Toc159 is the major TOC component cross-linked to the preprotein (Perry and Keegstra 1994; Ma et al. 1996; Kouranov and Schnell 1997). The blocking of early import intermediate formation by applying a Toc159 antibody also supports the primary receptor role of

Toc159 (Hirsch et al. 1994). Detection of an abundant form of atToc159 in the cytosol gave this model another dimension (Hiltbrunner et al. 2001b). This was seen in protoplasts by immunofluorescence microscopy, and supported by subfractionation studies in *Arabidopsis* and pea. It was proposed that Toc159 recognizes the preprotein in the cytosol, and then brings it to the TOC complex; in fact, soluble Toc159 was shown to interact exclusively with the TP of preproteins (Smith et al. 2004). This implies that Toc159 cycles between its soluble, cytosolic form and its membrane-integrated form (Fig. 1). The atypical hydrophilic M-domain of Toc159 may play a critical role in this integration/de-integration process. However, the relevance of cytosolic Toc159 has been questioned, and attributed to partial disruption of membranes due to the experimental procedures (Becker et al. 2004b). Whether this is true or not remains unanswered.

The "targeting model" is comparable to the co-translational translocation of proteins into the ER by the *s*ignal *r*ecognition *p*article (SRP) system. The latter starts with recognition of a nascent signal peptide, protruding from a ribosome, by the SRP (a GTPase). This complex carries the preprotein to the SRP-receptor (another GTPase) where GTP hydrolysis ensures preprotein transfer to the Sec translocase for transport over the membrane (Shan and Walter 2005; Bange et al. 2007). Toc159 is proposed to play a role analogous to that of SRP, while Toc34 may be analogous to the SRP-receptor. Intriguingly, a distant homology between TOC GTPases and chloroplastic SRP and SRP-receptors was recently presented (Hernández Torres et al. 2007). The SRP system can also act post-translationally, for example within chloroplasts for thylakoid targeting (Schuenemann 2004), which is more similar to the way in which chloroplast import occurs.

The Toc34 protein is believed to mediate the insertion of Toc159 in a GTPase-regulated fashion (Wallas et al. 2003). Hence, the initial binding to the membrane could occur by a heterodimerization between Toc34 and Toc159, both in the GTP state. That would perhaps induce GTP hydrolysis (each receptor acting as a GAP on the other), and thereby induce integration of the Toc159 M-domain. However, the role of heterodimerization in relation to GTP hydrolysis is debated (Sun et al. 2002; Weibel et al. 2003; Reddick et al. 2007; Yeh et al. 2007). Integration of a Toc159 receptor pre-bound to a preprotein client might simultaneously initiate the membrane translocation of the precursor (Fig. 1).

In the "motor model", where Toc34 is seen as the primary receptor (Sect. 4.1), Toc159 is permanently associated with the membrane and acts as a motor by driving the preprotein forward through the Toc75 channel (Fig. 1). The Toc159 motor action is powered by multiple cycles of GTP hydrolysis, each one pushing a new part of the preprotein into the channel (Schleiff et al. 2003a; Becker et al. 2004b). The model is based on several lines of evidence. Firstly, a minimal TOC complex consisting of a Toc159 fragment and Toc75 was able to mediate transport of preproteins into proteoliposomes at the expense of GTP hydrolysis (Schleiff et al. 2003a). Secondly, in isolated TOC core-complexes (whose stoichiometry was estimated to be 4–5:4:1 or 3:3:1 for Toc34:Toc75:Toc159), Toc34 and Toc75 were found in almost equal amounts (Schleiff et al. 2003b; Kikuchi et al. 2006). Also, crosslinking studies revealed that Toc159 is in close association with the preprotein throughout OEM translocation (Kouranov and Schnell 1997). Studies using proteoliposomes containing the TOC core-complex showed that precursor binding could only be inhibited when Toc34 was blocked using a competitive TP fragment; similar inhibition of Toc159 did not interfere with binding (Becker et al. 2004b). This again supported the proposed role of Toc34 as the primary receptor.

In this model, Toc159 remains in close association with Toc75 via an interaction that is nucleotide-insensitive (Becker et al. 2004b). Firstly, Toc34 in its GTP-bound state binds to the C-proximal domain of an incident TP. Then, the binding of the N-terminal domain of the same TP to Toc159 (also in its GTP-state) causes the two receptors to associate closely. Next, Toc34 hydrolyses GTP, and so releases its grip on the TP such that it is transferred entirely to Toc159. The latter process promotes GTP hydrolysis by Toc159, which in turn drives the preprotein into the Toc75 channel (Fig. 1) (Schleiff et al. 2003a; Becker et al. 2004b). Following GTP hydrolysis, the association of Toc34 with the TOC complex is weakened. Further experimentation is required to determine which elements of the "targeting" and "motor" models most closely reflect the in vivo situation.

The atToc159 null mutant (*ppi2*) is more severely sick than the atToc33 null mutant (*ppi1*); it has an albino phenotype and cannot produce normal chloroplasts (Bauer et al. 2000; Smith et al. 2002). Like *ppi1*, it has reduced accumulation of photosynthetic proteins and low expression of the corresponding genes. However, many housekeeping genes are not affected. These results strengthen the hypothesis that atToc159 and atToc33 are part of a specific import machinery preferentially used by photosynthetic proteins. In the *ppi2* mutant, a photosynthetic-TP-GFP fusion-protein was not targeted efficiently to plastids, whereas a housekeeping-TP-GFP fusion-protein was imported normally (Smith et al. 2004).

Other *Arabidopsis* Toc159 homologues are believed to mediate import of housekeeping proteins (Bauer et al. 2000), as revealed by analyses of corresponding null mutants (Hiltbrunner et al. 2004; Ivanova et al. 2004; Kubis et al. 2004; Hust and Gutensohn 2006). In contrast with *ppi2*, mutants of atToc120 and atToc90 (*ppi4*) did not display any visible phenotypes, while atToc132 mutants expressed only a moderate yellow–green, reticulate phenotype (Kubis et al. 2004). Hence, none of these atToc159 homologues is essential, individually, for plant viability. However, atToc132 and atToc120 are highly redundant, since a double-knockout mutant displayed a severe, near-albino phenotype (Ivanova et al. 2004; Kubis et al. 2004); this phenotype could be abolished by overexpression of either atToc132 or atToc120 (Kubis et al. 2004). Interestingly, atToc159 overexpression could not complement the atToc132/atToc120 double mutant, supporting the notion that two different, functionally specialized import pathways exist (Kubis et al. 2004). In contrast, the role of atToc90 is less clear, and it does not share significant redundancy with the other Toc159 homologues (Kubis et al. 2004). However, its expression pattern resembles that of atToc159, and so it has been proposed to assist the function of atToc159 (Bauer et al. 2000; Hiltbrunner et al. 2004).

4.3 Toc64/OEP64

Analysis of the TOC complex from pea chloroplasts identified an unknown, co-purifying 64-kDa protein, later referred to as Toc64 (Sohrt and Soll 2000). The N-terminus of Toc64 contains a transmembrane anchor, the central part shares homology with amidases, and the C-terminal region has three tetratricopeptide repeats (TPRs) exposed to the cytosol. Cross-linking studies revealed that psToc64 is in close proximity with several TOC and TIC components as well as precursor proteins (Sohrt and Soll 2000; Becker et al. 2004a). In *Arabidopsis*, there are three genes for Toc64-like proteins: *atTOC64-I*, *atTOC64-III* and *atTOC64-V* (roman numbers indicate chromosome location) (Table 1; Jackson-Constan and Keegstra 2001).

The likely orthologue of psToc64 is atToc64-III, which is localized in chloroplasts (Chew et al. 2004; Qbadou et al. 2007). The atToc64-III transmembrane domain, and its C-terminal flanking region, are sufficient to mediate targeting to the OEM (Lee et al. 2004). In pea, Toc64 has been proposed to mediate docking of the "guidance complex" (Sect. 3), and to act as a receptor for preproteins delivered by Hsp90 (Sohrt and Soll 2000, Qbadou et al. 2006); in the latter model, the Toc64 TPR domain binds to the chaperone rather than to the preprotein itself. Despite its similarity with psToc64, a clear role for atToc64-III in chloroplast protein import could not be established (Aronsson et al. 2007). The atToc64-V protein is also very similar to psToc64, but it is localized in mitochondria and so was renamed mtOM64 (*mi*tochondrial *O*uter *M*embrane protein, 64 kDa) (Chew et al. 2004). The atToc64-I protein lacks both the TPRs and the transmembrane anchor, and is essentially just the amidase domain; this isoform seems to be cytosolic (Chew et al. 2004; Pollmann et al. 2006). Interestingly, atToc64-I was shown to have indole-3-acetamide hydrolase activity, relating the protein to auxin biosynthesis, and so its name was changed to Amidase 1 (AMI1) (Pollmann et al. 2006). However, an atToc64-I null mutant did not show significant growth phenotypes, and so more studies are needed to resolve its role (Aronsson et al. 2007).

Two Toc64-like proteins were identified in the moss, *Physcomitrella patens* (Hofmann and Theg 2005b). A double null mutant affecting these components showed no obvious phenotypes (except a slight chloroplast shape alteration). Most importantly, experiments revealed no defects in chloroplast protein import efficiency in the double mutant (Hofmann and Theg 2005b). This argues against a crucial role for Toc64 in import. Even in the higher plant, *Arabidopsis*, Toc64 (atToc64-III) null mutants are indistinguishable from wild type (Aronsson et al. 2007). Absence of defects in atToc64-III mutants raised the possibility of redundancy, but even triple mutants lacking all three isoforms appeared normal. In the absence of a clearly defined role for Toc64, it was suggested that it should be renamed using the general designation, OEP64 (*O*uter *E*nvelope *P*rotein, 64 kDa). Data supporting a role for Toc64 in import were generated biochemically, whereas those arguing against this role were derived in vivo, in *Arabidopsis* and *P. patens*.

The topology of Toc64/OEP64 is also disputed. In addition to the originally defined, N-terminal transmembrane region, two additional membrane spans have been suggested to exist (Qbadou et al. 2007). However, contradicting data on the topology of Toc64 have been presented (Hofmann and Theg 2005c). If the more complex topology is correct, it may be that Toc64 acts on both sides of the OEM: in the cytosol, the TPR would facilitate docking of incident cargo proteins, which is consistent with the general role of TPRs as protein–protein interaction domains (Schlegel et al. 2007); in the intermembrane space (IMS), Toc64 would act as a "link" between TOC and TIC through interplay with Toc12, an IMS Hsp70 and Tic22 (Fig. 2a; Becker et al. 2004b; Qbadou et al. 2007).

However, preprotein translocation does not require a functional Toc64/OEP64 protein, since the TOC core-complex (Toc159, Toc75 and Toc34) was fully import-competent when reconstituted into proteoliposomes (Schleiff et al. 2003a). Moreover, knockout of a TOC protein involved in multifaceted activities on both sides of the OEM (as proposed for Toc64) would be predicted to have a clear phenotype (e.g., pale, albino, embryo-lethal), as found for other TOC components (Jarvis et al. 1998; Bauer et al. 2000; Baldwin et al. 2005). This is not the case for Toc64 (Hofmann and Theg 2005b; Aronsson et al. 2007), and so its role is far from clear. Nonetheless, the possibility remains that it acts as a "fine-tuner".

5 Channels and Intermediaries

Following recognition, preproteins are transferred to the OEM channel, of which Toc75 is the main component. Following OEM translocation, preproteins enter the IMS prior to their association with the IEM. Contact sites between the OEM

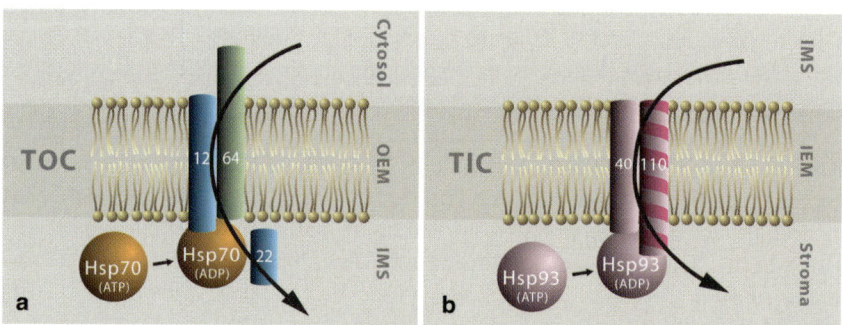

Fig. 2 *Putative motor complexes of the import machinery.* (**a**) Protein translocation across the outer membrane. The preprotein (*black line*) is recognized by Toc64 before being released to the TOC core-complex (not shown). Toc64 activates Toc12, which recruits Hsp70 in an ATP-bound state. Toc12 then stimulates ATP hydrolysis by Hsp70, enabling the chaperone to drive translocation of the preprotein in a ratchet mechanism. The preprotein is transferred to Tic22 for further movement to the TIC complex. (**b**) Protein translocation across the inner membrane. The preprotein (*black line*) moves through the TIC channel (not shown), binding at Tic110. This stimulates a Tic110–Tic40 interaction, which releases the transit peptide from Tic110 for association with Hsp93. The ATPase activity of Hsp93 is then stimulated by Tic40, which allows Hsp93 to complete import, possibly via a ratchet mechanism. OEM/IEM, outer/inner envelope membrane; TOC/TIC, translocon of the outer/inner envelope membrane of chloroplasts; IMS, intermembrane space

and the IEM are established to enable efficient translocation from TOC to TIC (Schnell and Blobel 1993; Perry and Keegstra 1994). Several components, including Toc12, an IMS Hsp70 and Tic22, are proposed to facilitate translocation across the IMS (Fig. 2a). On arrival at the IEM, the preprotein may contact Tic110 and/or Tic20, since both have been proposed to mediate channel formation in the IEM (Fig. 1).

5.1 Toc75

One of the first TOC components to be identified in pea was Toc75 (Waegemann and Soll 1991; Perry and Keegstra 1994; Schnell et al. 1994). Evidence for its role in import came from its proximity to preproteins engaged in import, as revealed by cross-linking (Perry and Keegstra 1994; Ma et al. 1996) and co-purification from solubilized envelopes (Schnell et al. 1994), and from the inhibition of import by Toc75 antibodies (Tranel et al. 1995). It is one of the most abundant proteins in the OEM (Cline et al. 1981), and can act as an aqueous ion channel in vitro, strongly suggesting that it forms the translocation pore (Fig. 1; Hinnah et al. 1997, 2002). As an integral membrane protein, Toc75 can withstand protease treatment and extraction by either salt or high pH (Schnell et al. 1994; Tranel et al. 1995). It is proposed to have either 16 or 18 amphiphilic β-strands that together make up a β-barrel domain (Sveshnikova et al. 2000a; Schleiff et al. 2003b), as often found in bacterial proteins (Gentle et al. 2005). The channel is estimated to be ~14-Å in diameter (Hinnah et al. 2002), which is sufficient only for largely unfolded proteins to pass. However, the import of a tightly folded substrate ~23-Å in diameter suggests a degree of elasticity (Clark and Theg 1997).

Interestingly, psToc75 shares 22% amino acid sequence identity with a cyanobacterial (*Synechocystis*) homologue, SynToc75 (Reumann et al. 1999). This homologue behaves as a channel protein with features resembling those of psToc75 (Bölter et al. 1998b). In pea, a second Toc75-related protein was identified on the basis of its similarity to SynToc75. This protein shares 31% identity with psToc75, is assumed to be a channel also, and was named psToc75-V due to the fact that the most similar *Arabidopsis* protein, atToc75-V, is encoded on chromosome 5 (Eckart et al. 2002). On the basis of phylogenetic studies, atToc75-V was proposed to be the most ancestral form of the Toc75 channel (Eckart et al. 2002). However, this was challenged by an idea that the two pea Toc75-like proteins each derived independently from cyanobacterial sequences (Inoue and Potter 2004).

Three Toc75-related proteins exist in *Arabidopsis*: atToc75-III, atToc75-IV and atToc75-V (Table 1; Jackson-Constan and Keegstra 2001; Eckart et al. 2002; Baldwin et al. 2005). The first of these, atToc75-III, is thought to be the true orthologue of psToc75, based partly on high sequence similarity (73% within the mature region) and similar developmental expression profiles (Tranel et al. 1995; Baldwin et al. 2005). The second protein, atToc75-IV, lacks a large N-terminal domain, is only expressed at very low levels, and most likely plays a rather minor role (Baldwin et al. 2005). Interestingly, atToc75-IV and atToc75-V both differ from atToc75-III in respect of their membrane insertion requirements: neither appears to be processed during insertion, whereas the latter has a bipartite targeting sequence, like psToc75, which directs the protein firstly towards the stroma, and then in a second step mediates insertion into the OEM (Inoue and Keegstra 2003; Inoue and Potter 2004; Baldwin et al. 2005). The size of fully assembled atToc75-V was reported to be 80 kDa, which is somewhat larger than originally proposed (Eckart et al. 2002; Inoue and Potter 2004). Since a role in protein import has not been demonstrated, the Toc75-V protein was renamed with the general designation, OEP80 (*Outer Envelope Protein, 80 kDa*) (Eckart et al. 2002; Inoue and Potter 2004). The role of OEP80 is unknown, but it has been proposed to mediate the translocation of highly hydrophobic α-helical solute transporters of the IEM, and the biogenesis β-barrel proteins of the OEM (Eckart et al. 2002; Inoue and Potter 2004).

Null mutants have been identified for atToc75-III and atToc75-IV. Those affecting the former are embryo-lethal, while atToc75-IV knockouts exhibit no obvious defects except for inefficient de-etiolation, indicating a possible role in etioplasts (Baldwin et al. 2005). These data support the notion that atToc75-III is the true orthologue of psToc75, and the main Toc75 channel in *Arabidopsis*.

5.2 Outer Envelope Translocation Models

Several lines of evidence indicate that there is a motor activity at the OEM, and that TOC translocation is not simply driven by TIC-associated machinery (Guera et al. 1993; Scott and Theg 1996; Kovacheva et al. 2007). Different hypotheses exist for the mechanism of translocation through Toc75. One of these is the Toc159 "motor model", in which the receptor acts like a sewing machine to pushing the preprotein through the channel in cycles of GTP hydrolysis (Sect. 4.2) (Schleiff et al. 2003a). However, this model is inconsistent with data indicating that import can still proceed in the presence of non-hydrolyzable GTP analogues, or following removal of the Toc159 G-domain (Kessler et al. 1994; Young et al. 1999, Chen et al. 2000). The G-domain may instead function to place the M-domain of Toc159 in a position suitable for translocation; this idea is supported by the partial complementation of *ppi2* using the M-domain only (Lee et al. 2003).

Another possibility involves ATP hydrolysis and may be called the "chaperone model". Initial studies on import showed that formation of early import intermediates is ATP dependent (Olsen et al. 1989; Olsen and Keegstra 1992). Moreover, an Hsp70 chaperone has been identified in the IMS in close association with the TOC complex (Marshall et al. 1990; Schnell et al. 1994). In this model, Hsp70 acts as a molecular ratchet to ensure the unidirectional movement of the preprotein (Neupert and Brunner 2002). The recently identified component, Toc12, may act as a co-chaperone by controlling the ATPase activity of the IMS Hsp70 (Sect. 5.3) (Fig. 2a; Becker et al. 2004a). In contrast with mitochondrial import, Hsp70 is proposed to mediate translocation over the OEM only; full transport into the stroma over the IEM seems to involve other chaperones (Sect. 6.2.3).

5.3 Intermediaries, with Emphasis on Toc12 And Tic22

Transport across the two envelope membranes takes place in concert (Schnell and Blobel 1993; Perry and Keegstra 1994; Schnell et al. 1994; Wu et al. 1994). Thus, the TOC and TIC translocons meet at so-called contact sites to enable efficient co-translocation (Schnell and Blobel 1993; Perry and Keegstra 1994). These contacts may be mediated by proteins present in the IMS (e.g., Toc12 and Tic22; Fig. 1).

Tic22 was identified by its interaction with preproteins arrested as early import intermediates (Ma et al. 1996; Kouranov and Schnell 1997). It was also demonstrated to lie mostly in the IMS and to be peripherally associated with the IEM. Its location led to the hypothesis that Tic22 is involved in the movement of preproteins from TOC to TIC (Kouranov et al. 1998, 1999). Tic22 has also been proposed to work in close association with Toc12 and Toc64 (Becker et al. 2004a).

The Toc12 protein is located in the OEM and is a DnaJ-like co-chaperone. Its C-terminal J-domain protrudes into the IMS, and is stabilized by an intramolecular disulphide bond, which might play a role in sensing the redox state of the chloroplast (Sect. 8) (Becker et al. 2004b). As mentioned earlier (Sect. 4.3), Toc12 is proposed to stimulate the ATPase activity of the IMS Hsp70, to assist the formation of early import intermediates. Together with Hsp70, Toc64 and Tic22, Toc12 was proposed to be part of an IMS translocase (Fig. 2a). In this model, Toc64 would coordinate preprotein arrival at the OEM with the readying or assembly of the IMS translocase for downstream steps. Tic22 would be the link to the TIC complex, so that upon preprotein arrival a contact site is induced (Becker et al. 2004a). This is an attractive hypothesis, but its relevance is unclear since the role of Toc64 is questionable (Hofmann and Theg 2005b; Aronsson et al. 2007). At present, no in vivo studies have been presented for Toc12 (Inoue 2007).

6 Translocation Events at the TIC Complex

Translocation over the IEM is less well characterized than OEM translocation. Nonetheless, IEM transport is proposed to involve: channel-forming protein(s); a motor complex that drives transport into the stroma; and, regulatory factors that control import by sensing plastid redox status (Fig. 1). Interestingly, there is considerable disagreement in the literature concerning the composition of the TIC machinery (e.g., the identity of the main channel component). This may reflect an inherent flexibility in the system, or indicate that multiple, different TIC complexes exist.

6.1 Candidate Channel Components of the TIC Complex

6.1.1 Tic110

As already mentioned, the identity of the TIC channel is uncertain, since three different components (Tic110, Tic20 and Tic21) have been proposed to perform this role (Fig. 1). The first of these, Tic110, was identified in pea by its close association with preproteins at a late stage of import (Schnell et al. 1994; Wu et al. 1994). Tic110 has an N-terminal membrane anchor of ~9 kDa comprising two helical membrane spans. The C-terminal bulk of the protein constitutes a ~98-kDa hydrophilic domain for which the location is debated. Initial reports that it protrudes into the IMS (Lübeck et al. 1996) have been corrected by more recent studies showing orientation towards the stroma (Kessler and Blobel 1996; Nielsen et al. 1997; Jackson et al. 1998; Inaba et al. 2003). Its location in the stroma is thought to be significant, enabling it to recruit stromal factors needed for import (Sect. 6.2).

When reconstituted into liposomes, the Tic110 C-terminal domain was reported to form a β-barrel structure; this was proposed to be part of the TIC channel, since it had a pore diameter of 15 Å and exhibited cation-selectivity (Heins et al. 2002). However, this model was disputed by others who proposed an alternative structure for the C-terminus. Upon overexpression in bacteria and plants, the domain was found to be soluble and to have an α-helical conformation (Inaba et al. 2003). Thus, any role of Tic110 as part of the TIC channel is most likely mediated by its N-terminal transmembrane spans, leaving the C-terminus free to manage later events of the import process; an important component of this role seems to be a domain that recognizes and binds to TPs (Inaba et al. 2003). It is clear that Tic110 is essential for the import mechanism, since reduced expression causes chlorosis and null mutations are embryo-lethal (Inaba et al. 2005; Kovacheva et al. 2005). Tic110 is found in most tissues and is distributed widely across different species (Davila-Aponte et al. 2003; Kovacheva et al. 2005).

6.1.2 Tic20

The Tic20 protein consists of four transmembrane α-helices and was first identified in pea by its association with arrested preproteins (Kouranov et al. 1998). In *Arabidopsis*, there are four paralogues: atTic20-I (the isoform most similar to psTic20; ~63% identity), atTic20-IV, atTic20-II and atTic20-V (Table 1). The latter three are progressively less similar to psTic20 (~20–30% identity). As mentioned earlier, Tic20 is thought to play a role in channel formation. Its weak homology and topological similarity with mitochondrial Tim23/22/17 preprotein translocase components support this notion (Rassow et al. 1999; Reumann et al. 1999). Expression of atTic20-I occurs in most tissues, just as for Tic110, and is highest in young tissues (Chen et al. 2002). By analyzing antisense lines, reduced expression of atTic20-I was correlated with a pale-yellow phenotype, which in turn was linked to a chloroplast import defect at the level of IEM translocation (Chen et al. 2002). These data support the role of Tic20 as a key player for IEM translocation.

6.1.3 Tic21

The most recently identified putative channel component is Tic21 (also called *C*hloroplast *I*mport *A*pparatus *5*; CIA5), which resembles Tic20 in terms of size and topology. This protein was identified in a forward-genetic screen for *Arabidopsis* mutants with defects in the chloroplast import of a selectable marker protein (Sun et al. 2001; Teng et al. 2006). The expression pattern of atTic21 is similar to those of Toc75 and Tic110, except for during germination where it is lower. Thus, it was suggested that it mainly functions at later developmental stages; since Tic20 has a more pronounced role during earlier stages, Tic21 might take over from Tic20 later on. Chloroplasts from mutant *cia5* plants display an IEM translocation deficiency, while binding to the OEM is unaffected. An atTic21 null mutant is albino, and accumulates unprocessed preproteins (Teng et al. 2006). However, the role of this component has been debated, since Duy et al. (2007) believe that it acts as an iron transporter and regulator of cellular metal homeostasis. Evidence for this comes from the accumulation of ferritin clusters and the differential regulation of genes involved in iron stress or transport in the mutant. An alternative name for the protein was therefore suggested: *P*ermease *I*n *C*hloroplasts *1* (PIC1). Further studies are needed to resolve these different ideas about Tic21.

6.1.4 PIRAC

Electrophysiological studies led to the identification of a *P*rotein *I*mport *R*elated *A*nion *C*hannel (PIRAC), and this has been proposed to be involved in preprotein translocation at the IEM (van den Wijngaard and Vredenberg 1997). The channel was blocked by preproteins and could also be inactivated by antibodies against Tic110, suggesting its close association with the TIC complex (van den Wijngaard and Vredenberg 1997; 1999). One problem with the putative involvement of PIRAC in preprotein translocation relates to the fact that TPs are positively charged, and so would have difficulty to pass through an anion channel. It was therefore proposed that PIRAC adapts to a cation channel structure upon preprotein arrival (van den Wijngaard et al. 2000). However, this idea, together with the molecular nature and significance of PIRAC needs to be established in future studies.

6.2 *A Motor Complex of the TIC Machinery*

Complete preprotein translocation over the IEM consumes large quantities of ATP in the stroma (Pain and Blobel 1987; Theg et al. 1989). This ATP consumption is attributed to stromal chaperones as part of a motor complex. Components of this putative motor include Tic110, Tic40 and Hsp93 (Fig. 2b). The first of these, Tic110 (Sect. 6.1.1), plays a dual role. First, it possesses a TP recognition site located close to the TIC channel exit; this is proposed to prevent preproteins from sliding back into the IMS after passing through the channel (Inaba et al. 2003). Second, the Tic110 stromal domain probably also acts to recruit molecular chaperones (Hsp93). The second component, Tic40, seems to act as a co-chaperone, and is proposed to regulate the activity of Hsp93. The close cooperation of these three components is supported by genetic interaction data (Kovacheva et al. 2005). The fact that they are expressed in all tissues suggests that they likely mediate import into all plastid types.

6.2.1 Tic40

The Tic40 protein is encoded by a single-copy gene in pea and *Arabidopsis* (Stahl et al. 1999; Chou et al. 2003). It was first characterized biochemically in pea, and shown to closely associate with preproteins at the IEM (Wu et al. 1994; Ko et al. 1995). It possesses one transmembrane α-helix at its N-terminus, while its large C-terminal, hydrophilic domain protrudes into the stroma (Chou et al. 2003). Tic40 and Tic110 can be cross-linked to each other implying that they work in close association (Stahl et al. 1999). A co-chaperone role of Tic40 was suggested by limited homology with *H*sp70-*i*nteracting *p*rotein (Hip) and *H*sp70/Hsp90-*o*rganizing *p*rotein (Hop) in a short C-terminal region, termed the Sti1 domain. Like Hip and Hop, Tic40 also seems to possess a TPR protein–protein interaction interface, upstream of the Sti1 domain (Stahl et al. 1999; Chou et al. 2003). Furthermore, a recent study showed that the putative Tic40 Sti1 domain could be functionally replaced with the equivalent region of human Hip, strongly supporting the notion of a co-chaperone role for Tic40 (Bédard et al. 2007). In contrast with Tic110, Tic40 null mutants in *Arabidopsis* are not lethal, but instead show a strongly pale phenotype that correlates with import defects (Chou et al. 2003; Kovacheva et al. 2005). This indicates that Tic40 function is not essential for the import mechanism, but rather serves to increase the efficiency of import.

6.2.2 Hsp93 (ClpC)

The Hsp93 chaperone is also known as ClpC, due to homology with the bacterial *C*aseino*l*ytic *p*rotease (Clp) ATPase (Shanklin et al. 1995). Like the bacterial Clp ATPase, Hsp93 is proposed to form a protease holocomplex with the proteolytic subunit, ClpP (Sokolenko et al. 1998). In addition to this stromal proteolytic role, Hsp93 associates with the import apparatus at the IEM. Initial support for its involvement in the TIC complex came from cross-linking studies (Akita et al. 1997). Interaction with the TIC complex occurred even in the absence of preprotein, and was destabilized by ATP (Nielsen et al. 1997; Kouranov et al. 1998).

In *Arabidopsis*, there are two Hsp93 genes: *atHSP93-V* and *atHSP93-III* (Table 1; Jackson-Constan and Keegstra 2001). These genes display slightly different expression patterns, but the mature proteins share ~92% identity and so are believed to act redundantly (Kovacheva et al. 2005). Null mutants of atHsp93-V and atHsp93-III have been characterized by several groups (Constan et al. 2004a; Park and Rodermel 2004; Sjögren et al. 2004; Kovacheva et al. 2005, 2007). The atHsp93-V mutants display a chlorotic phenotype and have underdeveloped chloroplasts. Interestingly, import efficiency in *hsp93-V* mutant chloroplasts was found to be reduced in two studies (Constan et al. 2004a; Kovacheva et al. 2005), but normal in another (Sjögren et al. 2004). Knockout mutants of atHsp93-III showed no phenotypic differences from wild-type plants, suggesting possible redundancy with atHsp93-V (Constan et al. 2004a; Kovacheva et al. 2005). This was confirmed by complementation analysis, and by the demonstration that the double-null genotype is embryo lethal (Kovacheva et al. 2007).

6.2.3 A Model for the Operation of the Motor Complex

At the early import intermediate stage, TOC–TIC supercomplexes are formed and the preprotein is in contact with the TIC machinery (Akita et al. 1997; Kouranov and Schnell 1997; Nielsen et al. 1997; Kouranov et al. 1998; Inaba et al. 2003). Such supercomplexes already contain Hsp93, so the preprotein is able to move quickly to the next stage of translocation. The IEM motor complex then comes into play, driven by stromal ATP. The close association of Tic110 and Tic40 has been firmly established (Stahl et al. 1999; Chou et al. 2003, 2006; Bédard et al. 2007). The Tic110–Tic40 interaction is stimulated when a preprotein is present in the Tic110 TP binding site, and is proposed to be mediated by the TPR domain of Tic40 (Chou et al. 2006). This interaction triggers the release of the TP from Tic110, and enables the preprotein to associate with Hsp93 (Fig. 2b). The Sti1 region of Tic40 then stimulates the ATPase activity of Hsp93, enabling the chaperone to complete the translocation process, presumably through a molecular ratchet mechanism (Neupert and Brunner 2002). Thus, in this model (Fig. 2b), the role of Tic40 is to organize the last steps of envelope translocation, by regulating the interaction of the preprotein with Tic110 and Hsp93, and by controlling the activity of Hsp93. The net result is to enhance the efficiency of translocation (Chou et al. 2006).

Some aspects of the model are unexpected. It is surprising that the TPR region of Tic40 interacts with Tic110, since the TPR domains of Hip and Hop interact with their chaperone partners (Frydman and Höhfeld 1997; Abbas-Terki et al. 2001). That the Sti1 region should stimulate ATPase activity of Hsp93 is also unexpected, since this is not one of the proposed functions of Hip/Hop Sti1 domains; Bédard et al. (2007) showed that the Sti1 region of Tic40 is functionally equivalent to the Hip Sti1 domain. Nonetheless, the model is a useful subject for future refinement.

7 SPP

Once inside the stroma, the TP of the preprotein is cleaved off by the *s*tromal *p*rocessing *p*eptidase (SPP) (Fig. 1; Richter and Lamppa 1999). This peptidase is important for plastid biogenesis, since its down regulation by antisense technology causes albino or lethal phenotypes. Chloroplast import efficiency is negatively affected by the reduction of SPP levels (Wan et al. 1998; Zhong et al. 2003). This may be an indirect effect due to inefficient processing and assembly of TOC/TIC components, or reflective of a closely integrated role for SPP at the TIC complex.

The SPP protein has a zinc-binding motif, found also in other pitrilysin metalloendopeptidases like the enzyme responsible for cleaving mitochondrial presequences (VanderVere et al. 1995; Roth 2004), and it is essential for catalytic activity (Richter and Lamppa 2003). The peptidase binds to the C-terminal ~10–15 residues of the TP. The consensus of the cleavage site is rather weak (Emanuelsson et al. 1999), and some data suggest that recognition of physicochemical properties is more important than a specific sequence of residues (Rudhe et al. 2004). After cleavage, the TP remains attached to SPP for another round of proteolysis, and is then released into the stroma to be further degraded by a presequence protease (Moberg et al. 2003; Bhushan et al. 2006). The TP can be accessed by SPP even while the preprotein's C-terminus is attached to the TOC complex, indicating that cleavage occurs soon after entry to the stroma (Schnell and Blobel 1993). Newly imported proteins are likely folded with help from the stromal chaperones, Hsp70 and Hsp60 (Tsugeki and Nishimura 1993; Jackson-Constan et al. 2001). The association of Hsp60 with the TIC complex in an ATP-dependent manner suggests that translocation and folding may take place in concert (Kessler and Blobel 1996).

8 Fine-Tuning the TIC Complex, with Emphasis on Redox

It is well documented that chloroplast gene expression is regulated by redox status, due to the importance of photosynthesis. Three putative TIC complex components, Tic55, Tic62 and Tic32, have been identified as possible sensors of chloroplast redox state, suggesting that import is subject to similar regulation (Fig. 1; Caliebe et al. 1997; Küchler et al. 2002; Hörmann et al. 2004). All three components were first identified in pea: Tic55 and Tic62 were detected through blue native PAGE analysis of the TIC complex (Caliebe et al. 1997; Küchler et al. 2002), whereas Tic32 was found to tightly associate with the N-terminal transmembrane part of Tic110 (Hörmann et al. 2004). Additionally, Toc12 has features that could respond to changes in redox state, since its IMS J-domain is stabilized by an intramolecular disulphide bond (Becker et al. 2004b). There is some evidence to suggest that redox changes do have an effect on import. In the presence of light, the non-photosynthetic ferredoxin III preprotein is translocated only as far as the IMS, whereas in the dark it enters the stroma (Hirohashi et al. 2001). However, so far no translocon component has been shown to be responsible for this "mistargeting" phenomenon. One possibility is that, under reductive conditions, Toc12 loses its disulphide bond, impairing its ability to stimulate Hsp70 in the IMS and leading to the disrupted passage of preproteins to the TIC complex.

8.1 Tic55

The Tic55 component is believed to be an integral IEM protein, and its characteristic features are a Rieske-type iron-sulphur centre (these normally function in electron transfer) and a mononuclear iron-binding site, which are also found in bacterial oxygenases (Caliebe et al. 1997). It was reported that *di*ethylpyrocarbonate (DEPC) treatment of pea chloroplasts inhibits protein import, and this was attributed to the disruption of histidine in the Tic55 Rieske centre, thereby supporting its role as part of the TIC complex (Caliebe et al. 1997). However, the possibility that other proteins were also affected by DEPC was not ruled out. To address this issue, we used null mutants of the *Arabidopsis* Tic55 orthologue (atTic55-II); when treated with DEPC, wild-type and *tic55* mutant chloroplasts were equally affected in terms of import capacity (Boij P, Patel R, Jarvis P, Aronsson H, unpublished data). Thus, the exact role of Tic55 is not clear. Furthermore, other groups have been unable to find Tic55 associated within TOC/TIC complexes (Kouranov et al. 1998, Reumann and Keegstra 1999). So far, no data have been reported for *Arabidopsis* Tic55 null mutants, but we observe that homozygous lines show no phenotypic deviance from the wild type (Boij P, Patel R, Jarvis P, Aronsson H, unpublished data). A weak homologue of Tic55 in *Synechocystis* was suggested to have a role as a cell death suppressor (Mason and Cammack 1992).

8.2 Tic62

Tic62 has been identified partly on the basis of its co-purification with Tic55. This component shares homology with eukaryotic NAD(P)H dehydrogenases and Ycf39-like proteins in cyanobacteria and non-green algae (Küchler et al. 2002). Circular dichroism measurements show that Tic62 is composed of two structurally different domains (Stengel et al. 2008). The NAD(P)H binding site lies in the N-terminal part, while the stromal-facing C-terminal part interacts with a *f*erredoxin-*N*AD(P)$^+$ oxido*r*eductase (FNR). FNR normally mediates electron transfer, during oxygenic photosynthesis, from ferredoxin to NADP$^+$ at the thylakoids. The FNR-binding domain of Tic62 proteins is thought to be a relatively recent development in vascular plants, with no sequence similarity to other known motifs (Balsera et al. 2007); it may be crucial for its function in the TIC complex. Reagents that interfere with either NAD binding or have an effect on the ratio of NAD(P)/NAD(P)H influence the import of leaf-specific FNR isoforms differently, suggesting that Tic62 regulates import through redox sensing (Küchler et al. 2002). Localization of

Tic62 (partitioning between the stroma and the IEM), and its association with the TIC complex and FNR, are all influenced by redox status, including the $NADP^+$/NADPH ratio in the stroma (Stengel et al. 2008).

8.3 Tic32

The Tic32 protein shares homology with *s*hort-chain *d*ehydrogenase/*r*eductase (SDR) proteins, and behaves as an integral membrane component. It was reported to associate with several other TIC components after co-immunoprecipitation (e.g., Tic22, Tic40, Tic62 and Tic110), and is proposed to play a role late in the translocation process. Null mutations for one of two *Arabidopsis* Tic32 homologues appear to be embryo lethal, suggesting an essential role for the protein (Hörmann et al. 2004). Interestingly, import of preproteins with a cleavable TP was reduced in the presence of calcium or calmodulin inhibitors (Chigri et al. 2005); in contrast, the import of other proteins with no cleavable TP (Sect. 9.3) was not affected. This calcium regulation was suggested to occur within the IMS or at the IEM, possibly involving TIC components and calmodulin. A subsequent study using affinity chromatography found Tic32 to be the predominant IEM protein bound to calmodulin, and this interaction was calcium dependent (Chigri et al. 2006). In the same study, Tic32 was shown to have NADPH-dependent dehydrogenase activity; moreover, NADPH (but not NADH or $NADP^+$) affected the interaction of Tic32 with Tic110. Binding of NADPH and calmodulin to Tic32 were mutually exclusive processes, suggesting that Tic32 is involved in sensing and integrating redox and calcium signals at the TIC complex (Chigri et al. 2006).

9 Targeting to the Envelope System

9.1 Outer Envelope Membrane Targeting

Several mechanisms exist for protein targeting to the OEM (Hofmann and Theg 2005c). Nonetheless, many OEM proteins have intrinsic, non-cleavable targeting information, consisting of a hydrophobic transmembrane span adjacent to a C-terminal positive region; the latter prevents such proteins from entering the endomembrane system, since the former are similar to signal peptides for ER translocation (Lee et al. 2001). Recently, an *a*n*k*yrin *r*epeat protein (AKR2) was identified as a cytosolic mediator of OEM targeting (Bae et al. 2008; Bédard and Jarvis 2008). It acts by binding to the targeting signals of client proteins, preventing their aggregation, and by docking at the OEM surface. Interestingly, AKR2-deficient mutants have reduced levels of many chloroplast proteins, not only OEM proteins. The internal chloroplast defects in these mutants may be indirectly related to defective OEM biogenesis, but it is interesting that AKR2 can bind to 14-3-3 proteins, since this raises the possibility that AKR2 is also a component of the "guidance complex" (May and Soll 2000). Another recent development indicates that the TOC channel protein, Toc75, is involved in OEM insertion (Tu et al. 2004).

Interestingly, Toc75 is unusual amongst OEM proteins, since it has a bipartite, cleavable targeting signal. The first part is a standard TP, while the second downstream part is an intraorganellar targeting signal that mediates "stop-transfer", release from the translocon, and membrane integration (Inoue and Keegstra 2003).

9.2 Inner Envelope Membrane and Intermembrane Space Targeting

Most proteins destined to the IMS or IEM have a cleavable TP. Information on targeting to the IMS is limited and based on MGD1 and Tic22, which are both located at the IEM and sit facing the IMS (Kouranov et al. 1999; Vojta et al. 2007). While MGD1 seems to use the TOC/TIC machinery for its translocation, it is less clear whether this is true for Tic22. Furthermore, the TP of Tic22, unlike that of MGD1, does not seem to be cleaved by SPP; instead, it is removed by an unidentified protease in the IMS. Thus, two pathways for IMS targeting seem plausible. Proteins of the IEM follow two routes: in the "stop-transfer" route, hydrophobic transmembrane domains induce lateral exit from the TIC machinery and membrane integration; in the "post-import" route, the preprotein first enters the stroma prior to second-step IEM integration (Li and Schnell 2006; Tripp et al. 2007).

9.3 Inner Membrane Targeting without a Transit Peptide

Unlike most proteins targeted to the IEM, Tic32 is imported without a cleavable TP. Instead, ten N-terminal amino acids hold the essential targeting information. Furthermore, neither recognition nor translocation of Tic32 involves the standard import components, Toc34, Toc159 and Toc75, and the energy requirement is low (<20-μM ATP), suggesting stromal chaperones are not involved. Crosslinking to Tic22 suggests that Tic32 may be assisted by Tic22 during its targeting through the IMS, prior to assembly into the IEM (Nada and Soll 2004).

Another protein targeted in a non-canonical manner is *c*hloroplast *e*nvelope *Q*uinone *O*xido*r*eductase *H*omologue (ceQORH). This protein does not carry a predicted TP, but still it is found in the IEM. In contrast with Tic32, the N-terminus of ceQORH is not needed for targeting. Instead, ~40 central residues are required for proper targeting. Like Tic32, its import is not mediated by the standard TOC components, but the energy requirement for import is higher for ceQORH than for Tic32 (Miras et al. 2007).

10 Alternative Import Pathways

Identification of multiple TOC receptor isoforms in *Arabidopsis* cast new light on the previous view of a standardized "general import pathway" (Sect. 4). Studies on *Arabidopsis* mutants for these receptors indicated the existence of different sub-pathways (e.g., photosynthetic versus non-photosynthetic, abundant versus non-abundant) (Bauer et al. 2000; Kubis et al. 2003), and it has even been proposed that cell-specific import pathways may operate (Yu and Li 2001). Whether different combinations of TOC components also attract different TIC complex combinations is an open question, although at least some key TIC components appear to be constitutive (Kovacheva et al. 2005). Bearing in mind that ~3,000 different proteins need to be imported, it is not surprising that variations on the basic import theme exist. Remarkably, it is now clear that chloroplast protein targeting is even more complex. Some proteins are targeted without cleavable TPs (Sect. 9), while others are transported via the endomembrane system.

10.1 Is There a Substrate-Dependent Import Pathway?

In chlorophyll biosynthesis, NADPH:protochlorophyllide oxidoreductase (POR) catalyzes the reduction of protochlorophyllide (Pchlide) to chlorophyllide in a strictly light-dependent manner. There are two main isoforms of POR in several species: PORA (responsible for reduction in illuminated, dark-grown material) and PORB (responsible for reduction in green material) (Aronsson et al. 2003b). It has been suggested that the import of PORA can only occur in the presence of its substrate, Pchlide. Pchlide accumulates in etioplasts but is found only in small amounts in chloroplasts. Thus, in barley, PORA was reported to be imported into etioplasts due to the presence of envelope-bound Pchlide, but not into chloroplasts (Reinbothe et al. 1995, 1997). This implies that a novel import pathway exists. However, this hypothesis has been challenged by several laboratories, since the import of PORA in various homologous systems, including *Arabidopsis* (Jarvis et al. 1998), pea (Aronsson et al. 2000, 2001) and wheat (Teakle and Griffiths 1993), was not dependent on Pchlide. Furthermore, barley PORA was also imported independently of Pchlide (Aronsson et al. 2000, Dahlin et al. 2000).

An explanation for these different results was offered by showing that the import of PORA in *Arabidopsis* was substrate-dependent in cotyledons only, and not in true leaves (Kim and Apel 2004). Such organ-specific import of PORA might help control the light-dependent transformation of a storage organ into a fully photosynthesizing leaf. More recently, putative components of the proposed PORA-specific import machinery have been identified (e.g., Toc33 and OEP16) (Reinbothe et al. 2004). However, these components have been demonstrated not to be of importance for PORA import in vivo (Kim et al. 2005; Philippar et al. 2007). Thus, the relevance of the substrate-dependent PORA import pathway model is unclear. Interestingly, the envelope translocation of a light-harvesting chlorophyll *a/b* binding protein

(LHCP) has also been proposed to be regulated by a pigment (chlorophyll) in *Chlamydomonas reinhardtii* (Eggink and Hoober 2000). Thus, pigments should not be excluded as possible factors for import.

10.2 Interactions with the Endomembrane System

Close associations between the *e*ndoplasmic *r*eticulum (ER) and the chloroplast envelope have been long established (Crotty and Ledbetter 1973). Recently, specific regions of the ER, termed the *pl*astid-*a*ssociated *m*embrane (PLAM), were identified as sites where a strong physical link exists (Andersson et al. 2007). Whether or not a vesicle transport system operates at or near these PLAM contacts remains to be established. Interestingly, indirect evidence for protein transport to chloroplasts through the ER has existed for some time, since glycoproteins and proteins with ER targeting signals have been identified inside plastids (Gaikwad et al. 1999; Chen et al. 2004; Asatsuma et al. 2005). While plastid protein transport through the ER is common in organisms with complex plastids with more than two envelope membranes (Nassoury and Morse 2005), it was only recently shown to exist in angiosperms.

The *c*arbonic *anh*ydrase 1 (CAH1) protein has an ER targeting signal and was detected in the chloroplast stroma. This protein cannot be imported directly by chloroplasts, but instead has to be imported into the ER first of all. Moreover, CAH1 is found in glycosylated form in the stroma, arguing that the transport pathway involves the Golgi apparatus where glycosylation occurs. Most likely, transport occurs through vesicles, since the inhibition of vesicle formation blocked further transport of CAH1 into chloroplasts (Villarejo et al. 2005). Like CAH1, the *n*ucleotide *p*yrophosphatase/*p*hosphodiesterase 1 (NPP1) protein has been shown to use an ER-to-chloroplast transport pathway (Nanjo et al. 2006). So far, no data on how these proteins enter the chloroplast exist. Perhaps vesicles fuse with the OEM to release their contents into the IMS, and then the proteins are translocated through the TIC complex or an unknown translocon; alternatively, onward transport may involve further vesicle formation at the IEM.

11 Conclusion

Substantial progress has been made in the last two decades. Most components of the canonical TOC/TIC machinery have been identified, and many of their individual roles have been at least partially characterized. Nevertheless, many questions still remain to be solved, as was discussed in each of the relevant sections above. Perhaps the most significant recent development has been the realization that the targeting of proteins to chloroplasts is not nearly as simple or as standardized as was once thought. It has become clear that multiple mechanisms operate to ensure that the many diverse proteins that must be transported to chloroplasts arrive safely and without undue delay.

Acknowledgement The authors thank Mats and Paula Töpel for art illustrations, the Swedish Research Council FORMAS (HA), the Royal Society Rosenheim Fellowship and the Biotechnology and Biological Sciences Research Council (PJ).

References

Abbas-Terki T, Briand PA, Donzé O, Picard D (2001) The Hsp90 co-chaperones Cdc37 and Sti1 interact physically and genetically. Biol Chem 383:1335–1342

Abdallah F, Salamini F, Leister D (2000) A prediction of the size and evolutionary origin of the proteome of chloroplasts of Arabidopsis. Trends Plant Sci 5:141–142

Akita M, Nielsen E, Keegstra K (1997) Identification of protein transport complexes in the chloroplastic envelope membranes via chemical cross-linking. J Cell Biol 136:983–994

Andersson MX, Goksör M, Sandelius AS (2007) Optical manipulation reveals strong attracting forces at membrane contact sites between endoplasmic reticulum and chloroplasts. J Biol Chem 282:1170–1174

Aronsson H, Jarvis P (2002) A simple method for isolating import-competent Arabidopsis chloroplasts. FEBS Lett 529:215–220

Aronsson H, Sohrt K, Soll J (2000) NADPH:protochlorophyllide oxidoreductase uses the general import pathway. Biol Chem 381:1263–1267

Aronsson H, Sundqvist C, Timko MP, Dahlin C (2001) Characterisation of the assembly pathway of the pea NADPH:protochlorophyllide (Pchlide) oxidoreductase (POR), with emphasis on the role of its substrate, Pchlide. Physiol Plant 111:239–244

Aronsson H, Combe J, Jarvis P (2003a) Unusual nucleotide-binding properties of the chloroplast protein import receptor, Toc33. FEBS Lett 544:79–85

Aronsson H, Sundvist C, Dahlin C (2003b) POR – import and membrane association of a key element in chloroplast development. Physiol Plant 118:1–9

Aronsson H, Combe J, Patel R, Jarvis P (2006) In vivo assessment of the significance of phosphorylation of the Arabidopsis chloroplast protein import receptor, atToc33. FEBS Lett 580:649–655

Aronsson H, Boij P, Patel R, Wardle A, Töpel M, Jarvis P (2007) Toc64/OEP64 is not essential for the efficient import of proteins into chloroplasts in *Arabidopsis thaliana*. Plant J 52:53–68

Aronsson H, Schüttler MA, Kelly AA, Sundqvist C, Dörmann P, Karim S, Jarvis P (2008) Monogalactosyldiacylglycerol deficiency in *Arabidopsis thaliana* affects pigment composition in the prolamellar body and impairs thylakoid membrane energization and photoprotection in leaves. Plant Physiol, 148:580–592

Asatsuma S, Sawada C, Itoh K, Okito M, Kitajima A, Mitsui T (2005) Involvement of alpha-amylase I-1 in starch degradation in rice chloroplasts. Plant Cell Physiol 46:858–869

Bae W, Lee YJ, Kim DH, Lee J, Kim S, Sohn EJ, Hwang I (2008) AKR2A-mediated import of chloroplast outer membrane proteins is essential for chloroplast biogenesis. Nat Cell Biol 10:220–227

Baldwin A, Wardle A, Patel R, Dudley P, Park SK, Twell D, Inoue K, Jarvis P (2005) A molecular-genetic study of the Arabidopsis Toc75 gene family. Plant Physiol 138:1–19

Balsera M, Stengel A, Soll J, Bölter B (2007) Tic62: a protein family from metabolism to protein translocation. BMC Evol Biol 7:43

Bange G, Wild K, Sinning I (2007) Protein translocation: checkpoint role for SRP GTPase activation. Curr Biol 17:R980–982

Bauer J, Chen KH, Hiltbunner A, Wehrli E, Eugster M, Schnell DJ, Kessler F (2000) The major protein import receptor of plastids is essential for chloroplast biogenesis. Nature 403:203–207

Becker T, Hritz J, Vogel M, Caliebe A, Bukau B, Soll J, Schleiff E (2004b) Toc12, a novel subunit of the intermembrane space preprotein translocon of chloroplasts. Mol Biol Cell 15:5130–5144

Becker T, Jelic M, Vojta A, Radunz A, Soll J, Schleiff E (2004a) Preprotein recognition by the Toc complex. EMBO J 23:520–530

Bédard J, Jarvis P (2005) Recognition and envelope translocation of chloroplast preproteins. J. Exp Bot 56:2287–2320

Bédard J, Jarvis P (2008) Green light for chloroplast outer-membrane proteins. Nat Cell Biol 10:120–122

Bédard J, Kubis S, Bimanadham S, Jarvis P (2007) Functional similarity between the chloroplast translocon component, Tic40, and the human co-chaperone, Hsp70-interacting protein (Hip). J Biol Chem 282:21404–21414

Bhushan S, Johnson KA, Eneqvist T, Glaser E (2006) Proteolytic mechanism of a novel mitochondrial and chloroplastic PreP peptidasome. Biol Chem 387:1087–1090

Bölter B, May T, Soll J (1998a) A protein import receptor in pea chloroplasts, Toc86, is only a proteolytic fragment of a larger polypeptide. FEBS Lett 441:59–62

Bölter B, Soll J, Schulz A, Hinnah S, Wagner R (1998b) Origin of a chloroplast protein importer. Proc Natl Acad Sci USA 95:15831–15836

Bos JL, Rehmann H, Wittinghofer A (2007) GEFs and GAPs: critical elements in the control of small G proteins. Cell 129:865–877; erratum Cell 130:385

Bruce BD (1998) The role of lipids in plastid protein transport. Plant Mol Biol 38:223–246

Bruce BD (2000) Chloroplast transit peptides: structure, function and evolution. Trends Cell Biol 10:440–447

Caliebe A, Grimm R, Kaiser G, Lübeck J, Soll J, Heins L (1997) The chloroplastic protein import machinery contains a Rieske-type iron–sulfur cluster and a mononuclear iron-binding protein. EMBO J 16:7342–7350

Chen K, Chen X, Schnell DJ (2000) Initial binding of preproteins involving the Toc159 receptor can be bypassed during protein import into chloroplasts. Plant Physiol 122:813–822

Chen L-J, Li H-M (1998) A mutant deficient in the plastid lipid DGD is defective in protein import into chloroplasts. Plant J 16:33–39

Chen MH, Huang LF, Li HM, Chen YR, Yu SM (2004) Signal peptide-dependent targeting of a rice alpha-amylase and cargo proteins to plastids and extracellular compartments of plant cells. Plant Physiol 135:1367–1377

Chen X, Smith MD, Fitzpatrick L, Schnell DJ (2002) In vivo analysis of the role of atTic20 in protein import into chloroplasts. Plant Cell 14:641–654

Chew O, Lister R, Qbadou S, Haezlewood JL, Soll J, Schleiff E, Millar AH, Whelan J (2004) A plant outer mitochondrial membrane protein with high amino acid sequence identity to a chloroplast protein import receptor. FEBS Lett 557:109–114

Chigri F, Soll J, Vothknecht UC (2005) Calcium regulation of chloroplast protein import. Plant J 42:821–831

Chigri F, Hörmann F, Stamp A, Stammers DK, Bölter B, Soll J, Vothknecht UC (2006) Calcium regulation of chloroplast protein translocation is mediated by calmodulin binding to Tic32. Proc Natl Acad Sci USA 103:16051–16056

Chou ML, Fitzpatrick LM, Tu SL, Budziszewski G, Potter-Lewis S, Akita M, Levin JZ, Keegstra K, Li HM (2003) Tic40, a membrane-anchored co-chaperone homolog in the chloroplast protein translocon. EMBO J 22:2970–2980

Chou ML, Chu CC, Chen LJ, Akita M, Li HM (2006) Stimulation of transit-peptide release and ATP hydrolysis by a cochaperone during protein import into chloroplasts. J Cell Biol 175:893–900

Chupin V, van't Hof R, de Kruijff B (1994) The transit sequence of a chloroplast precursor protein reorients the lipids in monogalactosyl diglyceride containing bilayers. FEBS Lett 350:104–108

Clark SA, Theg SM (1997) A folded protein can be transported across the chloroplast envelope and thylakoid membranes. Mol Biol Cell 8:923–934

Cline K, Andrews J, Mersey B, Newcomb EH, Keegstra K (1981) Separation and characterization of inner and outer envelope membranes of pea chloroplasts. Proc Natl Acad Sci USA 78:3595–3599

Constan D, Froehlich JE, Rangarajan S, Keegstra K (2004a) A stromal Hsp100 protein is required for normal chloroplast development and function in Arabidopsis. Plant Physiol 136:3605–3615

Constan D, Patel R, Keegstra K, Jarvis P (2004b) An outer envelope membrane component of the plastid protein import apparatus plays an essential role in Arabidopsis. Plant J 38:93–106

Crotty WJ, Ledbetter MC (1973) Membrane continuities involving chloroplasts and other organelles in plant cells. Science 182:839–841

Dabney-Smith C, van den Wijngaard PWJ, Treece Y, Vredenberg WJ, Bruce BD (1999) The C-terminus of a chloroplast precursor modulates its interaction with the translocation apparatus and PIRAC. J Biol Chem 274:32351–32359

Dahlin C, Cline K (1991) Developmental regulation of the plastid protein import apparatus. Plant Cell 3:1131–1140

Dahlin C, Aronsson H, Almkvist J, Sundqvist C (2000) Pchlide independent import of two NADPH:protochlorophyllide oxidoreductase proteins (PORA and PORB) from barley into isolated plastids. Physiol Plant 109:298–303

Davila-Aponte JA, Inoue K, Keegstra K (2003) Two chloroplastic protein translocation components, Tic110 and Toc75, are conserved in different plastid types from multiple plant species. Plant Mol Biol 51:175–181

Dörmann P, Hoffman-Benning S, Balbo I, Benning C (1995) Isolation of an *Arabidopsis* mutant deficient in the thylakoid lipid digalactosyldiacylglycerol. Plant Cell 7:1801–1810

Douce R, Joyard J (1990) Biochemistry and function of the plastid envelope. Annu Rev Cell Biol 6:173–216

Duy D, Wanner G, Meda AR, von Wirén N, Soll J, Philippar K (2007) PIC1, an ancient permease in Arabidopsis chloroplasts, mediates iron transport. Plant Cell 19:986–1006

Eckart K, Eichacker L, Sohrt K, Schleiff E, Heins L, Soll J (2002) A Toc75-like protein import channel is abundant in chloroplasts. EMBO Rep 3:557–562

Eggink LL, Hoober JK (2000) Chlorophyll binding to peptide maquettes containing a retention motif. J Biol Chem 275:9087–9090

Emanuelsson O, Nielsen H, von Heijne G (1999) ChloroP, a neural network-based method for predicting chloroplast transit peptides and their cleavage sites. Protein Sci 8:978–984

Emanuelsson O, Nielsen H, Brunak S, von Heijne G (2000) Predicting subcellular localization of proteins based on their N-terminal amino acid sequence. J Mol Biol 300:1005–1016

Frydman J, Höhfeld J (1997) Chaperones get in touch: the Hip-Hop connection. Trends Biochem Sci 22:87–92

Fulgosi H, Soll J (2002) The chloroplast protein import receptors Toc34 and Toc159 are phosphorylated by distinct protein kinases. J Biol Chem 277:8934–8940

Gaikwad A, Tewari KK, Kumar D, Chen W, Mukherjee SK (1999) Isolation and characterisation of the cDNA encoding a glycosylated accessory protein of pea chloroplast DNA polymerase. Nucleic Acids Res 27:3120–3129

Gentle IE, Burri L, Lithgow T (2005) Molecular architecture and function of the Omp85 family of proteins. Mol Microbiol 58:1216–1225

Guera A, America T, van Waas M, Weisbeek PJ (1993) A strong protein unfolding activity is associated with the binding of precursor chloroplast proteins to chloroplast envelopes. Plant Mol Biol 23:309–324

Goetze TA, Philippar K, Ilkavets I, Soll J, Wagner R (2006) OEP37 is a new member of the chloroplast outer membrane ion channels. J Biol Chem 281:17989–17998

Heins L, Mehrle A, Hemmler R, Wagner R, Küchler M, Hörmann F, Sveshnikov D, Soll J (2002) The preprotein conducting channel at the inner envelope membrane of plastids. EMBO J 21:2616–2625

Hernández Torres J, Maldonado MA, Chomilier J (2007) Tandem duplications of a degenerated GTP-binding domain at the origin of GTPase receptors Toc159 and thylakoidal SRP. Biochem Biophys Res Commun 364:325–331

Hiltbrunner A, Bauer J, Alvarez-Huerta M, Kessler F (2001a) Protein translocon at the *Arabidopsis* outer chloroplast membrane. Biochem Cell Biol 79:629–635

Hiltbrunner A, Bauer J, Vidi P-A, Infanger S, Weibel P, Hohwy M, Kessler F (2001b) Targeting of an abundant cytosolic form of the protein import receptor at Toc159 to the outer chloroplast membrane. J Cell Biol 154:309–316

Hiltbrunner A, Grünig K, Alvarez-Huerta M, Infanger S, Bauer J, Kessler F (2004) AtToc90, a new GTP-binding component of the Arabidopsis chloroplast protein import machinery. Plant Mol Biol 54:427–440

Hinnah SC, Hill K, Wagner R, Schlicher T, Soll J (1997) Reconstitution of a chloroplast protein import channel. EMBO J 16:7351–60

Hinnah SC, Wagner R, Sveshnikova N, Harrer R, Soll J (2002) The chloroplast protein import channel Toc75: pore properties and interaction with transit peptides. Biophys J 83:899–911

Hirohashi T, Hase T, Nakai M (2001) Maize non-photosynthetic ferredoxin precursor is missorted to the intermembrane space of chloroplasts in the presence of light. Plant Physiol 125:2154–2163

Hirsch S, Muckel E, Heemeyer F, von Heijne G, Soll J (1994) A receptor component of the chloroplast protein translocation machinery. Science 266:1989–1992

Hofmann NR, Theg SM (2005a) Chloroplast outer membrane protein targeting and insertion. Trends Plant Sci 10:450–457

Hofmann NR, Theg SM (2005b) Toc64 is not required for import of proteins into chloroplasts in the moss *Physcomitrella patens*. Plant J 43:675–687

Hofmann NR, Theg SM (2005c) Protein- and energy-mediated targeting of chloroplast outer envelope membrane proteins. Plant J 44:917–927

Hörmann F, Küchler M, Sveshnikov D, Oppermann U, Li Y, Soll J (2004) Tic32, an essential component in chloroplast biogenesis. J Biol Chem 279:34756–34762

Hust B, Gutensohn M (2006) Deletion of core components of the plastid protein import machinery causes differential arrest of embryo development in *Arabidopsis thaliana*. Plant Biol 8:18–30

Inaba T, Li M, Alvarez-Huerta M, Kessler F, Schnell DJ (2003) atTic110 functions as a scaffold for coordinating the stromal events of protein import into chloroplasts. J Biol Chem 278:38617–38627

Inaba T, Alvarez-Huerta M, Li M, Bauer J, Ewers C, Kessler F, Schnell DJ (2005) Arabidopsis tic110 is essential for the assembly and function of the protein import machinery of plastids. Plant Cell 17:1482–1496

Inoue K (2007) The chloroplast outer envelope membrane: the edge of light and excitement. J Integr Plant Biol 49:1100–1111

Inoue K, Keegstra K (2003) A polyglycine stretch is necessary for proper targeting of the protein translocation channel precursor to the outer envelope membrane of chloroplasts. Plant J 34:661–669

Inoue K, Potter D (2004) The chloroplastic protein translocation channel Toc75 and its paralog OEP80 represent two distinct protein families and are targeted to the chloroplastic outer envelope by different mechanisms. Plant J 39:354–65

Inoue K, Demel R, de Kruijff B, Keegstra K (2001) The N-terminal portion of the preToc75 transit peptide interacts with membrane lipids and inhibits binding and import of precursor proteins into isolated chloroplasts. Eur J Biochem 268:4036–4043

Ivanova Y, Smith MD, Chen K, Schnell DJ (2004) Members of the Toc159 import receptor family represent distinct pathways for protein targeting to plastids. Mol Biol Cell 15:3379–3392

Jackson DT, Froehlich JE, Keegstra K (1998) The hydrophilic domain of Tic110, an inner envelope membrane component of the chloroplastic protein translocation apparatus, faces the stromal compartment. J Biol Chem 273:16583–16588

Jackson-Constan D, Keegstra K (2001) Arabidopsis genes encoding components of the chloroplastic protein import apparatus. Plant Physiol 125:1567–1576

Jackson-Constan D, Akita M, Keegstra K (2001) Molecular chaperones involved in chloroplast protein import. Biochim Biophys Acta 1541:102–113

Jarvis P (2008) The targeting of nucleus-encoded proteins to chloroplasts in plants (Tansley Review). New Phytol 179:257–285

Jarvis P, Chen LJ, Li H, Peto CA, Fankhauser C, Chory J (1998) An *Arabidopsis* mutant defective in the plastid general protein import apparatus. Science 282:100–103

Jarvis P, Dörmann P, Peto CA, Lutes J, Benning C, Chory J (2000) Galactolipid deficiency and abnormal chloroplast development in the Arabidopsis *MGD synthase 1* mutant. Proc Natl Acad Sci USA 97:8175–8179

Jelic M, Sveshnikova N, Motzkus M, Hörth P, Soll J, Schleiff E (2002) The chloroplast import receptor Toc34 functions as preprotein-regulated GTPase. Biol Chem 383:1875–1883

Jelic M, Soll J, Schleiff E (2003) Two Toc34 homologues with different properties. Biochemistry 42:5906–5916

Kalanon M, McFadden GI (2008) The chloroplast protein translocation complexes of *Chlamydomonas reinhardtii*: a bioinformatic comparison of Toc and Tic components in plants, green algae and red algae. Genetics 179:95–112

Kessler F, Blobel G (1996) Interaction of the protein import and folding machineries in the chloroplast. Proc Natl Acad Sci USA 93:7684–7689

Kessler F, Schnell DJ (2002) A GTPase gate for protein import into chloroplasts. Nat Struct Biol 9:81–83

Kessler F, Schnell DJ (2004) Chloroplast protein import: solve the GTPase riddle for entry. Trends Cell Biol 14:334–338

Kessler F, Schnell DJ (2006) The function and diversity of plastid protein import pathways: a multilane GTPase highway into plastids. Traffic 7:248–257

Kessler F, Blobel G, Patel HA, Schnell DJ (1994) Identification of two GTP-binding proteins in the chloroplast protein import machinery. Science 266:1035–1039

Kikuchi S, Hirohashi T, Nakai M (2006) Characterization of the preprotein translocon at the outer envelope membrane of chloroplasts by blue native PAGE. Plant Cell Physiol 47:363–371

Kim C, Apel K (2004) Substrate-dependent and organ-specific chloroplast protein import in planta. Plant Cell 16:88–98

Kim C, Ham H, Apel K (2005) Multiplicity of different cell- and organ-specific import routes for the NADPH-protochlorophyllide oxidoreductases A and B in plastids of Arabidopsis seedlings. Plant J 42:329–340

Ko K, Budd D, Wu C, Seibert F, Kourtz L, Ko ZW (1995) Isolation and characterization of a cDNA clone encoding a member of the Com44/Cim44 envelope components of the chloroplast protein import apparatus. J Biol Chem 270:28601–28608

Kobayashi K, Kondo M, Fukuda H, Nishimura M, Ohta H (2007) Galactolipd synthesis in chloroplast inner envelope is essential for proper thylakoid biogenesis, photosynthesis, and embryogenesis. Proc Natl Acad Sci USA 104:17216–17221

Kouranov A, Schnell DJ (1997) Analysis of the interactions of preproteins with the import machinery over the course of protein import into chloroplasts. J Cell Biol 139:1677–1685

Kouranov A, Chen X, Fuks B, Schnell DJ (1998) Tic20 and Tic22 are new components of the protein import apparatus at the chloroplast inner envelope membrane. J Cell Biol 143:991–1002

Kouranov A, Wang H, Schnell DJ (1999) Tic22 is targeted to the intermembrane space of chloroplasts by a novel pathway. J Biol Chem 274:25181–25186

Kovacheva S, Bédard J, Patel R, Dudley P, Twell D, Ríos G, Koncz C, Jarvis P (2005) In vivo studies on the roles of Tic110, Tic40 and Hsp93 during chloroplast protein import. Plant J 41:412–428

Kovacheva S, Bédard J, Wardle A, Patel R, Jarvis P (2007) Further in vivo studies on the role of the molecular chaperone, Hsp93, in plastid protein import. Plant J 50:364–379

Kubis S, Baldwin A, Patel R, Razzaq A, Dupree P, Lilley K, Kurth J, Leister D, Jarvis P (2003) The Arabidopsis ppi1 mutant is specifically defective in the expression, chloroplast import, and accumulation of photosynthetic proteins. Plant Cell 15:1859–1871

Kubis S, Patel R, Combe J, Bédard J, Kovacheva S, Lilley K, Biehl A, Leister D, Ríos G, Koncz C, Jarvis P (2004) Functional specialization amongst the Arabidopsis Toc159 family of chloroplast protein import receptors. Plant Cell 16:2059–2077

Küchler M, Decker S, Hörmann F, Soll J, Heins L (2002) Protein import into chloroplasts involves redox-regulated proteins. EMBO J 21:6136–6145

Lee YJ, Kim DH, Kim YW, Hwang I (2001) Identification of a signal that distinguishes between the chloroplast outer envelope membrane and the endomembrane system in vivo. Plant Cell 13:2175–2190

Lee KH, Kim SJ, Lee YJ, Jin JB, Hwang I (2003) The M domain of atToc159 plays an essential role in the import of proteins into chloroplasts and chloroplast biogenesis. J Biol Chem 278:36794–36805

Lee YJ, Sohn EJ, Lee KH, Lee DW, Hwang I (2004) The transmembrane domain of AtToc64 and its C-terminal lysine-rich flanking region are targeting signals to the chloroplast outer envelope membrane. Mol Cells 17:281–291

Li H-M, Schnell DJ (2006) Reconstitution of protein targeting to the inner envelope membrane of chloroplasts. J Cell Biol 175:249–259

Li H-M, Kesavulu MM, Su P-H, Yeh Y-H, Hsiao C-D (2007) Toc GTPases. J Biomed Sci 14:505–508

Lübeck J, Soll J, Akita M, Nielsen E, Keegstra K (1996) Topology of IEP110, a component of the chloroplastic protein import machinery present in the inner envelope membrane. EMBO J 15:4230–4238

Lynch M, Blanchard JL (1998) Deleterious mutation accumulation in organelle genomes. Genetica 102–103:29–39

Ma Y, Kouranov A, LaSala SE, Schnell DJ (1996) Two components of the chloroplast protein import apparatus, IAP86 and IAP75, interact with the transit sequence during the recognition and translocation of precursor proteins at the outer envelope. J Cell Biol 134:315–327

Marshall JS, DeRocher AE, Keegstra K, Vierling E (1990) Identification of heat shock protein hsp70 homologues in chloroplasts. Proc Natl Acad Sci USA 87:374–378

Martin W, Herrman RG (1998) Gene transfer from organelles to the nucleus: how much, what happens, and why? Plant Physiol 118:9–17

Mason JR, Cammack R (1992) The electron-transport proteins of hydroxylating bacterial dioxygenases. Annu Rev Microbiol 46:277–305

May T, Soll J (2000) 14-3-3 proteins form a guidance complex with chloroplast precursor proteins in plants. Plant Cell 12:53–63

Miras S, Salvi D, Piette L, Seigneurin-Berny D, Grunwald D, Reinbothe C, Joyard J, Reinbothe S, Rolland N (2007) Toc159- and Toc75-independent import of a transit sequence-less precursor into the inner envelope of chloroplasts. J Biol Chem 282:29482–29492

Moberg P, Ståhl A, Bhushan S, Wright SJ, Eriksson A, Bruce BD, Glaser E (2003) Characterization of a novel zinc metalloprotease involved in degrading targeting peptides in mitochondria and chloroplasts. Plant J 36:616–628

Nada A, Soll J (2004) Inner envelope protein 32 is imported into chloroplasts by a novel pathway. J Cell Sci 117:3975–3982

Nakai K, Horton P (1999) PSORT: a program for detecting sorting signals in proteins and predicting their subcellular localization. Trends Biochem Sci 24:34–35

Nakrieko K-A, Mould RM, Smith AG (2004) Fidelity of targeting to chloroplasts is not affected by removal of the phosphorylation site from the transit peptide. Eur J Biochem 271:509–516

Nanjo Y, Oka H, Ikarashi N, Kaneko K, Kitajima A, Mitsui T, Muñoz FJ, Rodríguez-López M, Baroja-Fernández E, Pozueta-Romero J (2006) Rice plastidial N-glycosylated nucleotide pyrophosphatase/phosphodiesterase is transported from the ER-golgi to the chloroplast through the secretory pathway. Plant Cell 18:2582–25892

Nassoury N, Morse D (2005) Protein targeting to the chloroplasts of photosynthetic eukaryotes: getting there is half the fun. Biochim Biophys Acta 1743:5–19

Neupert W, Brunner M (2002) The protein import motor of mitochondria. Nat Rev Mol Biol Cell 3:555–565

Nielsen E, Akita M, Davila-Aponte J, Keegstra K (1997) Stable association of chloroplastic precursors with protein-translocation complexes that contain proteins from both envelope membranes and a stromal Hsp1000 molecular chaperone. EMBO J 16:935–946

Olsen LJ, Keegstra K (1992) The binding of precursor proteins to chloroplasts requires nucleoside triphosphates in the intermembrane space. J Biol Chem 267:433–439

Olsen LJ, Theg SM, Selman BR, Keegstra K (1989) ATP is required for the binding of precursor proteins to chloroplasts. J Biol Chem 264:6724–6729

Olson JM (2006) Photosynthesis in the Archean era. Photosynth Res 88:109–117

Oreb M, Zoryan M, Vojta A, Maier UG, Eichacker LA, Schleiff E (2007) Phospho-mimicry mutant of atToc33 affects early development of *Arabidopsis thaliana*. FEBS Lett 581:5945–5951

Pain D, Blobel G (1987) Protein import into chloroplasts requires a chloroplast ATPase. Proc Natl Acad Sci USA 84:3288–3292

Park S, Rodermel SR (2004) Mutations in ClpC2/Hsp100 suppress the requirement for FtsH in thylakoid membrane biogenesis. Proc Natl Acad Sci USA 101:12765–12770

Perry SE, Keegstra K (1994) Envelope membrane proteins that interact with chloroplastic precursor proteins. Plant Cell 6:93–105

Philippar K, Geis T, Ilkavets I, Oster U, Schwenkert S, Meurer J, Soll J (2007) Chloroplast biogenesis: the use of mutants to study the etioplast-chloroplast transition. Proc Natl Acad Sci USA 104:678–683

Pilon M, Wienk H, Sips W, de Swaaf M, Talboom I, van't Hof R, de Korte-Kool G, Demel R, Weisbeek P, de Kruijff B (1995) Functional domains of the ferredoxin transit sequence involved in chloroplast import. J Biol Chem 270:3882–3893

Pinnaduwage P, Bruce B (1996) In vitro interaction between a chloroplast transit peptide and chloroplast outer envelope lipids is sequence-specific and lipid class-dependent. J Biol Chem 271:32907–32915

Pollmann S, Neu D, Lehmann T, Berkowitz O, Schafer T, Weiler EW (2006) Subcellular localization and tissue specific expression of amidase 1 from *Arabidopsis thaliana*. Planta 224:1241–1253

Qbadou S, Becker T, Mirus O, Tews I, Soll J, Schleiff E (2006) The molecular chaperone Hsp90 delivers precursor proteins to the chloroplast import receptor Toc64. EMBO J 25:1836–1847

Qbadou S, Becker T, Bionda T, Reger K, Ruprecht M, Soll J, Schleiff E (2007) Toc64 – a preprotein-receptor at the outer membrane with bipartide function. J Mol Biol 367:1330–1346

Rassow J, Dekker PJ, van Wilpe S, Meijer M, Soll J (1999) The preprotein translocase of the mitochondrial inner membrane: function and evolution. J Mol Biol 286:105–120

Reddick LE, Vaughn MD, Wright SJ, Campbell IM, Bruce BD (2007) In vitro comparative kinetic analysis of the chloroplast Toc GTPases. J Biol Chem 282:11410–11426

Reinbothe S, Runge S, Reinbothe C, van Cleve B, Apel K (1995) Substrate-dependent transport of the NADPH:protochlorophyllide oxidoreductase into isolated plastids. Plant Cell 7:161–172

Reinbothe C, Lebedev N, Apel K, Reinbothe S (1997) Regulation of chloroplast protein import through a protochlorophyllide-responsive transit peptide. Proc Natl Acad Sci USA 94:8890–8894

Reinbothe S, Quigley F, Springer A, Schemenewitz A, Reinbothe C (2004) The outer plastid envelope protein Oep16: role as precursor translocase in import of protochlorophyllide oxidoreductase A. Proc Natl Acad Sci USA 101:2203–2208

Rensink WA, Pilon M, Weisbeek PJ (1998) Domains of a transit sequence required for in vivo import in Arabidopsis chloroplasts. Plant Physiol 118:691–699

Reumann S, Keegstra K (1999) The endosymbiotic origin of the protein import machinery of chloroplastic envelope membranes. Trends Plant Sci 4:302–307

Reumann S, Davila-Aponte J, Keegstra K (1999) The evolutionary origin of the protein-translocating channel of chloroplastic envelope membranes: identification of a cyanobacterial homolog. Proc Natl Acad Sci USA 96:784–789

Reumann S, Inoue K, Keegstra K (2005) Evolution of the general protein import pathway. Mol Membr Biol 22:73–86

Richter S, Lamppa GK (1999) Stromal processing peptidase binds transit peptides and initiates their ATP-dependent turnover in chloroplasts. J Biol Chem 147:33–43

Richter S, Lamppa GK (2003) Structural properties of the chloroplast stromal processing peptidase required for its function in transit peptide removal. J Biol Chem 278:39497–39502

Roth RA (2004) Pitrilysin. In: Handbook of proteolytic enzymes, 2 edn. Barrett AJ, Rawlings ND and Woessner JF (eds).Elsevier, London, pp 868–871

Rudhe C, Clifton R, Chew O, Zemam K, Richter S, Lamppa G, Whelan J, Glaser E (2004) Processing of the dual targeted precursor protein of glutathione reductase in mitochondria and chloroplasts. J Mol Biol 343:639–647

Schleiff E, Tien R, Salomon M, Soll J (2001) Lipid composition of outer leaflet of chloroplast outer envelope determines topology of OEP7. Mol Biol Cell 12:4090–4102

Schleiff E, Jelic M, Soll J (2003a) A GTP-driven motor moves proteins across the outer envelope of chloroplasts. Proc Natl Acad Sci USA 100:4604–4609

Schleiff E, Soll J, Küchler M, Kühlbrandt W, Harrer R (2003b) Characterization of the translocon of the outer envelope of chloroplasts. J Cell Biol 160:541–551

Schlegel T, Mirus O, von Haeseler A, Schleiff E (2007) The tetratricopeptide repeats of receptors involved in protein translocation across membranes. Mol Biol Evol 24:2763–2774

Schnell DJ, Blobel G (1993) Identification of intermediates in the pathway of protein import into chloroplasts and their localization to envelope contact sites. J Cell Biol 120:103–115

Schnell DJ, Kessler F, Blobel G (1994) Isolation of components of the chloroplast protein import machinery. Science 266:1007–1012

Schnell DJ, Blobel G, Kessler F, Keegstra K, Ko K, Soll J (1997) A consensus nomenclature for the protein-import components of the chloroplast envelope. Trends Cell Biol 7:303–304

Schuenemann D (2004) Structure and function of the chloroplast signal recognition particle. Curr Genet 44:295–304

Scott SV, Theg SM (1996) A new chloroplast protein import intermediate reveals distinct translocation machineries in the two envelope membranes: energetics and mechanistic implications. J Cell Biol 132:63–75

Seedorf M, Waegemann K, Soll S (1995) A constituent of the chloroplast import complex represents a new type of GTP-binding protein. Plant J 7:401–411

Shan SO, Walter P (2005) Co-translational protein targeting by the signal recognition particle. FEBS Lett 579:921–926

Shanklin J, DeWitt ND, Flanagan JM (1995) The stroma of higher plant plastids contain ClpP and ClpC, functional homologs of *Escherichia coli* ClpP and ClpA: an archetypal two-component ATP-dependent protease. Plant Cell 7:1713–1722

Sjögren LL, MacDonald TM, Sutinen S, Clarke AK (2004) Inactivation of the clpC1 gene encoding a chloroplast Hsp100 molecular chaperone causes growth retardation, leaf chlorosis, lower photosynthetic activity, and a specific reduction in photosystem content. Plant Physiol 136:4114–4126

Small I, Peeters N, Legeai F, Lurin C (2004) Predotar: a tool for rapidly screening proteomes for N-terminal targeting sequences. Proteomics 4:1581–1590

Smeekens S, Bauerle C, Hageman J, Keegstra K, Weisbeek P (1986) The role of the transit peptide in the routing of precursors toward different chloroplast compartments. Cell 46:365–375

Smith MD, Hiltbrunner A, Kessler F, Schnell DJ (2002) The targeting of the at Toc159 preprotein receptor to the chloroplast outer membrane is mediated by its GTPase domain and is regulated by GTP. J Cell Biol 159:833–843

Smith MD, Rounds CM, Wang F, Chen K, Afitlhile M, Schnell DJ (2004) atToc159 is a selective transit peptide receptor for the import of nucleus-encoded chloroplast proteins. J Cell Biol 165:323–334

Sohrt K, Soll J (2000) Toc64, a new component of the protein translocon of chloroplasts. J Cell Biol 148:1213–1221

Sokolenko A Lerbs-Mache S Altschmied L, Herrmann RG (1998) Clp protease complexes and their diversity in chloroplasts. Planta 207:286–295

Soll J, Schleiff E (2004) Protein import into chloroplasts. Nat Rev Mol Cell Biol 5:198–208

Stahl T, Glockman C, Soll J, Heins L (1999) Tic40, a new "old"subunit of the chloroplast protein import translocon. J Biol Chem 274:37467–37472

Stengel A, Benz P, Balsera M, Soll J, Bölter B (2008) TIC62 – redox-regulated translocon composition and dynamics. J Biol Chem /doi/10.1074/jbc.M706719200

Sun CW, Chen LJ, Lin LC, Li HM (2001) Leaf-specific upregulation of chloroplast translocon genes by a CCT motif-containing protein, CIA2. Plant Cell 13:2053–2061

Sun Y-J, Forouhar F, Li H-M, Tu S-L, Yeh Y-H, Kao S, Shr H-L, Chou C-C, Chen C, Hsiao C-D (2002) Crystal structure of pea Toc34, a novel GTPase of the chloroplast protein translocation. Nat Struct Biol 9:95–100

Sveshnikova N, Grimm R, Soll J, Schleiff E (2000a) Topology studies of the chloroplast protein import channel Toc75. Biol Chem 381:687–693

Sveshnikova N, Soll J, Schleiff E (2000b) Toc34 is a preprotein receptor regulated by GTP and phosphorylation. Proc Natl Acad Sci USA 97:4973–4978

Teakle GR, Griffiths WT (1993) Cloning, characterization and import studies on protochlorophyllide reductase from wheat (Triticum aestivum). Biochem J 296:225–230

Teng YS, Su YS, Chen LJ, Lee YJ, Hwang I, Li HM (2006) Tic21 is an essential translocon component for protein translocation across the chloroplast inner envelope membrane. Plant Cell 18:2247–2257

Theg S, Bauerle C, Olsen L, Selman B, Keegstra K (1989) Internal ATP is the only energy requirement for the translocation of precursor. J Biol Chem 264:6730–6736

Tranel PJ, Froehlich J, Goyal A, Keegstra K (1995) A component of the chloroplastic protein import apparatus is targeted to the outer envelope membrane via a novel pathway. EMBO J 14:2436–2446

Tripp J, Inoue K, Keegstra K, Froehlich J (2007) A novel serine/proline-rich domain in combination with a transmembrane domain is required for the insertion of AtTic40 into the inner envelope membrane of chloroplasts. Plant J 52:824–838

Tsugeki R, Nishimura M (1993) Interaction of homologues of Hsp70 and Cpn60 with ferredoxin-NADP$^+$ reductase upon its import into chloroplasts. FEBS Lett 320:198–202

Tu SL, Chen LJ, Smith MD, Su YS, Schnell DJ, Li HM (2004) Import pathways of chloroplast interior proteins and the outer-membrane protein OEP14 converge at Toc75. Plant Cell 16:2078–2088

van den Wijngaard PWJ, Vredenberg WJ (1997) A 50-pico-siemens anion channel of the chloroplast envelope is involved in chloroplast import. J Biol Chem 272:29430–29433

van den Wijngaard PWJ, Vredenberg WJ (1999) The envelope anion channel involved in chloroplast protein import is associated with Tic110. J Biol Chem 274:25201–25204

van den Wijngaard PWJ, Demmers JAA, Thompson SJ, Wienk HLJ, de Kruijff BD, Vredenberg WJ (2000) Further analysis of the involvement of the envelope anion channel PIRAC in chloroplast protein import. Eur J Biochem 267:3812–3817

VanderVere PS, Bennett TM, Oblong JE, Lamppa GK (1995) A chloroplast processing enzyme involved in precursor maturation shares a zinc-binding motif with a recently recognized family of metalloendopeptidases. Proc Natl Acad Sci USA 92:7177–7181

van't Hof R, Demel RA, Keegstra K, de Kruijff B (1991) Lipid–peptide interactions between fragments of the transit peptide of ribulose-1,5-bisphosphate carboxylase/oxygenase and chloroplast membrane lipids. FEBS Lett 291:350–354

van't Hof R, van Klompenburg W, Pilon M, Kozubek A, de Korte-Kool G, Demel RA, Weisbeek PJ, de Kruijff B (1993) The transit sequence mediates the specific interaction of the precursor of ferredoxin with chloroplast envelope membrane. J Biol Chem 268:4037–4042

Villarejo A, Burén S, Larsson S, Déjardin A, Monné M, Rudhe C, Karlsson J, Jansson S, Lerouge P, Rolland N, von Heijne G, Grebe M, Bako L, Samuelsson G (2005) Evidence for a protein transported through the secretory pathway en route to the higher plant chloroplast. Nat Cell Biol 7:1224–1231

Voigt A, Jakob M, Klösgen RB, Gutensohn M (2005) At least two Toc34 protein import receptors with different specificities are also present in spinach chloroplasts. FEBS Lett 579:1343–1349

Vojta L, Soll J, Bölter B (2007) Protein transport in chloroplasts – targeting to the intermembrane space. FEBS J 274:5043–5054

von Heijne G (1995) Membrane protein assembly: rules of the game. Bioessays 17:25–30

von Heijne G, Steppuhn J, Herrmann RG (1989) Domain structure of mitochondrial and chloroplast targeting peptides. Eur J Biochem 180:535–545

Waegemann K, Soll J (1991) Characterization of the protein import apparatus in isolated outer envelopes of chloroplasts. Plant J 1:149–158

Waegemann K, Soll J (1996) Phosphorylation of the transit sequence of chloroplast precursor proteins. J Biol Chem 271:6545–6554

Wallas TR, Smith MD, Sanchez-Nieto S, Schnell DJ (2003) The roles of toc34 and toc75 in targeting the toc159 preprotein receptor to chloroplasts. J Biol Chem 278:44289–44297

Wan J, Bringloe D, Lamppa GK (1998) Disruption of chloroplast biogenesis and plant development upon down-regulation of a chloroplast processing enzyme involved in the import pathway. Plant J 15:459–468

Weibel P, Hiltbrunner A, Brand L, Kessler F (2003) Dimerization of Toc-GTPases at the chloroplast protein import machinery. J Biol Chem 278:37321–37329

Wu C, Seibert FS, Ko K (1994) Identification of chloroplast envelope proteins in close physical proximity to a partially translocated chimeric precursor protein. J Biol Chem 269:32264–32271

Yeh YH, Kesavulu MM, Li HM, Wu SZ, Sun YJ, Konozy EH, Hsiao CD (2007) Dimerization is important for the GTPase activity of chloroplast translocon components atToc33 and psToc159. J Biol Chem 282:13845–13853

Young ME, Keegstra K, Froehlich JE (1999) GTP promotes the formation of early-import intermediates but is not required during the translocation step of protein import into chloroplasts. Plant Physiol 121:237–243

Yu T-S, Li H-M (2001) Chloroplast protein translocation components atToc159 and atToc33 are not essential for chloroplast biogenesis in guard cells and root cells. Plant Physiol 127:90–96

Zhong R, Wan J, Jin R, Lamppa G (2003) A pea antisense gene for the chloroplast stromal processing peptidase yields seedling lethals in Arabidopsis: survivors show defective GFP import in vivo. Plant J 34:802–812

Chloroplast Membrane Lipid Biosynthesis and Transport

M.X. Andersson and P. Dörmann (✉)

Abstract The photosynthetic membranes of chloroplasts are characterized by a high abundance of glycolipids. The two galactolipids monogalactosyldiacylglycerol and digalactosyldiacylglycerol (DGDG) are the predominant constituents of thylakoid membranes, while phospholipids (phosphatidylcholine, phosphatidylglycerol) and a sulfolipid (sulfoquinovosyldiacylglycerol) are minor components. Galactolipids are synthesized in the envelope membranes from precursors originating from the chloroplast or from the endoplasmic reticulum (ER). Direct contact sites ("plastid-associated membranes") between the ER and the chloroplast may be involved in the transport of lipid precursors to the envelope membranes. During chloroplast development, thylakoids are established from invaginations of the inner envelope, whereas in mature chloroplasts, a vesicle-based transport system was suggested to supply galactolipids to the thylakoids. During phosphate limitation, phospholipids are replaced with glycolipids in plastidial and extraplastidial membranes. DGDG produced in the chloroplast was suggested to be transferred to the mitochondria via direct contact sites. The transport of DGDG to the plasma membrane and tonoplast is believed to be mediated via the ER and Golgi vesicle trafficking system.

1 Introduction

Oxygenic photosynthesis in its very essence is a membrane-bound process. The primary light absorption is mediated by membrane-bound protein–pigment complexes, the electron transport chain consists of membrane-bound components, and the proton gradient that drives ATP synthase is built across the thylakoid membrane. It is thus not surprising that the higher-plant chloroplast is a very membrane rich organelle.

P. Dörmann
Institute of Molecuar Physiology and Biotechnology of Plants, University of Bonn,
Karlrobert-Kreiten-Str. 13, 53115 Bonn, Germany
e-mail: Doermann@uni-bonn.de

The chloroplast internal thylakoid membranes are the site of the photosynthetic electron transport chain and constitute an extremely large surface area in green plant tissues. In fact, the bulk of the membrane lipids in a green leaf are situated in the thylakoid membranes. The chloroplast is delimited from the cytosol by the double-envelope membrane, and all molecules exchanged between the chloroplast and other cellular compartments must at some point pass across the envelope. The envelope is also the site for the final assembly of the major chloroplast lipids.

A consensus on the overall lipid composition of the different chloroplast membranes was reached in the mid-1980s. At the same time, many of the major membrane lipid biosynthetic pathways were worked out. The 1990s and the beginning of the postgenomic era marked the cloning of a number of important chloroplast lipid biosynthesis genes from the model plant *Arabidopsis thaliana* and other species. Membrane lipid biosynthesis mutants generated by forward or reverse genetic approaches shed more light on the in vivo function(s) of individual membrane lipid classes. In the last decade it has become clear that a particular chloroplast galactolipid, digalactosyldiacylglycerol (DGDG), can play a major role in extraplastidial membranes during phosphate-limited growth conditions. Thus, galactolipid synthesis under normal and phosphate-limited conditions requires the involvement of several cellular compartments. The remaining questions are concerned mainly with the regulation of the synthesis pathways and transport of chloroplast lipids and precursors. In the present chapter, we will first give a brief overview of the chloroplast lipids and their particular functions, and then summarize the current understanding of biosynthesis and transport of chloroplast membrane lipids. We have tried to include as many as possible of the primary literature sources, but owing to space limitations, it was impossible to cover all relevant aspects. We apologize to all whose excellent contributions we have not been able to refer to.

2 The Chloroplast Lipidome

The lipid composition of the chloroplast membranes differs in several ways from that of other membranes in the plant cell and seems to largely reflect the cyanobacterial origin of the organelle. Phosphoglycerolipids constitute only a minor proportion of the chloroplast membranes; instead the chloroplast membranes are highly enriched in galactoglycerolipids. Sterols and sphingolipids, which are important constituents of the plant plasma membrane, Golgi apparatus and tonoplast are completely absent from the chloroplast membranes. The chloroplasts are the only plant organelles that contain the anionic sulfur-containing lipid sulfoquinovosyldiacylglycerol (SQDG). The thylakoids, which constitute the bulk of the membrane surface in a mature chloroplast (approximately 90 mol% of the membrane lipids), are, in principle, composed of four different lipid classes, the galactolipids monogalactosyldiacylglycerol (MGDG) and DGDG and the two anionic lipids phosphatidylglycerol (PG) and SQDG. The structures of the most common molecular species of the chloroplast membrane lipids are shown in Fig. 1 and the membrane lipid composition of chloroplasts and chloroplast subfractions is shown in Table 1.

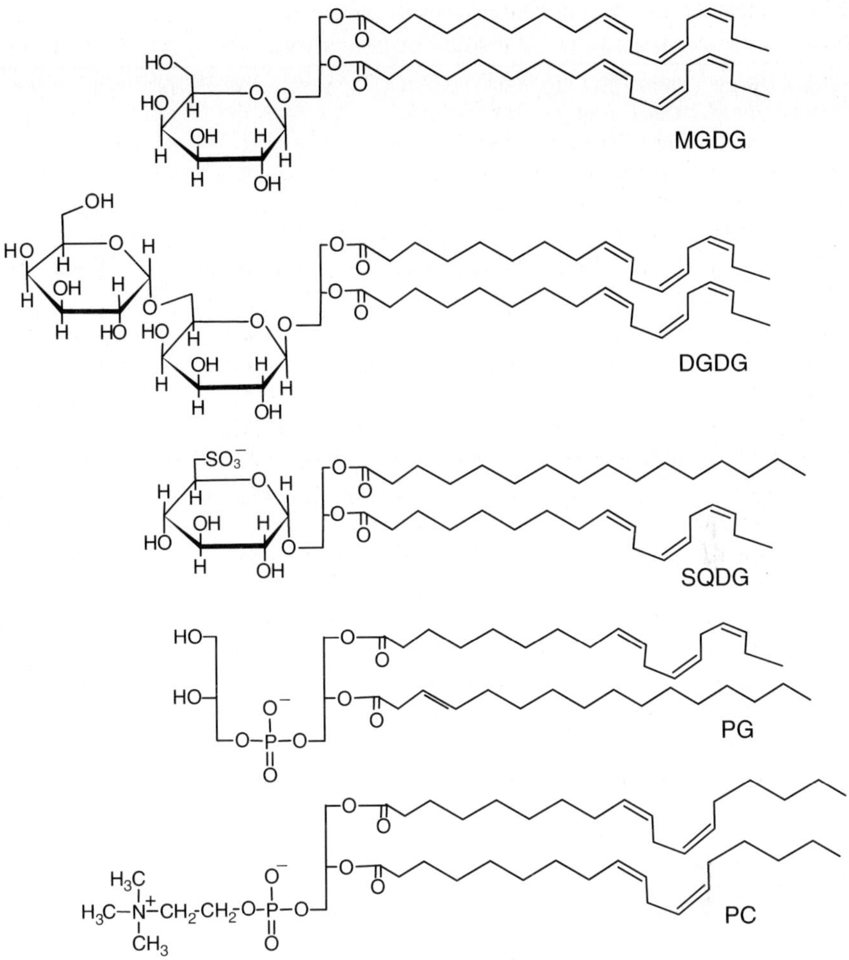

Chloroplast Membrane Lipid Biosynthesis and Transport

Fig. 1 Glycerolipids found in chloroplast membranes. Chloroplasts contain large amounts of the two galactolipids monogalactosyldiacylglycerol (*MGDG*) and digalactosyldiacylglycerol (*DGDG*). Furthermore, two anionic lipids are found in chloroplasts, the sulfolipid sulfoquinovosyldiacylglycerol (*SQDG*) and the phospholipid phosphatidylglycerol (*PG*). The zwitterionic phospholipid phosphatidylcholine (*PC*) is found primarily in the outer-envelope membrane

The galactose head group in MGDG and DGDG is directly linked through a β-glycosidic bond to the glycerol backbone, and the second galactose in the DGDG head group is linked by an α-glycosidic bond to the first galactose (Fig. 1). The fatty acids found in the chloroplast membrane lipids are highly unsaturated. Trienoic acids usually constitute more than 80–90% of the acyl groups found in MGDG and DGDG. SQDG and PG, on the other hand, contain a relatively larger proportion of more saturated fatty acids. The most common trienoic acid found in plants is linolenic acid

Table 1 Lipid composition (mol%) of chloroplast membranes

Plant	Membrane	MGDG	DGDG	SQDG	PG	PC	PI	PE
Spinach (Block et al., 1983b)	Total envelope	36	29	6	9	18	2	ND
	Outer envelope	17	29	6	10	32	5	ND
	Inner envelope	49	30	5	8	6	1	ND
	Thylakoid	52	26	6.5	9.5	4.5	1.5	ND
Pea (Andersson et al., 2001)	Intact chloroplasts	46	32	7	6	7	1	ND
	Thylakoid	51	33	8	5	2	0	ND
Pea (Cline et al., 1981)	Outer envelope	6	33	3	6	44	5	2
	Inner envelope	45	31	2	7	10	2	1
Wheat (Bahl et al., 1976)	Total envelope	22	44	10	9	14	0	0
	Lamellar thylakoid	42	37	9	10	2	0	0
	Grana thylakoid	47	36	7	9	1	0	0
Broad bean (Mackender and Leech, 1974)	Total envelope	29	32	ND	9	30	ND	0
	Thylakoid	65	26	ND	6	3	ND	0
Acer hippocastanum (Chapman et al., 1986)	Thylakoid	43	31	5	15	–[a]	–[a]	–[a]
Pea (Chapman et al., 1986)	Thylakoid	42	29	8	11	–[a]	–[a]	–[a]
Dark grown wheat (Selstam and Sandelius, 1984)	Envelope	44	37	6	6	5	2	ND
	Prothylakoid	45	40	8	7	–	–	ND
	Prolamellar body	52	32	8	8	–	–	ND
Consensus	Outer envelope	6–17	30	3–6	6–10	32–44	5	0
	Inner envelope	45–49	30	2–5	6–8	6–10	1–2	0
	Thylakoid	42–65	26–33	5–8	5–15	2–4	0–1	0

MGDG monogalactosyldiacylglycerol, *DGDG* digalactosyldiacylglycerol, *SQDG* sulfoquinovosyldiacylglycerol, *PG* phosphatidylglycerol, *PC* phosphatidylcholine, *PI* phosphatidylinositol, *PE* phosphatidylethanolamine
ND not detectable, – not reported
[a] Not reported, but "minor phospholipids" present in low amounts of 6–8 mol%

(18:3). Hexadecatrienoic acid (16:3) is restricted to the *sn*-2 position of MGDG and occurs in significant amounts only in some plant species, the so-called 16:3 plants (for example *Arabidopsis* and spinach). The ratio of 18:3 to 16:3 fatty acids esterified to chloroplast galactolipids varies considerably among different plant species (Mongrand et al., 1998). The majority of plant species are devoid of 16:3 and are referred to as "18:3 plants". Chloroplast PG in both 16:3 and 18:3 plants contains only molecular lipid species synthesized by the prokaryotic pathway (see Sect. 5) (Roughan, 1985; Dorne and Heinz, 1989). Chloroplast PG contains approximately 20% of the unusual monene fatty acid *trans*-3-hexadecenoic acid, which is strictly associated with the *sn*-2 position (Dubacq and Tremolieres, 1983; Browse et al., 1986).

Methods to isolate the outer and inner chloroplast envelopes in purity and quantity sufficient for reliable biochemical characterization were developed in the early 1980s (Cline et al., 1981; Block et al., 1983a,b). The inner envelope is in terms of lipid composition closely related to the thylakoids, whereas the outer-envelope lipid composition is more related to that of other extraplastidial membranes (Table 1). Notably, the outer-envelope membrane contains a substantial amount of the zwitterionic phospholipid phosphatidylcholine (PC; Fig. 1). The amount of PC in the inner envelope and the thylakoids is very low. A few reports indeed suggest that the occurrence of PC in all other chloroplast compartments except the outer leaflet of the outer envelope is entirely artefactual (Dorne et al., 1985; Dorne et al., 1990). Nevertheless, the majority of studies report a PC content in thylakoids and inner envelope of 1–4 mol%. The ratio of DGDG to MGDG is much higher in the outer envelope than in the inner envelope and the thylakoids. The lipid-to-protein ratio changes from the outer envelope to the thylakoid. The outer envelope is lipid-rich, with a ratio of lipid to protein of about 2.5, whereas the inner envelope and the thylakoids have lipid-to-protein ratios of about 1 and 0.4, respectively (Block et al., 1983b).

3 Chloroplast Membrane Lipid Function

The primary function of the chloroplast membrane lipids is to provide a structural environment for the photosynthetic membrane protein complexes and a barrier for the different solutes present in the thylakoid lumen and chloroplast stroma, which is also a prerequisite for the establishment of the photosynthetic proton gradient. In addition, particular lipids are also found in very close association with or embedded into the photosynthetic membrane protein complexes and linked to specific biochemical functions. The major thylakoid lipid, polyunsaturated MGDG, has a small polar head group but at the same time bulky fatty acid chains. The other major chloroplast lipid, DGDG, has a much larger head group and thus an almost cylindrical geometry. DGDG by itself in excess water forms a stable bilayer, whereas MGDG favours the formation of inverted hexagonal or cubic phases. Pure lipid mixtures resembling the composition of the thylakoid membrane do not form stable bilayers on their own, but rather complex mixtures of inverted hexagonal and cubic phases (Brentel et al., 1985). The non-bilayer-forming tendency is highly dependent on the degree of desaturation of the membrane lipids (Gounaris et al., 1983). The reason why the non-bilayer lipid mixture still forms a stable thylakoid membrane is probably the very high content of bilayer-spanning proteins. In addition to the bilayer-spanning proteins, carotenoids may also contribute to the stabilization of the inner envelope and the thylakoid membrane. Several studies indicated that carotenoids contribute to membrane stability of lipid bilayers (Gruszecki and Sielewiesiuk, 1991; Gabrielska and Gruszecki, 1996; Wisniewska and Subczynski, 1998; Berglund et al., 1999; Munné-Bosch and Alegre, 2002; Szilágyi et al., 2008). The exact amount of carotenoids freely dissolved in the thylakoid lipid bilayer is not known and the contribution of the carotenoids to the thylakoid membrane stability remains an open question. It has, however, been suggested that the xanthophyll cycle participates in regulating thylakoid membrane stability (Gruszecki and Strzalka, 1991; Szilágyi et al., 2008).

Mutations which cause a substantial loss of MGDG or DGDG lead to severely compromised chloroplast function in *Arabidopsis* (Dörmann et al., 1995; Härtel et al., 1997; Jarvis et al., 2000; Kelly et al., 2003; Kobayashi et al., 2007); however, the chonsequences of a loss of DGDG are not as strong as a loss of MGDG. DGDG deficiency has severe effects on photosynthetic performance and growth (Kelly et al., 2003), whereas the loss of MGDG in the *mgd1* mutant leads to albino plants with a strong decrease in thylakoid membrane abundance and photosynthetic capacity (Kobayashi et al., 2007). A complete loss of chloroplast DGDG has severe effects on photosynthetic performance and growth (Kelly et al., 2003), whereas the loss of MGDG in the *mgd1* mutant leads to albino plants with a strong decrease in thylakoid membranes and photosynthetic capacity (Kobayashi et al., 2007). *Arabidopsis* is able to cope with a total loss of SQDG, at least under standard conditions (Essigmann et al., 1998; Yu et al., 2002). Although it seems that the two anionic thylakoid lipids can compensate for each other to some extent, it is clear that any substantial loss of chloroplast PG strongly affects photosynthetic activity, growth and thylakoid development (Hagio et al., 2002; Xu et al., 2002; Babiychuk et al., 2003).

An important question is what the function of the very high content of trienoic acids in the chloroplast membrane lipids might be? In fact, an *Arabidopsis* fatty acid desaturase tripple mutant which completely lacks trienoic fatty acids has a very mild phenotype with regard to growth, photosynthesis and chloroplast ultrastructure under standard growth conditions (Routaboul et al., 2000); however, the mutant reveals severe defects in growth and photosynthesis at low temperatures. Thus, the high amount of trienoic fatty acids in chloroplast membrane lipids appears to be important for maintaining membrane functionality at low temperatures. It was recently shown that the loss of trienoic fatty acids from chloroplast membrane lipids had a detrimental effect on one of the pathways for translocation of proteins across the thylakoid membrane to the thylakoid lumen (Ma and Browse, 2006). Certain lipids also appear to have other more specific roles beside their general structural contribution to the chloroplast membranes. Several chloroplast lipids have been shown to be intrinsic components of the photosystems (Jordan et al., 2001; Loll et al., 2005, 2007). MGDG in the outer envelope was suggested to be required for protein recognition and targeting to the chloroplast (Bruce, 1998). The trienoic acids in the chloroplast lipids serve as substrates for the chloroplast lipoxygenase pathway (Feussner and Wasternack, 2002). This pathway provides several potent signalling compounds to the plant; the best characterized is jasmonic acid. The jasmonic acid which is responsible for anther dehiscence in *Arabidopsis* is synthesized from 18:3 released from PC by an envelope-localized acyl hydrolase (Ishiguro et al., 2001), whereas wound-induced jasmonic acid is predominantly produced from 18:3 released by a galactolipase (Hyun et al., 2008).

4 Minor Chloroplast Lipids

In addition to the major lipid classes found in the chloroplast membranes, a number of minor lipid species have also been reported. The chloroplast membranes contain a small proportion of intermediates of the biosynthetic pathways for chloroplast lipids; these will be discussed in the context of the respective biosynthesis pathways. Other minor lipid classes reported in chloroplasts or chloroplast subfractions include inositol phospholipids (Siegenthaler et al., 1997; Bovet et al., 2001), oligogalactolipids, phosphorylated galactolipids (Müller et al., 2000) and acylated galactolipids (Heinz, 1967; Heinz and Tulloch, 1969; Heinz et al., 1978). In addition, a number of oxygenated species of chloroplast galactolipids have been described in plant tissues, but never actually isolated from a purified chloroplast fraction. Trigalactosyldiacylglycerols (TGDG) and tetragalactosyldiacylglycerols are not normally found in plant lipid extracts, but have been described as minor constituents in isolated chloroplasts (Cline et al., 1981; Wintermans et al., 1981).

In general, a small proportion of phosphatidylinositol (PI; Table 1) was reported to be present in chloroplast envelopes. In other cellular membranes, the mono-, di- and triphosphorylated analogues of PI provide important functions in intracellular signalling and cytoskeletal organization (Mueller-Roeber and Pical, 2002). An ATP-dependent, wortmannin-sensitive PI kinase activity has been reported to occur in outer envelope from spinach chloroplasts (Siegenthaler et al., 1997; Bovet et al., 2001). Thus, it is possible that phosphatidylinositide-dependent signalling or cytoskeletal reorganization can also emanate from the chloroplast. In addition to phosphatidylinositolphosphates, Bovet and colleagues (Müller et al., 2000; Bovet et al., 2001) reported CTP-dependent phosphorylation of MGDG and lyso-MGDG in spinach chloroplast envelopes. These phosphorylated lipids have so far only been reported as radiolabelled products after in vitro feeding with nucleotides. Steady-state concentrations and the physiological relevance of the phosphorylated chloroplast lipids remain open questions.

The chloroplast lipids are rich in polyunsaturated fatty acids and thus sensitive to chemical or enzymatic oxidation. The higher-plant chloroplast contains lipoxygenases with specificity for the 13-position of C_{18} fatty acids as well as several other enzymes which catalyse downstream reaction of oxygenated fatty acids (Feussner and Wasternack, 2002). Several different chloroplast lipid species containing oxygenated fatty acids have been described in extracts from *Arabidopsis*. Information on other plant species, however, remains very scarce. The oxygenated galactolipids described so far include MGDG and DGDG containing keto fatty acids (Buseman et al., 2006) or oxo-phytodienoic acid at the *sn*-1 and/or the *sn*-2 position (Stelmach et al., 2001; Hisamatsu et al., 2003; Stenzel et al., 2003; Hisamatsu et al., 2005; Ohashi et al., 2005; Buseman et al., 2006; Nakajyo et al., 2006; Böttcher and Weiler, 2007). MGDG esterified with oxo-phytodienoic acid has been demonstrated to be present in thylakoid and envelope fractions as well as detergent-solubilized thylakoid pigment protein complexes (Böttcher and Weiler, 2007). In addition, MGDG carrying an extra oxo-phytodienoic acid on the glycerol backbone as well as on the C-6 position of the galactose head group has been found in *Arabidopsis* (Andersson et al., 2006; Kourtchenko et al., 2007). The latter substances were found to accumulate in response to wounding and in the hypersensitive response induced by avirulence peptides. The other oxo-phytodienoic acid containing galactolipids were also induced by wounding of *Arabidopsis* tissues (Buseman et al., 2006; Böttcher and Weiler, 2007; Kourtchenko et al., 2007). Oxo-phytodienoic acid containing galactolipids have so far only been found in extracts from *Arabidopsis thaliana*, *Arabidopsis arenosa* (Böttcher and Weiler, 2007) and *Ipomoea tricolor* (Ohashi et al., 2005). The evidence for a physiological function of these particular oxylipin-containing lipids remains rather scarce, but senescence-promoting effects, antipathogenic properties, stomatal closing and a role as precursors for free oxylipins have been reported. In addition to these particular molecular lipid species, it seems likely that the everyday wear and tear of the photosynthesis machinery would cause some damage to unsaturated galactolipids. Thus, it seems likely that a steady-state concentration of oxygenated fatty acids is present in the thylakoid lipid pool and that these fatty acids are continuously removed from the complex lipid pool. However, this remains poorly explored territory.

5   Origin of Chloroplast Membrane Lipid Acyl Groups

The bulk of the fatty acid synthesis in plant cells takes place in the chloroplast stroma. Plant mitochondria also contribute to fatty acid synthesis, but only in a very minor way. The fatty acid synthesis machinery inside the plastid is related to that of bacteria rather than to cytosolic fatty acid synthesis in other eukaryotic organisms. The substrate for fatty acid synthesis in the stroma was traditionally thought to be acetate, but this has been questioned and the topic is not completely settled (Bao et al., 2000; Rawsthorne, 2002). Regardless of the actual identity of the fatty acid synthesis substrate, exogenous acetate is efficiently channelled into the pathway. This has been instrumental in many acyl labelling studies where radiolabelled acetate has been fed to isolated chloroplast or intact plant tissue. During the fatty acid synthesis cycle, the growing acyl chain is attached via a thioester bond to the small (approximately 9 kDa) acidic acyl carrier protein (ACP). For each turn of the cycle, the fatty acid grows by two carbon atoms. Two steps in the cycle require reducing power which is derived from NAD(P)H. A special stroma-localized desaturase accepts 18:0-ACP as a substrate and introduces a *cis* double bond at the $\Delta 9$-position, yielding oleic acid (18:1). Thus, 18:1 and hexadecanoic acid (16:0) are the major fatty acids synthesized in the stroma (Ohlrogge and Browse, 1995; Rawsthorne, 2002). The fatty acids are then either used directly in the chloroplast for glycerolipid synthesis (prokaryotic pathway), or exported to the endoplasmic reticulum (ER) for lipid synthesis (eukaryotic pathway). In 18:3 plants all the fatty acids found in the chloroplast glycolipids take the detour through the ER, whereas in 16:3 plants a portion of the fatty acids are assembled into galactolipids without ever having to leave the chloroplast (Fig. 2). The reason for the difference between the fatty acid composition of chloroplast lipids in 18:3 and 16:3 plants is the different specifici-

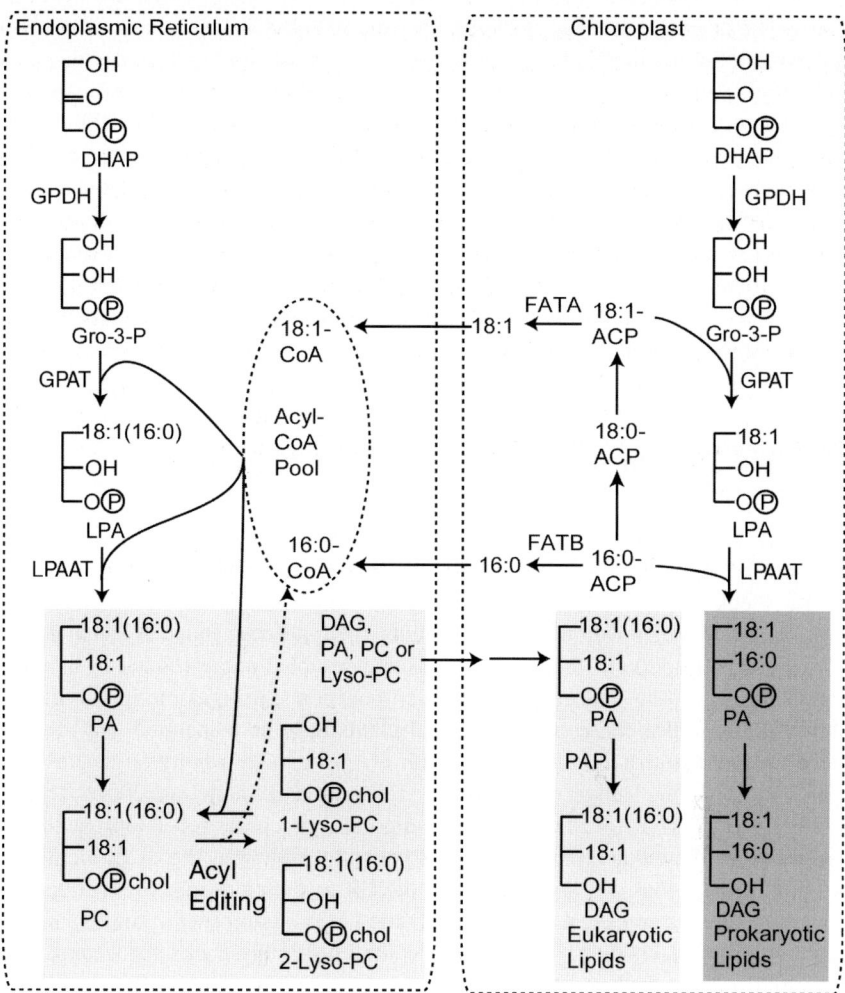

Fig. 2 Dihydroxyacetonephosphate serves as precursor for the synthesis of glycerol-3-phosphate, which is sequentially acylated leading to lyso-phosphatidic acid and phosphatidic acid (*PA*) production in chloroplasts and at the endoplasmic reticulum (ER). Acyl editing represents an alternative pathway for the incorporation of plastid-derived acyl groups into the cytosolic PC pool. PA (or another PC-derived lipid metabolite, DAG, PC or lyso-PC) is transported from the ER to the chloroplast. While dephosphorylation of ER-derived PA results in the synthesis of glycerolipids with eukaryotic structure (C_{16} and C_{18} at *sn*-1, C_{18} at *sn*-2, *light grey*), plastidial PA is the precursor for glycerolipids with prokaryotic structure (C_{18} at *sn*-1, C_{16} and C_{18} at *sn*-2, *dark grey*). The distinct distribution of acyl groups to the glycerol backbone is based on the substrate specificity of ER-localized, acylcoenzyme A (*acyl-CoA*) dependent acyltransferases, and plastidial, acyl carrier protein (*ACP*) dependent acyltransferases

ties of the acyltransferases which assemble phosphatidic acid (PA) in the ER and the chloroplast (Roughan and Slack, 1982; Heinz and Roughan, 1983; Browse et al., 1986). The basis for the loss of the prokaryotic pathway in 18:3 plants is not well understood, but the fact that several independent groups of plants are devoid of the prokaryotic pathway suggests that the loss of the capability to synthesize prokaryotic lipids has occurred several times during evolution (Mongrand et al., 1998). Fatty acids that are exported from the chloroplast are cleaved from ACP by stromal ACP thioesterases. The fatty acids are transported by some unknown mechanism to the outer envelope, where acylcoenzyme A (acyl-CoA) synthases reactivate the fatty acids and render them soluble in the cytosol. There is experimental evidence for a direct channelling from the stromal fatty acid synthesis to the acyl-CoA synthesis (Koo et al., 2004). The acyl-CoAs are then utilized by acyltransferases for phospholipid synthesis in the ER.

6 Lipid Transport from the ER to the Chloroplast

All plants rely on the import of diacylglycerol (DAG) backbones assembled in the ER to the chloroplast for galactolipid synthesis (Fig. 2). DAG units derived from the ER ("eukaryotic lipids") are devoid of 16:3 and contain mostly C_{18} fatty acids at the *sn*-1 and *sn*-2 positions (minor amounts of C_{16} are also found at the *sn*-1 position). A large fraction of the lipid precursors assembled at the ER has to be transported back to the chloroplast for incorporation into galactolipids. In situ radiolabelling pulse chase studies on 18:3 plants have demonstrated that the radiolabel becomes transiently associated with PC prior to incorporation into MGDG (Slack et al., 1977; Hellgren et al., 1995; Hellgren and Sandelius, 2001). Thus, it seems likely that ER-localized PC is an important precursor for chloroplast lipids derived from the eukaryotic biosynthesis pathway. However, the exact identity of the lipid moiety that is transported from the ER to the chloroplast is still unclear. PC might seem to represent a likely candidate since it is present in the ER and the outer chloroplast envelope. In fact, since the chloroplast lacks the capacity for assembling the phosphorylcholine head group, some kind of transport mechanism for PC between the ER and the chloroplast must exist. In addition to PC (Oursel et al., 1987; Andersson et al., 2004), lyso-PC (Mongrand et al., 1997; Mongrand et al., 2000), DAG (Williams et al., 2000) and most recently PA (Xu et al., 2005; Awai et al., 2006; Lu et al., 2007) have been suggested to represent the lipid molecules transferred from the ER to the chloroplast.

Possible modes of transport include the diffusion of water-soluble molecules (e.g. lyso-PC), protein-mediated transport, vesicle transport or lipid transfer via ER–chloroplast contact sites (Moreau et al., 1998). Transport of proteins and membrane material synthesized at the ER and destined for the secretory pathway is known to be mediated via vesicles. The vesicles are derived from ER membranes and after budding off the ER system they fuse with the *cis* Golgi membranes, and their cargo is subsequently sorted to the vacuole or plasma membrane. A detailed overview of the vesicle-mediated lipid transport system in plants was recently included in Jouhet et al. (2007). However, the existence of an analogous, vesicle-based transport system for the transfer of proteins or lipids to the plastids is unclear. Specific contact sites of the ER with other cellular membranes have previously been described (Staehelin, 1997). In yeast, special regions of the ER are believed to be involved in lipid and protein transfer to the mitochondria (mitochondria-associated membranes, MAM) (Gaigg et al., 1995; Achleitner et al., 1999) or the plasma membrane (plasma-membrane-associated membranes, PAM) (Pichler et al., 2001). These specialized regions were found to be closely associated with their respective organelles and could be co-isolated with these. Contact sites between the ER and the plastid outer envelopes have previously been observed by electron microscopy of various tissues (Wooding and Northcote, 1965; Schlötz, 1975; Whatley et al., 1991; Kaneko and Keegstra, 1996). Apparent contacts between the ER and chloroplasts were also observed by confocal microscopy in *Arabidopsis* expressing green fluorescent protein (GFP) targeted to the ER lumen (Hanson and Kohler, 2001; Andersson et al., 2007). Fluorescent pieces of ER remained associated with chloroplasts isolated from GFP-expressing protoplasts (Andersson et al., 2007). Furthermore, use of optical tweezers demonstrated that chloroplast-associated ER was firmly attached to the chloroplast surface, indicating that tight connections, possibly based on protein–protein interactions, exist between the outer envelope of chloroplasts and the ER (Andersson et al., 2007). An ER-derived fraction could also be isolated from intact pea chloroplast (Andersson et al., 2007). This fraction was by analogy to MAM and PAM dubbed PLAM for plastid-associated membranes. Although direct evidence for lipid transfer between the PLAM and the chloroplast has not yet been obtained, further characterization of this fraction seems promising. The contact sites between ER and chloroplast envelopes could also be related to the so-called stromules, tubular stroma-containing extensions of the envelope membranes that interconnect plastids in plants (Kwok and Hanson, 2003). Interestingly, such tubular connections of plastid envelopes were also observed with nuclear and cell membranes, suggesting that they could in general be involved in connecting the plastids with the cellular membrane system (Kwok and Hanson, 2004). In conclusion, the current understanding points towards a scenario where the transfer of eukaryotic lipid precursors to the chloroplasts is mediated via ER–outer envelope contact sites. Similarly, such contact sites could be involved in the export of galactolipids observed during phosphate deprivation (see below).

Protein factors involved in ER to chloroplast lipid transfer have for a long time remained enigmatic. Analysis of *Arabidopsis* mutants affected in galactolipid metabolism suggested that lipid transport is mediated via an ATP-binding cassette (ABC) transporter complex in the envelope membranes (Xu et al., 2003). ABC transporters are membrane proteins that in general transport small molecules, i.e. phytohormones, peptides or sugars, across membranes, accompanied by the hydrolysis of ATP. The *Arabidopsis* genome contains more than 100 genes with sequence similarities to ABC transporters (Sánchez-Fernández et al., 2001). In addition to the nuclear binding fold domain, ABC transporters are characterized by the presence of one or more transmembrane domains. ABC transporters can be encoded by one multifunctional protein, or by several genes encoding polypeptides that assemble into a functional transporter complex. With employment of a genetic screening strategy, *Arabidopsis* mutants were isolated with alterations in regulation and transport of galactolipids (Xu et al., 2003). The *tgd1* mutant (trigalactosyldiacylglycerol 1) is characterized by the accumulation of triacylglycerol and the unusual oligogalactolipid TGDG. *TGD1* encodes a permease-like protein which constitutes one of the subunits of an ABC transporter. The TGD transporter represented the first ABC transport complex for lipid molecules discovered in higher plants. Biochemical and radioactive labelling experiments suggest that PA is the lipid molecule that is transported through the TGD transporter complex (Xu et al., 2005). *TGD1* localizes to the inner chloroplast envelope membrane. The *tgd2* (Awai et al., 2006) and *tgd3* (Lu et al., 2007) mutants of *Arabidopsis* show the same biochemical phenotype as *tgd1*, e.g. they also accumulate triacylglycerol and TGDG. *TGD2* encodes a substrate binding domain subunit of the lipid ABC transporter which binds PA with high affinity, and *TGD3* encodes a small ATPase associated with the transporter complex. Taken together, this suggests that the three TGD proteins form a lipid transport complex in the envelope membrane (Lu et al., 2007). These data imply that the TGD lipid transporter is involved in the transfer of PA from the ER to the envelopes, where it is hydrolyzed by a PA phosphatase (PAP), yielding DAG, the substrate for galactolipid synthesis. However, there is clearly much less PAP activity in the chloroplast envelope in 18:3 plants than 16:3 plants (see further below). Thus, PA transfer from the ER to the chloroplast might be favoured in 16:3 plants. However, there must be additional ER–chloroplast transport mechanisms at least for PC given the fact that PC is present in both the chloroplast envelope and the ER, and the chloroplast envelope is devoid of PC synthases. The published data regarding 18:3 plants do seem to favour the hypothesis that PC and/or lyso-PC are the chloroplast galactolipid precursors that are transported from the ER to the chloroplast envelope (Oursel et al., 1987; Andersson et al., 2004; Mongrand et al., 1997; Mongrand et al., 2000). On the other hand, sequences with high similarity to *TGD* genes from *Arabidopsis* are present in the genome of rice, suggesting that a TGD/PA-mediated transport mechanism is also operating in 18:3 plants. Further research is clearly needed to completely resolve the issue of which lipid(s) is transported from the ER to the chloroplast. The *Arabidopsis* inventory of putative lipid ABC transporters (Jouhet et al., 2007) is certainly large enough to provide more than one chloroplast-localized transporter. Still, the TGD complex must be of high importance for chloroplast biogenesis, since effective silencing of the *TGD1* gene leads to embryo lethality (Xu et al., 2005). One interesting feature of the TGD proteins is that GFP fusions of all three proteins seem to be localized in dense patches on the chloroplast surface (Xu et al., 2005; Awai et al., 2006; Lu et al., 2007). It is extremely tempting to speculate that these patches might correspond to the sites on the chloroplast surface interacting with the PLAM.

7 Phospholipid Metabolism in the Chloroplast Envelope

The higher-plant chloroplast contains a complete glycerolipid synthesis machinery, the prokaryotic pathway; however, only in the so-called 16:3 plants does this pathway contribute DAG backbones for galactolipid synthesis (Fig. 2). In all plants, the plastidial pathway contributes to PG synthesis in the chloroplast. A soluble, stroma-localized enzyme transfers fatty acids from ACP to glycerol-3-phosphate to yield inner-envelope-localized lyso-PA (Douce and Joyard, 1977). The lyso-PA is then acylated at the *sn*-2 position by an inner-envelope-localized enzyme to yield PA (Andrews et al., 1985). The chloroplast acyltransferases specifically transfer C_{18} fatty acids to the *sn*-1 position of the glycerol; the *sn*-2 acyltransferase is less specific and also accepts C_{16} fatty acids (Frentzen et al., 1983). This pathway seems to be essentially conserved between cyanobacteria and higher-plant chloroplasts (Murata and Nishida, 1987). The genes encoding the two acyltransferases required for PA synthesis in the inner chloroplast envelope have been identified in *Arabidopsis* and knockout mutations have been investigated. Surprisingly, the two different mutations were reported to cause rather different phenotypes. Mutations in the first acyltransferase gene cause a rather subtle phenotype (Kunst et al., 1988, 1989), whereas a knockout mutation of the other enzyme is embryo-lethal (Bin et al., 2004; Kim and Huang, 2004). These conflicting data were recently resolved by the demonstration that the previously described mutants in the first acyltransferase were leaky alleles and therefore could sustain a low rate of prokaryotic lipid synthesis in the chloroplast (Xu et al., 2006). The important products of the prokaryotic lipid pathway are probably not the galactolipids, but rather PG. Apparently the small amount of plastidially produced PA is more efficiently directed into PG synthesis than into galactolipid synthesis (Xu et al., 2006). The recently described transgenic approach of channelling ER-derived lipid backbones into plastidial PG (Fritz et al., 2007) might be a suitable way to rescue a complete knockout mutation of the prokaryotic pathway.

PA produced by the plastidial pathway is converted to PG by three inner-envelope-localized enzymes (Fig. 2) (Andrews and Mudd, 1985). PA is activated to CDP-DAG by CDP-DAG synthase, and a PG-phosphate synthase transfers glycerol-3-phosphate to the lipid. The resulting PG-phosphate is finally dephosphorylated to PG by PG-phosphate phosphatase. Interestingly, the PG-phosphate synthase protein is targeted to both the plastid and the mitochondria (Muller and Frentzen, 2001; Babiychuk et al., 2003). The protein was found to be essential for chloroplast function, but dispensable for mitochondrial function, indicating that mitochondria but not chloroplasts can import PG from the ER (Babiychuk et al., 2003). There are three candidate genes in the *Arabidopsis* genome for the plastid CDP-DAG synthase, but experimental data regarding this activity are missing. No genes have been identified for the plastidial PG-phosphate phosphatase, although candidate genes have been suggested on the basis of sequence similarity and phylogeny (Lykidis, 2007).

PA of eukaryotic fatty acid composition in the inner envelope seems to be capable of entering the PG biosynthesis pathway (Fritz et al., 2007). Therefore, there is currently no good explanation for the question how eukaryotic lipid species are kept out of the chloroplast PG pool, given the finding that PA might be a major lipid precursor transported from the ER to the inner envelope (Xu et al., 2005; Awai et al., 2006; Lu et al., 2007). A possible solution for this dilemma would be a very tight channelling of different PA pools in the inner envelope. One suggestion is that the TGD transporter complex binds PA in the inner envelope and shields the imported PA from the CDP-DAG synthase activity (Fritz et al., 2007).

In 16:3 plants, the PA produced in the chloroplast envelope is utilized for galactolipid synthesis. This requires PAP activity to form DAG. This activity was demonstrated in envelope preparations from spinach more than 30 years ago (Douce and Joyard, 1977). The PAP activity in pea chloroplasts is localized to the inner envelope (Andrews et al., 1985). Only very recently, a small gene family that might encode the chloroplast-localized PAPs was identified in *Arabidopsis* (Nakamura et al., 2007). A knockout mutation of one of the particular isoforms in *Arabidopsis* caused reduced pollen fertility, but no definite link could be made to plastidial galactolipid synthesis. The envelope PAP activity has been proposed to be the metabolic point of divergence between 16:3 and 18:3 plants (Heinz and Roughan, 1983; Gardiner et al., 1984). This is also supported by the observation that chloroplast PAP activity is much lower (more than 10 times lower) in 18:3 than in 16:3 plants (Gardiner et al., 1984). However, this suggestion fits poorly with the hypothesis that direct import of PA from the ER to the inner envelope is the basis for formation of eukaryotic chloroplast lipids (Xu et al., 2005; Awai et al., 2006). Again, channelling and/or regulation of chloroplast PAP activity might resolve this apparent discrepancy.

A chloroplast-envelope-localized activity which transfers fatty acids from acyl-CoA to lyso-PC has been described in leek and pea seedlings (Bessoule et al., 1995; Kjellberg et al., 2000; Mongrand et al., 2000). This has led to the suggestion that lyso-PC might be the precursor translocated from the ER to the chloroplast (Bessoule et al., 1995; Mongrand et al., 1997; Mongrand et al., 2000). The chloroplast-localized lyso-PC acyltransferase was found to be protected from the protease thermolysine and co-fractionated with the inner envelope in pea (Kjellberg et al., 2000). In general, very little is known about the roles of similar acyltransferase activities in plant membranes. It was, however, recently reported that acyl editing of PC in the ER might be more important than previously recognized for de novo synthesis of phospholipids in the ER (Bates et al., 2007).

If PC delivered directly from the ER or assembled from lyso-PC in the envelope represents a precursor for galactolipid synthesis in the chloroplast, the phosphorylcholine head group needs first to be cleaved off to generate DAG required for galactolipid synthesis. This could be accomplished in two ways. The phosphorylcholine could be cleaved in one piece by the action of a phospholipase C (PLC). Alternatively, the head group could be cleaved off in two steps. First a phospholipase D (PLD) could cleave choline, and then PAP activity could remove the phosphate group. *In vitro* experiments clearly demonstrated that eukaryotic PC in the chloroplast envelope can serve as a precursor for galactolipid synthesis, provided that the PC is degraded by exogenously added PLC (Miquel et al., 1988). However, no PLC or PLD activity has ever been demonstrated to be present in an isolated chloroplast fraction. Thus, if PC in the envelope is a precursor for galactolipid synthesis in the chloroplast, the required phospholipase(s) must be a soluble cytosolic enzyme which only transiently associates with the chloroplast. *In vitro* experiments with isolated pea chloroplasts and envelopes demonstrated that PC synthesized in the chloroplast envelope from lyso-PC and acyl-CoA could be used as a substrate for galactolipid synthesis provided that cytosolic proteins were added to the reaction mixture (Andersson et al., 2004). Furthermore, the sensitivity to specific inhibitors suggested that the cytosolic enzymes required were PLD and PAP. The same set of enzyme activities could also be involved if PA is the major eukaryotic precursor transported from the ER to the chloroplast. In this case, the PLD activity could produce PA in the ER and a soluble PAP would provide the activity required for DAG production in the envelope. This was also supported by *in vitro* transfer experiments of PA from liposomes to chloroplasts isolated from wild-type *Arabidopsis* and the *tgd1* mutant (Xu et al., 2005). Information on soluble PAP activities in plants is rather scarce and most of the attention has focused on the envelope-localized PAP and the plasma-membrane-bound PAPs most probably involved in signalling rather than bulk lipid metabolism. Soluble PAP activity has been described in *Vicia faba* leaves (Königs and Heinz, 1974) and developing seeds of *Brassica napus* (Kocsis et al., 1996; Furukawa-Stoffer et al., 1998). No candidate genes have been identified, but the data presented by Andersson et al. (2004) suggest that the sought-after activity is a soluble PAP that is insensitive to *N*-ethylmaleimide and has a native size exceeding 100 kDa.

8 Glycolipid Biosynthesis in the Chloroplast Envelope

Glycolipids are assembled from DAG and UDP-sugars in the envelope membranes of chloroplasts (Fig. 3) (Neufeld and Hall, 1964). *Arabidopsis* contains three MGDG synthases: MGD1, which localizes to the inner envelope, is involved in the synthesis of the largest proportion of MGDG. MGD2 and MGD3, which are found in the outer envelope, are only active in some tissues or under specific conditions, e.g. phosphate deprivation (Awai et al., 2001). The MGDG synthase transfers a galactose moiety from UDP-galactose onto DAG under inversion of the anomeric configuration. Therefore, the galactose is linked in β-anomeric configuration to DAG. Two DGDG synthases are present in *Arabidopsis*, DGD1 and DGD2 that galactosylate the C-6 position of the galactose moiety in MGDG (Dörmann et al., 1999; Kelly et al., 2003). Since the reaction follows a retaining mechanism, the outermost galactose residue in DGDG has α-configuration. DGD1 is responsible for the production of the largest proportion of DGDG in leaves, while both enzymes contribute to DGDG production under phosphate deprivation (Kelly et al., 2003). The two DGDG synthases localize to the outer envelope of chloroplasts (Froehlich et al., 2001; Kelly et al., 2003).

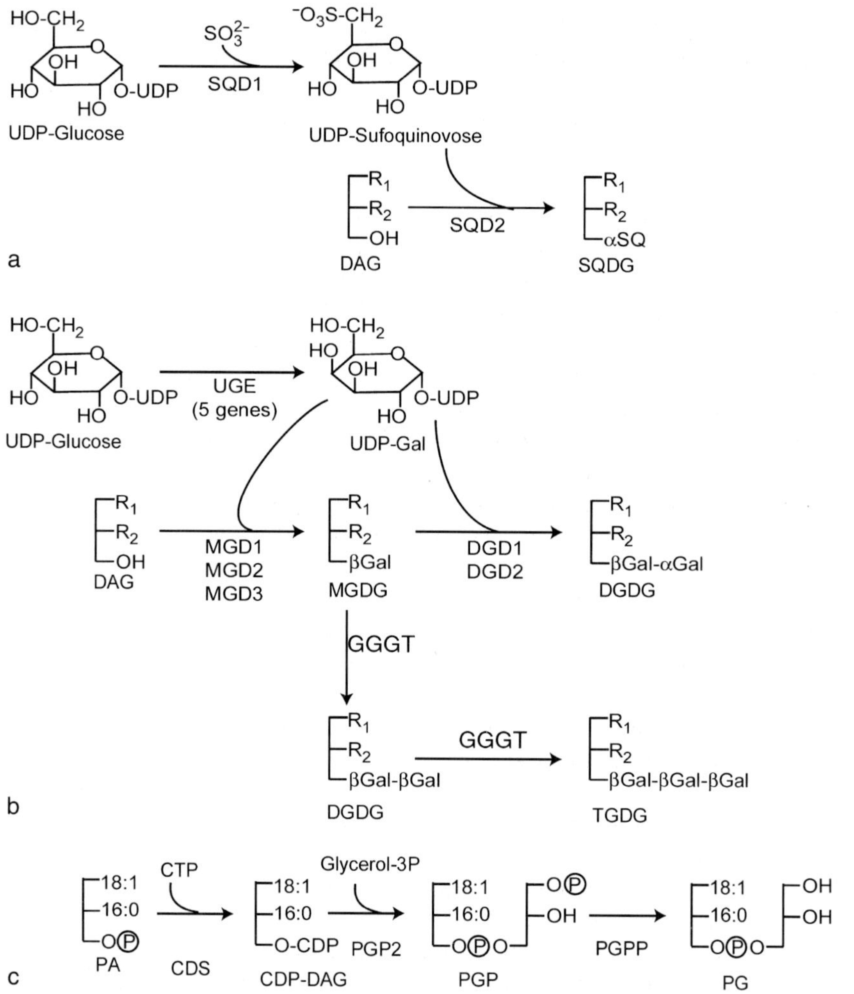

Fig. 3 Synthesis of glycoglycerolipids in plants. UDP-sulfoquinovose, the donor for the head group of sulfolipid (SQDG), is synthesized from UDP-glucose and sulfite by SQD1 (*top part*). Subsequently, SQD2 produces SQDG from diacylglycerol and UDP-sulfoquinovose. The bottom part shows galactolipid synthesis. UDP-glucose is converted into UDP-galactose by one of the five UDP-glucose epimerases present in *Arabidopsis*. UDP-galactose is the head group donor for MGDG and DGDG synthesis by one of the MGDG synthases (MGD1, MGD2 and MGD3) or DGDG synthases (DGD1 and DGD2). The galactolipid:galactolipid galactosyltransferase (*GGGT*), a processive oligogalactolipid synthase with unknown identity and unknown function, produces small amounts of trigalactosyldiacylglycerol (*TGDG*). Note that the glycosidic linkages derived from the MGDG synthase reaction and from the GGGT reaction are β, while DGDG synthases produce α-glycosidic bonds

In addition to the two DGDG synthases described above, *Arabidopsis* contains a third activity, the galactolipid:galactolipid galactosyltransferase (GGGT), capable of synthesizing DGDG and oligogalactolipids with three (TGDG) or more galactose units in the head group (van Besouw and Wintermans, 1978; Heemskerk et al., 1990). The GGGT activity localizes to the outer chloroplast envelope (Douce, 1974; Cline and Keegstra, 1983). The amount of oligogalactolipids, including TGDG, in leaves is usually very low, but it can increase to high amounts in stressed leaves (Sakaki et al., 1990). The GGGT activity was described as an enzyme transferring a galactose unit from one MGDG to the other, thereby producing DAG and DGDG (van Besouw and Wintermans, 1978). GGGT is independent of the DGDG synthases DGD1 and DGD2 since GGGT activity is still detectable in the *dgd1 dgd2* double mutant (Kelly et al., 2003). As indicated above, TGDG accumulates in the mutants *tgd1*, *tgd2* and *tgd3* (Xu et al., 2003). The reason why mutations in the TGD lipid transporter complex stimulate TGDG synthesis in *Arabidopsis* remains unknown. It is possible that the TGDG-synthesizing activity from *tgd1* is identical or related to the GGGT activity described above. The fact that the galactose moieties in TGDG of the *tgd1* mutant are all in β-linkage indicates that this enzyme is specific for β-anomeric bonds, in contrast to the two DGDG synthases DGD1 and DGD2, which are specific for α-glycosidic bonds (Xu et al., 2003). However, the identity of GGGT or the TGDG-synthesizing activity in the *tgd* mutants, the function of TGDG synthesis and the relation to the known enzymatic steps of galactolipid synthesis remain enigmatic.

The sulfolipid SQDG contains a modified glucose head group carrying a sulfonic acid moiety at the C-6 position (Benson et al., 1959). The head group for SQDG synthesis, UDP-sulfoquinovose, is produced from UDP-glucose and sulfite (SO_3^{2-}) in the stroma by the SQD1 gene product in *Arabidopsis*. SQD1 shows sequence similarity to UDP-sugar epimerases (Essigmann et al., 1998). Crystallization of the protein revealed that SQD1 binds NADH as a cofactor (Mulichak et al., 1999). The reaction mechanism for the conversion of UDP-glucose into UDP-sulfoquinovose proceeds through a reduced "glucosene" form which serves as the acceptor for sulfite addition. Subsequently, the sulfoquinovose is transferred onto DAG in the outer-envelope membrane of chloroplasts by SQDG synthase (SQD2) (Yu et al., 2002). The reaction follows a retaining mechanism resulting in α-glycosidic linkage between the sugar and the glycerol backbone.

9 Chloroplast Lipid Fatty Acid Desaturation

The introduction of further double bonds beyond the Δ9-double bond in 18:1 in chloroplast membrane lipids is dependent on membrane-bound desaturases which accept intact glycerolipids as substrates (Schmidt and Heinz, 1990a,b, 1993; Sperling et al., 1993). Two ER-localized membrane lipid desaturases were cloned by forward genetic screens for *Arabidopsis* mutants with deficiencies in unsaturated fatty acids (Arondel et al., 1992; Yadav et al., 1993; Okuley et al., 1994). The sequences for the two ER desaturases FAD2 and FAD3 were then used to identify three chloroplast-localized lipid desaturases FAD6, FAD7 and FAD8 (Iba et al., 1993; Falcone et al., 1994; Gibson et al., 1994; Hitz et al., 1994). The ER-localized desaturases FAD2 and FAD3 produce linoleic acid (18:2) and 18:3, which are esterified to phospholipids in the ER. Thus, these two enzymes contribute to desaturation of chloroplast lipid species assembled through the eukaryotic pathway. A specific envelope-localized desaturase introduces the Δ7-double bond in 16:0 esterified to MGDG assembled by the prokaryote pathway. This activity is encoded by a single nuclear gene, *FAD5*, in *Arabidopsis* (Mekhedov et al., 2000; Heilmann et al., 2004). No candidate gene for the FAD4 desaturase which introduces the *trans*-3 double bond in palmitic acid esterified to chloroplast PG has yet been found. In *Arabidopsis*, three chloroplast-localized desaturases catalyse the production of dienoic and trienoic acids on galactolipids. FAD6 introduces a second double bond in palmitoleic acid (16:1) and 18:1 and FAD7 and FAD8 introduce the third. The ER-localized lipid desaturases use cytochrome b_5 as an electron donor, whereas the chloroplast-localized desaturases use reduced ferredoxin.

10 Lipid Transport from the Envelope to the Thylakoids

Glycolipids (MGDG, DGDG and SQDG) are synthesized in the envelope membranes, but accumulate in the thylakoid membranes in high amounts. Similarly, PG and other lipids of the photosynthetic membranes (phylloquinone, plastoquinone, tocopherol, carotenoids) are also derived from the envelope membranes. Thus, massive transport of membrane material from the envelope to the thylakoids is required for the establishment of thylakoid membranes during chloroplast development (Fig. 4). Contact sites between the inner envelope and the thylakoid are quite frequently observed in developing chloroplasts (Carde et al., 1982). It has thus been proposed that the thylakoid membranes originate from invaginations of the inner-envelope membrane (Mühlethaler and Frey-Wyssling, 1959; Vothknecht and Westhoff, 2001). Recent data obtained from the analysis of an *Arabidopsis mgd1* mutant support this hypothesis (Kobayashi et al., 2007). This mutant, which is completely devoid of MGDG, develops large invaginations from the chloroplast envelope, suggesting that the loss of MGDG prevents the separation of regular thylakoids derived from the inner envelope.

In mature chloroplasts, contacts between the inner envelope and the thylakoid are only very rarely observed. The inner envelope and the thylakoids are usually separated by at least 50–100 nm of aqueous stroma (Morré et al., 1991b; Ryberg et al., 1993). Even though mature chloroplasts in comparison to developing chloroplasts do not require extensive transport of lipids from the envelope to the thylakoids, there is still a certain turnover of thylakoid lipids in a mature chloroplast. Thus, there is always a need for a certain flow of membrane material from the inner envelope to the thylakoid. A vesicle transport system has been proposed as a model for lipid transfer from the inner envelope to the thylakoids. Electron microscopy of leaves exposed

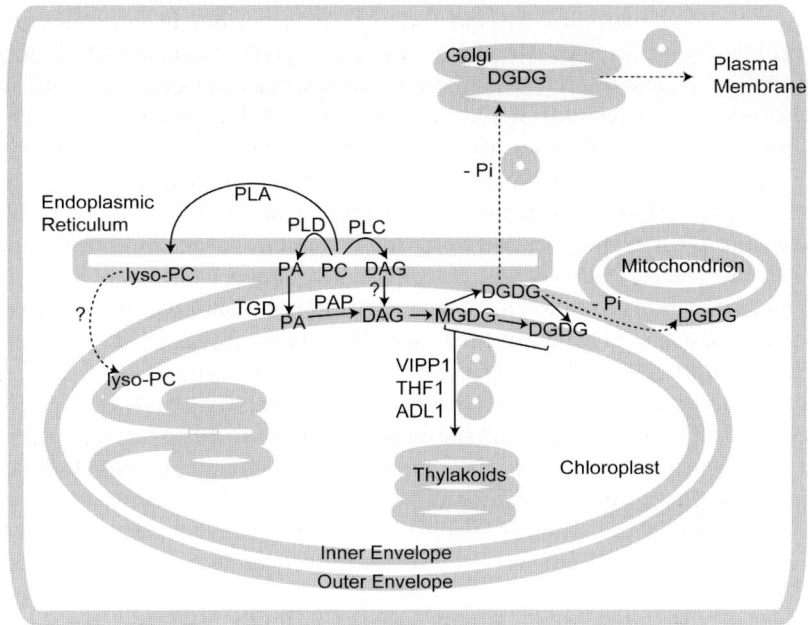

Fig. 4 Intraplastidial and extraplastidial lipid transport in *Arabidopsis*. Thylakoid lipids are derived from the inner-envelope membrane by invagination or from vesicle transport. Eukaryotic lipid precursors assembled at the ER are transported to the chloroplast, presumably via the TGD transporter complex. During phosphate deprivation, DGDG is exported from chloroplasts to extraplastidial membranes, i.e. plasma membrane and mitochondria

to low temperature revealed the accumulation of vesicles in the stroma close to the envelope membranes (Morré et al., 1991b). While vesicles are usually absent from *Arabidopsis* chloroplasts raised at normal growth conditions, vesicle formation and fusion can be blocked by inhibitors that affect cytosolic vesicle transport systems (Westphal et al., 2001b). Furthermore, contact sites between the two envelope membranes, and vesicles originating from the envelope have been observed in osmotically stressed chloroplasts, indicating that lipids and proteins might move between the two envelope membranes via contact sites, and that vesicles might be involved in the transfer to the thylakoids (Cremers et al., 1988). Finally, *in organello* assays also demonstrate that the transport of newly formed galactolipids to the thylakoid in intact chloroplasts is inhibited at the same temperature that results in the accumulation of vesicles in the stroma (Andersson et al., 2001).

The *Arabidopsis* genome codes for a number of proteins with putative chloroplast localization signals and that show sequence similarity to factors of the yeast cytosolic/ER-localized vesicle transport system, e.g. COPII coat proteins and small GTPases (Andersson and Sandelius, 2004). This suggests that a transport mechanism related to ER–Golgi transport in the cytosol might also operate within the plastid. It is known that galactolipids can be released from isolated envelopes in an

ATP- and stroma-dependent fashion (Räntfors et al., 2000). In vitro transfer of galactolipids from envelope to thylakoid seems to depend on similar factors (Morré et al., 1991a). A stromal protein, plastid fusion/transformation factor (Pftf), showing sequence similarity to yeast and mammalian NSF and bacterial FtsH proteins is required for fusion of vesicles derived from chromoplasts of pepper (Hugueney et al., 1995). The *Arabidopsis* dynamin-like 1 (ADL1) protein shows sequence similarity to mammalian dynamin proteins (Park et al., 1998). *Arabidopsis* lines deficient in ADL1 expression show yellowish leaves and contain fewer chloroplasts with strongly reduced thylakoids. Dynamin-like proteins harbour an N-terminal GTPase activity and are known to be involved in vesicle trafficking in the eukaryotic cells. The fact that ADL1 localizes to the chloroplasts suggests that this protein might participate in vesicle trafficking in the plastid.

A vesicle-based mechanism for lipid transport from the envelope to the thylakoids is also supported by data obtained from additional *Arabidopsis* mutant analysis. The thylakoid formation 1 (*THF1*) gene controls thylakoid development in *Arabidopsis* (Wang et al., 2004). THF1 localizes to the chloroplast, and the corresponding mutant shows disturbed thylakoid development and the accumulation of vesicles in the stroma. The gene is present in all organisms performing oxidative photosynthesis, but shows no sequence similarity to other genes in the databases. A mutant in vesicle inducing in plastid protein 1 (VIPP1) shows the accumulation of vesicles and a strongly reduced thylakoid system in the chloroplasts of *Arabidopsis* and also in cyanobacteria (Kroll et al., 2001; Westphal et al., 2001a). VIPP1 was previously discovered as a protein associated with thylakoids and the inner envelope (Li et al., 1994). VIPP1 shows sequence similarity with the phage shock protein A (PspA) from bacteria (Westphal et al., 2001a). VIPP1, similar to its ancestor protein PspA, forms large homo-oligomeric rings which localize to the inner envelope of chloroplasts (Aseeva et al., 2004). VIPP1 rings assemble into larger, rod-like filaments which could be the basis for the microtubule-like structures observed previously in chloroplasts (Liu et al., 2007). Plastidial chaperons and co-chaperons (CDJ2, CGE2 and the heat shock protein HSP70B) are involved in the ATP-dependent assembly and disassembly of VIPP1-containing complexes, thereby modulating thylakoid development (Liu et al., 2007). Thus, while thylakoid biogenesis in developing chloroplasts seems to be based on invaginations of the inner envelope, lipid transfer to thylakoids in mature leaves might be mediated via vesicle transport. This vesicle transport system appears to have features and components derived from both the endosymbiont cyanobacteria and the host eukaryote.

11 Chloroplast Lipid Metabolism and Galactolipid Export During Phosphate-Limited Growth

Approximately one third of all organically bound phosphate in an *Arabidopsis* leaf is bound to phospholipids (Poirier et al., 1991). Since galactolipids are phosphate-free, they largely contribute to phosphate homeostasis in the plant. Under normal growth conditions, galactolipids accumulate predominately in chloroplast membranes. The amount of galactolipids detected in extraplastidial membranes is in general very low, and it is has been a matter of debate as to whether they are authentic components or mere contaminations derived from plastidial membranes during isolation. The first indication that growth under phosphate limitation results in the accumulation of glycolipids as surrogates for phospholipids came from the analysis of bacterial lipid metabolism (Minnikin et al., 1974). In *Rhodobacter* and in cyanobacteria, the amount of sulfolipid increases strongly at the expense of phospholipids during phosphate-limited conditions (Benning et al., 1993; Güler et al., 1996). Similarly, the sulfolipid content in chloroplasts of *Arabidopsis* increases upon phosphate deprivation at the expense of phospholipids (Essigmann et al., 1998). Furthermore, the non-ionic galactolipid DGDG also accumulates in cyanobacteria and in *Arabidopsis* under phosphate limitation (Güler et al., 1996; Härtel et al., 2000). In contrast to SQDG, which remains localized to the chloroplast membranes, DGDG accumulates in the chloroplast as well as in extraplastidial membranes (Härtel et al., 2000).

The increase in SQDG and DGDG observed during phosphate deprivation represents an active process involving the induction of gene expression. Under phosphate limitation, the expression of the *SQD1* and *SQD2* messenger RNAs is induced, resulting in increased SQDG production (Essigmann et al., 1998; Yu et al., 2002). Furthermore, the expression of MGD2 and MGD3, but not of MGD1, is induced upon phosphate deprivation (Awai et al., 2001). Therefore, it is believed that MGD2 and MGD3 provide the substrate for the extra amount of DGDG synthesized during phosphate deprivation. Induction of DGD1 and DGD2 expression during phosphate limitation provides the means for increased DGDG production (Kelly and Dörmann, 2002; Kelly et al., 2003). While DGD1 is believed to synthesize DGDG for the chloroplasts, DGD2, which is only induced during phosphate deprivation, produces DGDG exported to extraplastidial membranes.

The first indication that DGDG might substitute for phospholipids in extraplastidial membranes during phosphate-limited growth was provided by Härtel et al. (2000). Phosphate limitation was shown to cause accumulation of DGDG in microsome fractions isolated from *Arabidopsis*. The first unambiguous demonstration of extraplastidial phospholipid replacement by DGDG was based on the isolation of plasma membranes in very high purity from roots and shoots of phosphate-starved oat (Andersson et al., 2003). Subsequently, the same phenomenon was shown to occur in oat root tonoplasts (Andersson et al., 2005) and in mitochondria in *Arabidopsis* cell suspension (Jouhet et al., 2004). The degree of phospholipid replacement by DGDG is quite impressive. In plasma membranes isolated from oat roots cultivated without phosphate for 4 weeks, the phospholipid content decreased by about 70% (Andersson et al., 2003; Andersson et al., 2005). Most of the phospholipids were replaced by DGDG, albeit there was also a minor contribution from glucosylceramide and sterolglucosides (Andersson et al., 2005). The phospholipid to DGDG replacement has been shown to occur in *Arabidopsis* (Härtel et al., 2000, 2001), oat (Andersson et al., 2003), sycamore maple (Jouhet et al., 2003), soybean (Gaude et al., 2004), rice, maize, radish, garden nasturtium and sunflower (H. Tjellström, M.A. Andersson, K.E. Larsson and A.S. Sandelius, unpublished results).

Since the phosphate-starvation-induced DGDG synthase machinery is clearly localized to the chloroplast envelope, a mechanism for the export of DGDG from the chloroplast must exist. Jouhet et al. (2004) suggested that DGDG is directly transferred from the chloroplast to the mitochondria. This is supported by electron micrographs showing that phosphate limitation causes an increase in close contacts between chloroplasts and mitochondria. For export of DGDG to the plasma membrane and the tonoplast, it seems reasonable to assume that the lipid is exported to the ER and from there sorted into the secretory pathway towards the tonoplast or the plasma membrane. In the *Arabidopsis fad2* mutant, which lacks ER localized oleoyl-PC desaturase, 18:2 and 18:3 fatty acids are still found in the extraplastidial phospholipids (Miquel and Browse, 1992); thus, lipid-bound fatty acids are exported from the chloroplast under other conditions as well. However, the export of DGDG so far is the only certain example for the export of intact glycerolipids from the chloroplast. The molecular details of the export of DGDG to mitochondria or the ER are at this point obscure.

If the replacement of phospholipids with DGDG takes place in fully expanded cells, the phospholipids first have to be degraded. In rapidly expanding tissues, on the other hand, phospholipid degradation would be less important than de novo synthesis of DGDG, which would quickly dilute the pre-existing phospholipids. In either case, DAG must be made available for galactolipid synthesis in the chloroplast envelope. *Arabidopsis* and *Acer pseudoplatanus* suspension cells respond to removal of phosphate from the medium with a transient increase in PC prior to accumulation of DGDG (Jouhet et al., 2003). While the PC content increased, there was a sharp decline in the content of other phospholipids. This underlines the role of PC as a precursor for galactolipid synthesis. It was suggested that hydrolysis of phospholipids plays a major role for the liberation of phosphate and the rerouting of DAG backbones from phospholipids to galactolipids. The current knowledge of phospholipase activities involved in these processes is very limited. A small family of genes with similarities to bacterial PC-specific PLC was identified in the *Arabidopsis* genome. One of the phospholipase genes, *NPC4*, was also found to be upregulated by phosphate starvation and the protein accumulated in the plasma membrane (Nakamura et al., 2005). The recombinant NPC4 protein exhibits a calcium-independent PLC activity towards PC, phosphatidylethanolamine and PA. However, inactivation of the *NPC4* gene in the *npc4* mutant had no effect on the phospholipid to DGDG exchange during phosphate starvation. The NPC4 protein was also found to accumulate in plasma membranes isolated from phosphate-starved oat roots (Andersson et al., 2005); however, in the latter case the major lipase activity detected in the isolated plasma membranes was PLD rather than PLC. The two *Arabidopsis* PLD isoforms PLDζ1 and PLDζ2 were also found to be induced by phosphate starvation (Cruz-Ramirez et al., 2006; Li et al., 2006a). A knockout mutation of PLDζ2 caused a reduction in phospholipid replacement in phosphate-starved roots, and the simultaneous knockout of both isoforms caused a change in the morphological response of the roots to low phosphate levels (Li et al., 2006b). It is thus clear that the phospholipid to DGDG replacement is accompanied by induction of phospholipase genes. However, an alternative function of these phospholipases could be the production of intracellular signals for phosphate homeostasis or for bulk phospholipid degradation as a prerequisite for DGDG production.

12 Chloroplast Galactolipid Degradation Pathways

Chloroplast lipids are subject to constant degradation and resynthesis. Thus, glycerolipids in the membranes are hydrolysed by phospholipases, glycolipases and glycosidases, and replaced with newly assembled lipids. The identity of the enzymes involved in this process, however, is to a large extent unknown.

Two galactosidase activities were described with specificities for a and b glycosidic bonds (Sastry and Kates, 1964; Helmsing, 1967). While α-galactosidase activity is required for the hydrolysis of the outermost galactose residue in DGDG, the inner galactose of DGDG and that of MGDG are cleaved by β-galactolipase activity. Furthermore, an α-glycosidase activity is involved in sulfolipid breakdown. The identities of the genes involved remain unknown. *Arabidopsis* contains almost 400 glycosylhydrolase sequences (Henrissat et al., 2001) which can be organized into α and β specific enzymes. However, the large number of candidate genes prevents a straightforward sequence-homology-based approach of identifying lipid glycosylhydrolase genes. Furthermore, it is still unclear whether galactolipids are first cleaved by galactolipases (i.e. removing the acyl groups), and the remaining digalactosylglycerol or galactosylglycerol moieties cleaved by galactosidases, or whether galactosidases directly act on the two galactolipids MGDG and DGDG.

Fatty acids are cleaved from glycolipids by the action of glycolipases, in analogy to the phospholipases A1 and A2 which are well known from the animal and yeast field. Galactolipases have been characterized in plant extracts (Sastry and Kates, 1964; Helmsing, 1967). The enzymes are specific for hydrolysis of acyl groups from *sn*-1 and *sn*-2 positions of MGDG and DGDG. Two genes encoding chloroplast-localized galactolipases and phospholipases involved in releasing 18:3 destined for the oxylipin pathway (i.e. jasmonate production) have been identified in *Arabidopsis* (Ishiguro et al., 2001; Hyun et al., 2008). A sulfolipase activity was described for *Scenedesmus*, but similar to the galactolipases, the identity of the corresponding gene remains unknown (Yagi and Benson, 1962). Labelling experiments with $H_2^{18}O$ suggested that the rate of lipid deacylation/acylation ("retailoring") reactions in non-stressed leaves is quite high, similar to the de novo fatty acid synthesis rate (Pollard and Ohlrogge, 1999). Bao et al. (2000) showed that the rate of fatty acid breakdown amounts to about 4% of total leaf fatty acid per day; therefore, the entire acyl matrix in membranes is turned over in about 25 days. It is possible that acyl retailoring is involved in channelling different fatty acids to specific membrane lipids. For example, some of the ER-derived fatty acids bound to PC seem to be the precursors for galactolipids in the chloroplast (Roughan, 1970). Removal of polyunsaturated fatty acids damaged by reactive oxygen species might be essential for maintaining a functional thylakoid lipid matrix.

The amount of free fatty acids in leaves is generally very low. During senescence or abiotic stress, large amounts of lipids are broken down, resulting in the release of free fatty acids (Conconi et al., 1996). A certain proportion of the fatty acids derived from galactolipid breakdown are exported from the plastid. Free fatty acids in *Arabidopsis* are converted into their acyl-CoA derivatives by one of the nine long-chain acyl-CoA synthetases (Fulda et al., 2002; Hayashi et al., 2002; Schnurr et al., 2002; Shockey et al., 2002). Acyl-CoA esters serve as substrates for the further degradation via β-oxidation in the peroxisome. Furthermore, ACP synthases were identified in *Arabidopsis* which are possibly involved in plastidial fatty acid remodelling (Koo et al., 2005).

In addition to galactolipid-derived free fatty acids, free phytol and chlorophyllide from chlorophyll are released by action of chlorophyllases in photosynthetic membranes during stress or senescence. While the pathway of chlorophyllide degradation has been studied in detail, not much is known about the fate of phytol in plant metabolism. Phytol represents a C_{20} alcohol derived from the isoprenoid pathway. It is known that phytol and free fatty acids have detergent-like characteristics and, therefore, their accumulation is toxic for the membranes. During senescence or nitrogen deprivation, large amounts of fatty acid phytyl esters accumulate in the chloroplasts (Ischebeck et al., 2006; Gaude et al., 2007). The phytyl esters are mostly localized to plastoglobules, small lipid protein particles in the chloroplasts. Phytyl esters have previously been identified in algae, mosses and in higher plants subjected to abiotic stress or senescence (Csupor, 1971; Gellerman et al., 1975; Cranwell et al., 1985; Patterson et al., 1993). The identity of the protein involved in fatty acid phytyl ester synthesis and its role during senescence remains enigmatic. It is possible that phytyl esters represent a transient sink for free fatty acids and phytol released from galactolipids and chlorophyll, respectively, and that they are finally channelled into degradation by β-oxidation in the peroxisome. In this regard it is interesting to note that plastoglobules contain a number of additional non-polar lipids that accumulate during stress or senescence, including phylloquinone, fatty acid phytyl ester and tocopherol (Lohmann et al., 2006; Vidi et al., 2006).

13 Outlook

The last 30 years has seen tremendous advances in our understanding of the biosynthesis and function of the chloroplast membrane lipids. Nearly all the important genes involved in synthesis of the major chloroplast lipids have been cloned from *Arabidopsis*. Taking advantage of the emerging plant genome projects, one can now employ these sequences to identify orthologues in other plants, including the major crop species. Figures 2–4 show tentative models for the different biochemical pathways presented here. Glycerolipids are synthesized in only a few membranes of the plant cells, in particular the chloroplast envelopes, the ER and, to a lesser extent, the mitochondrion. Therefore, lipid trafficking is required to transfer precursors or entire lipid molecules to other organelles. Direct contact sites seem to be involved in the transport of membrane material between the ER and the envelope. The first genes encoding an ABC transporter presumably involved in the transfer of PA from the ER to the chloroplast have been cloned. The molecular basis for DGDG export from the plastids during phosphate deprivation is not understood. Within young chloroplasts, thylakoids seem to be derived from inner-envelope invaginations, while a vesicle-based system seems to be operating in older chloroplasts. However, the molecular basis for intraplastidial lipid trafficking is unclear. Future research will have to focus on the missing factors involved in intra- and extraplastidial lipid homeostastis and lipid trafficking.

References

Achleitner G, Gaigg B, Krasser A, Kainersdorfer E, Kohlwein SD, Perktold A, Zellnig G, Daum G (1999) Association between the endoplasmic reticulum and mitochondria of yeast facilitates interorganelle transport of phospholipids through membrane contact. *Eur J Biochem* 264:545–553

Andersson MX, Sandelius AS (2004) A chloroplast-localized vesicular transport system: A bio-informatics approach. BMC Genomics 5

Andersson MX, Kjellberg JM, Sandelius AS (2001) Chloroplast biogenesis. *Regulation of lipid transport to the thylakoid in chloroplasts isolated from expanding and fully expanded leaves of pea. Plant Physiol* 127:184–193

Andersson MX, Stridh MH, Larsson KE, Liljenberg C, Sandelius AS (2003) Phosphate-deficient oat replaces a major portion of the plasma membrane phospholipids with the galactolipid digalactosyldiacylglycerol. *FEBS Lett* 537:128–132

Andersson MX, Kjellberg JM, Sandelius AS (2004) The involvement of cytosolic lipases in converting phosphatidyl choline to substrate for galactolipid synthesis in the chloroplast envelope. *Biochim Biophys Acta* 1684:46–53

Andersson MX, Larsson KE, Tjellstrom H, Liljenberg C, Sandelius AS (2005) Phosphate-limited oat: The plasma membrane and the tonoplast as major targets for phospholipid-to-glycolipid replacement and stimulation of phospholipases in the plasma membrane. *J Biol Chem* 280:27578–27586

Andersson MX, Hamberg M, Kourtchenko O, Brunnström Å, McPhail KL, Gerwick WH, Göbel C, Feussner I, Ellerstrom M (2006) Oxylipin profiling of the hypersensitive response in *Arabidopsis thaliana*: Formation of a novel oxo-phytodienoic acid-containing galactolipid, Arabidopside E. *J Biol Chem* 281:31528–31537

Andersson MX, Goksor M, Sandelius AS (2007) Optical manipulation reveals strong attracting forces at membrane contact sites between endoplasmic reticulum and chloroplasts. *J Biol Chem* 282:1170–1174

Andrews J, Mudd JB (1985) Phosphatidylglycerol synthesis in pea (*Pisum sativum* cultivar Laxton's Progress no 9) chloroplasts: Pathway and localization. *Plant Physiol* 79:259–265

Andrews J, Ohlrogge JB, Keegstra K (1985) Final step of phosphatidic acid synthesis in pea chloroplasts occurs in the inner envelope membrane. *Plant Physiol* 78:459–465

Arondel V, Lemieux B, Hwang I, Gibson S, Goodman HM, Somerville CR (1992) Map-based cloning of a gene controlling omega-3-fatty-acid desaturation in *Arabidopsis*. *Science* 258:1353–1355

Aseeva E, Ossenbuhl F, Eichacker LA, Wanner G, Soll J, Vothknecht UC (2004) Complex formation of Vipp1 depends on its alpha-helical PspA-like domain. *J Biol Chem* 279:35535–35541

Awai K, Maréchal E, Block MA, Brun D, Masuda T, Shimada H, Takamiya K, Ohta H, Joyard J (2001) Two types of MGDG synthase genes, found widely in both 16:3 and 18:3 plants, differentially mediate galactolipid synthesis in photosynthetic and nonphotosynthetic tissues in *Arabidopsis thaliana*. *Proc Natl Acad Sci USA* 98:10960–10965

Awai K, Xu CC, Tamot B, Benning C (2006) A phosphatidic acid-binding protein of the chloroplast inner envelope membrane involved in lipid trafficking. *Proc Natl Acad Sci USA* 103:10817–10822

Babiychuk E, Muller F, Eubel H, Braun HP, Frentzen M, Kushnir S (2003) Arabidopsis phosphatidylglycerophosphate synthase 1 is essential for chloroplast differentiation, but is dispensable for mitochondrial function. *Plant J* 33:899–909

Bahl J, Francke B, Monéger R (1976) Lipid composition of envelopes, prolamellar bodies and other plastid membranes in etiolated, green and greening wheat leaves. *Planta* 129:193–201

Bao XM, Focke M, Pollard M, Ohlrogge J (2000) Understanding in vivo carbon precursor supply for fatty acid synthesis in leaf tissue. *Plant J* 22:39–50

Bates PD, Ohlrogge JB, Pollard M (2007) Incorporation of newly synthesized fatty acids into cytosolic glycerolipids in pea leaves occurs via acyl editing. *J Biol Chem* 282:31206–31216

Benning C, Beatty JT, Prince RC, Somerville CR (1993) The sulfolipid sulfoquinovosyldiacylglycerol is not required for photosynthetic electron-transport in *Rhodobacter sphaeroides* but enhances growth under phosphate limitation. *Proc Natl Acad Sci USA* 90:1561–1565

Benson AA, Daniel H, Wiser R (1959) A sulfolipid in plants. *Proc Natl Acad Sci USA* 45:1582–1587

Berglund AH, Nilsson R, Liljenberg C (1999) Permeability of large unilamellar digalactosyldiacylglycerol vesicles for protons and glucose – influence of α-tocopherol, β-carotene, zeaxanthin and cholesterol. *Plant Physiol Biochem* 37:179–186

Bessoule JJ, Testet E, Cassagne C (1995) Synthesis of phosphatidylcholine in the chloroplast envelope after import of lysophosphatidylcholine from endoplasmic-reticulum membranes. *Eur J Biochem* 228:490–497

Bin Y, Wakao S, Fan JL, Benning C (2004) Loss of plastidic lysophosphatidic acid acyltransferase causes embryo-lethality in *Arabidopsis*. *Plant Cell Physiol* 45:503–510

Block MA, Dorne AJ, Joyard J, Douce R (1983a) Preparation and characterization of membrane fractions enriched in outer and inner envelope membranes from spinach chloroplasts: 1 Electrophoretic and immunochemical analyses. *J Biol Chem* 258:13273–13280

Block MA, Dorne AJ, Joyard J, Douce R (1983b) Preparation and characterization of membrane fractions enriched in outer and inner envelope membranes from spinach chloroplasts: 2 Biochemical characterization. *J Biol Chem* 258:13281–13286

Böttcher C, Weiler EW (2007) cyclo-Oxylipin-galactolipids in plants: Occurrence and dynamics. *Planta* 226:629–637

Bovet L, Müller MO, Siegenthaler PA (2001) Three distinct lipid kinase activities are present in spinach chloroplast envelope membranes: Phosphatidylinosotol phosphorylation is sensitive to wortmannin and not dependent on chloroplast ATP. *Biochem Biophys Res Comm* 289:269–275

Brentel I, Selstam E, Lindblom G (1985) Phase equilibria of mixtures of plant galactolipids. The formation of a bicontinuous phase. *Biochim Biophys Acta* 812:816–826

Browse J, Warwick N, Somerville CR, Slack CR (1986) Fluxes through the prokaryotic and eukaryotic pathways of lipid synthesis in the "16:3" plant *Arabidopsis thaliana*. *Biochem J* 235:25–31

Bruce BD (1998) The role of lipids in plastid protein transport. *Plant Mol Biol* 38:223–246

Buseman CM, Tamura P, Sparks AA, Baughman EJ, Maatta S, Zhao J, Roth MR, Esch SW, Shah J, Williams TD, Welti R (2006) Wounding stimulates the accumulation of glycerolipids containing oxophytodienoic acid and dinor-oxophytodienoic acid in Arabidopsis leaves. *Plant Physiol* 142:28–39

Carde JP, Joyard J, Douce R (1982) Electron microscopic studies of envelope membranes from spinach chloroplasts. *Biol Cell* 36c:62–70

Chapman DJ, Defelice J, Barber J (1986) Polar lipid-composition of chloroplast thylakoids isolated from leaves grown under different lighting conditions. *Photosynth Res* 8:257–265

Cline K, Keegstra K (1983) Galactosyltransferases involved in galactolipid biosynthesis are located in the outer membrane of pea (*Pisum sativum* cultivar Laxtons Progress No 9) chloroplast envelopes. *Plant Physiol* 71:366–372

Cline K, Andrews J, Mersey B, Newcomb EH, Keegstra K (1981) Separation and characterization of inner and outer envelope membranes of pea (*Pisum sativum* cultivar Laxtons Progress 9) chloroplasts. *Proc Natl Acad Sci USA* 78:3595–3599

Conconi A, Miquel M, Browse J, Ryan CA (1996) Intracellular levels of free linolenic and linoleic acids increase in tomato leaves in response to wounding. *Plant Physiol* 111:797–803

Cranwell PA, Robinson N, Eglinton G (1985) Esterified lipids of the freshwater dinoflagellate *Peridinium lomnickii*. *Lipids* 20:645–651

Cremers FFM, Voorhout WF, van der Krift TP, Leunissen-Bijvelt JJM, Verkleij AJ (1988) Visualisation of contact sites between outer and inner envelope membranes in isolated chloroplasts. *Biochim Biophys Acta* 933:334–340

Cruz-Ramirez A, Oropeza-Aburto A, Razo-Hernandez F, Ramirez-Chavez E, Herrera-Estrella L (2006) Phospholipase Dζ2 plays an important role in extraplastidic galactolipid biosynthesis and phosphate recycling in *Arabidopsis* roots. *Proc Natl Acad Sci USA* 103:6765–6770

Csupor L (1971) Das Phytol in vergilbten Blättern. *Planta Med* 19:37–40

Dörmann P, Hoffmann S, Balbo I, Benning C (1995) Isolation and characterization of an Arabidopsis mutant deficient in the thylakoid lipid digalactosyl diacylglycerol. *Plant Cell* 7:1801–1810

Dörmann P, Balbo I, Benning C (1999) Arabidopsis galactolipid biosynthesis and lipid trafficking mediated by DGD1. *Science* 284:2181–2184

Dorne AJ, Heinz E (1989) Position and pairing of fatty acids in phosphatidylglycerol from pea leaf chloroplasts and mitochondria. *Plant Sci* 60:39–46

Dorne AJ, Joyard J, Block MA, Douce R (1985) Localization of phosphatidylcholine in outer envelope membrane of spinach chloroplasts. *J Cell Biol* 100:1690–1697

Dorne AJ, Joyard J, Douce R (1990) Do thylakoids really contain phosphatidylcholine. *Proc Natl Acad Sci USA* 87:71–74

Douce R (1974) Site of biosynthesis of galactolipids in spinach chloroplasts. *Science* 183:852–853

Douce R, Joyard J (1977) Site of synthesis of phosphatidic acid and diacyglycerol in spinach chloroplasts. *Biochim Biophys Acta* 486:273–285

Dubacq JP, Tremolieres A (1983) Occurrence and function of phosphatidylglycerol containing Δ3-*trans*-hexadecenoic acid in photosynthetic lamellae. *Physiol Vég* 21:293–312

Essigmann B, Güler S, Narang RA, Linke D, Benning C (1998) Phosphate availability affects the thylakoid lipid composition and the expression of *SQD1*, a gene required for sulfolipid biosynthesis in *Arabidopsis thaliana*. *Proc Natl Acad Sci USA* 95:1950–1955

Falcone DL, Gibson S, Lemieux B, Somerville C (1994) Identification of a gene that complements an Arabidopsis mutant deficient in chloroplast omega-6 desaturase activity. *Plant Physiol* 106:1453–1459

Feussner I, Wasternack C (2002) The lipoxygenase pathway. *Annu Rev Plant Biol* 53:275–297

Frentzen M, Heinz E, McKeon TA, Stumpf PK (1983) Specificities and selectivities of glycerol-3-phosphate acyltransferase and monoacylglycerol-3-phosphate acyltransferase (EC 23151) from pea (*Pisum sativum*) and spinach (*Spinacia oleracea*) chloroplasts. *Eur J Biochem* 129:629–636

Fritz M, Lokstein H, Hackenberg D, Welti R, Roth M, Zähringer U, Fulda M, Hellmeyer W, Ott C, Wolter FP, Heinz E (2007) Channeling of eukaryotic diacylglycerol into the biosynthesis of plastidial phosphatidylglycerol. *J Biol Chem* 282:4613–4625

Froehlich JE, Benning C, Dörmann P (2001) The digalactosyldiacylglycerol (DGDG) synthase DGD1 is inserted into the outer envelope membrane of chloroplasts in a manner independent of the general import pathway and does not depend on direct interaction with monogalactosyldiacylglycerol synthase for DGDG biosynthesis. *J Biol Chem* 276:31806–31812

Fulda M, Shockey J, Werber M, Wolter FP, Heinz E (2002) Two long-chain acyl-CoA synthetases from *Arabidopsis thaliana* involved in peroxisomal fatty acid β-oxidation. *Plant J* 32:93–103

Furukawa-Stoffer TL, Byers SD, Hodges DM, Laroche A, Weselake RJ (1998) Identification of N-ethylmaleimide-sensitive and -insensitive phosphatidate phosphatase activity in microspore-derived cultures of oilseed rape. *Plant Sci* 131:139–147

Gabrielska J, Gruszecki WI (1996) Zeaxanthin (dihydroxy-β-carotene) but not β-carotene rigidifies lipid membranes: A ^{1}H NMR study of carotenoid-egg phosphatidylcholine liposomes. *Biochim Biophys Acta* 1285:167–174

Gaigg B, Simbeni R, Hrastnik C, Paltauf F, Daum G (1995) Characterization of a microsomal subfraction associated with mitochondria of the yeast, *Saccharomyces cerevisiae* Involvement in synthesis and import of phospholipids into mitochondria. *Biochim Biophys Acta* 1234:214–220

Gardiner SE, Heinz E, Roughan PG (1984) Rates and products of long-chain fatty-acid synthesis from carbon-14-labeled acetate in chloroplasts isolated from leaves of 16:3 and 18:3 plants. *Plant Physiol* 74:890–896

Gaude N, Tippmann H, Flemetakis E, Katinakis P, Udvardi M, Dörmann P (2004) The galactolipid digalactosyldiacylglycerol accumulates in the peribacteroid membrane of nitrogen-fixing nodules of soybean and Lotus. *J Biol Chem* 279:34624–34630

Gaude N, Bréhélin C, Tischendorf G, Kessler F, Dörmann P (2007) Nitrogen deficiency in Arabidopsis affects galactolipid composition and gene expression and results in accumulation of fatty acid phytyl esters. *Plant J* 49:729–739

Gellerman JL, Anderson WH, Schlenk H (1975) Synthesis and analysis of phytyl and phytenoyl wax esters. *Lipids* 10:656–661

Gibson S, Arondel V, Iba K, Somerville C (1994) Cloning of a temperature-regulated gene encoding a chloroplast omega-3 desaturase from *Arabidopsis thaliana*. *Plant Physiol* 106:1615–1621

Gounaris K, Mannock DA, Sen A, Brain APR, Williams WP, Quinn PJ (1983) Polyunsaturated fatty acyl residues of galactolipids are involved in the control of bilayer/non-bilayer lipid transitions in higher plant chloroplasts. *Biochim Biophys Acta* 732:229–242

Gruszecki WI, Sielewiesiuk J (1991) Galactolipid multibilayers modified with xanthophylls – orientational and diffractometric studies. *Biochim Biophys Acta* 1069:21–26

Gruszecki WI, Strzalka K (1991) Does the xanthophyll cycle take part in the regulation of fluidity of the thylakoid membrane. *Biochim Biophys Acta* 1060:310–314

Güler S, Seeliger A, Härtel H, Renger G, Benning C (1996) A null mutant of *Synechococcus* sp PCC7942 deficient in the sulfolipid sulfoquinovosyl diacylglycerol. *J Biol Chem* 271:7501–7507

Hagio M, Sakurai I, Sato S, Kato T, Tabata S, Wada H (2002) Phosphatidylglycerol is essential for the development of thylakoid membranes in *Arabidopsis thaliana*. *Plant Cell Physiol* 43:1456–1464

Hanson MR, Kohler RH (2001) GFP imaging: methodology and application to investigatecellular compartmentation in plants. *J Exp Bot* 52:529–539

Härtel H, Lokstein H, Dörmann P, Grimm B, Benning C (1997) Changes in the composition of the photosynthetic apparatus in the galactolipid-deficient *dgd1* mutant of *Arabidopsis thaliana*. *Plant Physiol* 115:1175–1184

Härtel H, Dörmann P, Benning C (2000) DGD1-independent biosynthesis of extraplastidic galactolipids after phosphate deprivation in *Arabidopsis*. *Proc Nat Acad Sci USA* 97:10649–10654

Härtel H, Dörmann P, Benning C (2001) Galactolipids not associated with the photosynthetic apparatus in phosphate-deprived plants. *J Photochem Photobiol* 61:46–51

Hayashi H, De Bellis L, Hayashi Y, Nito K, Kato A, Hayashi M, Hara-Nishimura I, Nishimura M (2002) Molecular characterization of an Arabidopsis acyl-coenzyme A synthetase localized on glyoxysomal membranes. *Plant Physiol* 130:2019–2026

Heemskerk JWM, Storz T, Schmidt RR, Heinz E (1990) Biosynthesis of digalactosyldiacylglycerol in plastids from 16:3 and 18:3 plants. *Plant Physiol* 93:1286–1294

Heilmann I, Mekhedov S, King B, Browse J, Shanklin J (2004) Identification of the Arabidopsis palmitoyl-monogalactosyldiacylglycerol Delta 7-desaturase gene *FAD5*, and effects of plastidial retargeting of Arabidopsis desaturases on the *fad5* mutant phenotype. *Plant Physiol* 136:4237–4245

Heinz E (1967) Acylgalaktosyldiglycerid aus Blatthomogenaten. *Biochim Biophys Acta* 144:321–332

Heinz E, Roughan P (1983) Similarities and differences in lipid metabolism of chloroplasts isolated from 18:3 and 16:3 plants. *Plant Physiol* 72:273–279

Heinz E, Tulloch AP (1969) Reinvestigation of the structure of acyl galactosyl diglyceride from spinach leaves. *Hoppe Seylers Z Physiol Chem* 350:493–498

Heinz E, Bertrams M, Joyard J, Douce R (1978) Demonstration of an acyltransferase activity in chloroplast envelopes. *Z Pflanzenphysiol* 87:325–331

Hellgren LI, Sandelius AS (2001) Age dependent variation in membrane lipid synthesis in leaves of garden pea (*Pisum sativum* L). *J Exp Bot* 52:2275–2282

Hellgren LI, Carlsson AS, Sellden G, Sandelius S (1995) In situ leaf lipid metabolism in garden pea (*Pisum sativum* L) exposed to moderately enhanced levels of ozone. *J Exp Bot* 46: 221–230

Helmsing PJ (1967) Hydrolysis of galactolipids by enzymes in spinach leaves. *Biochim Biophys Acta* 144:470–472

Henrissat B, Coutinho PM, Davies GJ (2001) A census of carbohydrate-active enzymes in the genome of *Arabidopsis thaliana*. *Plant Mol Biol* 47:55–72

Hisamatsu Y, Goto N, Hasegawa K, Shigemori H (2003) Arabidopsides A and B, two new oxylipins from *Arabidopsis thaliana*. *Tetrahedron Lett* 44:5553–5556

Hisamatsu Y, Goto N, Sekiguchi M, Hasegawa K, Shigemori H (2005) Oxylipins arabidopsides C and D from *Arabidopsis thaliana*. *J Nat Prod* 68:600–603

Hitz WD, Carlson TJ, Booth JR, Kinney AJ, Stecca KL, Yadav NS (1994) Cloning of a higher-plant plastid ω-6 fatty-acid desaturase cDNA and its expression in a cyanobacterium. *Plant Physiol* 105:635–641

Hugueney P, Bouvier F, Badillo A, Dharlingue A, Kuntz M, Camara B (1995) Identification of a plastid protein involved in vesicle fusion and/or membrane-protein translocation. *Proc Natl Acad Sci USA* 92:5630–5634

Hyun Y, Choi S, Hwang HJ, Yu J, Nam SJ, Ko J, Park JY, Seo YS, Kim EY, Ryu SB, Kim WT, Lee YH, Kang H, Lee I (2008) Cooperation and functional diversification of two closely related galactolipase genes for jasmonate biosynthesis. *Dev Cell* 14:183–192

Iba K, Gibson S, Nishiuchi T, Fuse T, Nishimura M, Arondel V, Hugly S, Somerville C (1993) A gene encoding a chloroplast omega-3-fatty-acid desaturase complements alterations in fatty-acid desaturation and chloroplast copy number of the *fad7* mutant of *Arabidopsis thaliana*. *J Biol Chem* 268:24099–24105

Ischebeck T, Zbierzak AM, Kanwischer M, Dörmann P (2006) A salvage pathway for phytol metabolism in *Arabidopsis*. *J Biol Chem* 281:2470–2477

Ishiguro S, Kawai-Oda A, Ueda J, Nishida I, Okada K (2001) The *DEFECTIVE IN ANTHER DEHISCENCE1* gene encodes a novel phospholipase A1 catalyzing the initial step of jasmonic acid biosynthesis, which synchronizes pollen maturation, anther dehiscence, and flower opening in *Arabidopsis*. *Plant Cell* 13:2191–2209

Jarvis P, Dörmann P, Peto CA, Lutes J, Benning C, Chory J (2000) Galactolipid deficiency and abnormal chloroplast development in the *Arabidopsis MGD synthase 1* mutant. *Proc Natl Acad Sci USA* 97:8175–8179

Jordan P, Fromme P, Witt HT, Klukas O, Saenger W, Krauss N (2001) Three-dimensional structure of cyanobacterial photosystem I at 25 Å resolution. *Nature* 411:909–917

Jouhet J, Maréchal E, Bligny R, Joyard J, Block MA (2003) Transient increase of phosphatidylcholine in plant cells in response to phosphate deprivation. *FEBS Lett* 544:63–68

Jouhet J, Maréchal E, Baldan B, Bligny R, Joyard J, Block MA (2004) Phosphate deprivation induces transfer of DGDG galactolipid from chloroplast to mitochondria. *J Cell Biol* 167:863–874

Jouhet J, Maréchal E, Block MA (2007) Glycerolipid transfer for the building of membranes in plant cells. *Prog Lipid Res* 46:37–55

Kaneko Y, Keegstra K (1996) Plastid biogenesis in embryonic pea leaf cells during early germination. *Protoplasma* 195:59–67

Kelly AA, Dörmann P (2002) *DGD2*, an *Arabidopsis* gene encoding a UDP-galactose-dependent digalactosyldiacylglycerol synthase is expressed during growth under phosphate-limiting conditions. *J Biol Chem* 277:1166–1173

Kelly AA, Froehlich JE, Dörmann P (2003) Disruption of the two digalactosyldiacylglycerol synthase genes *DGD1* and *DGD2* in *Arabidopsis* reveals the existence of an additional enzyme of galactolipid synthesis. *Plant Cell* 15:2694–2706

Kim HU, Huang AHC (2004) Plastid lysophosphatidyl acyltransferase is essential for embryo development in *Arabidopsis*. *Plant Physiol* 134:1206–1216

Kjellberg JM, Trimborn M, Andersson M, Sandelius AS (2000) Acyl-CoA dependent acylation of phospholipids in the chloroplast envelope. *Biochim Biophys Acta* 1485:100–110

Kobayashi K, Kondo M, Fukuda H, Nishimura M, Ohta H (2007) Galactolipid synthesis in chloroplast inner envelope is essential for proper thylakoid biogenesis, photosynthesis, and embryogenesis. *Proc Natl Acad Sci USA* 104:17216–17221

Kocsis MG, Weselake RJ, Eng JA, Furukawa-Stoffer TL, Pomeroy MK (1996) Phosphatidate phosphatase from developing seeds and microspore-derived cultures of *Brassica napus*. *Phytochemistry* 41:353–363

Königs B, Heinz E (1974) Investigation of some enzymatic activities contributing to the biosynthesis of galactolipid precursors in *Vicia faba*. *Planta* 118:159–169

Koo AJK, Ohlrogge JB, Pollard M (2004) On the export of fatty acids from the chloroplast. *J Biol Chem* 279:16101–16110

Koo AJK, Fulda M, Browse J, Ohlrogge JB (2005) Identification of a plastid acyl-acyl carrier protein synthetase in *Arabidopsis* and its role in the activation and elongation of exogenous fatty acids. *Plant J* 44:620–632

Kourtchenko O, Andersson MX, Hamberg M, Brunnström A, Göbel C, McPhail KL, Gerwick WH, Feussner I, Ellerström M (2007) Oxo-phytodienoic acid-containing galactolipids in *Arabidopsis*: jasmonate signaling dependence. *Plant Physiol* 145:1658–1669

Kroll D, Meierhoff K, Bechtold N, Kinoshita M, Westphal S, Vothknecht UC, Soll J, Westhoff P (2001) *VIPP1*, a nuclear gene of *Arabidopsis thaliana* essential for thylakoid membrane formation. *Proc Natl Acad Sci USA* 98:4238–4242

Kunst L, Browse J, Somerville C (1988) Altered regulation of lipid biosynthesis in a mutant of *Arabidopsis* deficient in chloroplast glycerol-3-phosphate acyltransferase activity. *Proc Natl Acad Sci USA* 85:4143–4147

Kunst L, Browse J, Somerville C (1989) Altered chloroplast structure and function in a mutant of *Arabidopsis* deficient in plastid glycerol-3-phosphate acyltransferase activity. *Plant Physiol* 90:846–853

Kwok EY, Hanson MR (2003) Microfilaments and microtubules control the morphology and movement of non-green plastids and stromules in *Nicotiana tabacum*. *Plant J* 35:16–26

Kwok EY, Hanson MR (2004) Stromules and the dynamic nature of plastid morphology. *J Microsc* 214:124–137

Li HM, Kaneko Y, Keegstra K (1994) Molecular-cloning of a chloroplastic protein associated with both the envelope and thylakoid membranes. *Plant Mol Biol* 25:619–632

Li MY, Welti R, Wang XM (2006a) Quantitative profiling of *Arabidopsis*. polar glycerolipids in response to phosphorus starvation*Roles of phospholipases Dζ1 and Dζ2 in phosphatidylcholine hydrolysis and digalactosyldiacylglycerol accumulation in phosphorus-starved plants. Plant Physiol* 142:750–761

Li MY, Qin CB, Welti R, Wang XM (2006b) Double knockouts of phospholipases D zeta 1 and D zeta 2 in *Arabidopsis* affect root elongation during phosphate-limited growth but do not affect root hair patterning. *Plant Physiol* 140:761–770

Liu C, Willmund F, Golecki JR, Cacace S, Heß B, Markert C, Schroda M (2007) The chloroplast HSP70B-CDJ2-CGE1 chaperones catalyse assembly and disassembly of VIPP1 oligomers in *Chlamydomonas*. *Plant J* 50:265277

Lohmann A, Schöttler MA, Bréhélin C, Kessler F, Bock R, Cahoon EB, Dörmann P (2006) Deficiency in phylloquinone (vitamin K_1) methylation affects prenyl quinone distribution,

photosystem I abundance and anthocyanin accumulation in the *Arabidopsis AtmenG* mutant. *J Biol Chem* 281:40461–40472
Loll B, Kern J, Saenger W, Zouni A, Biesiadka J (2005) Towards complete cofactor arrangement in the 30 Å resolution structure of photosystem II. *Nature* 438:1040–1044
Loll B, Kern J, Saenger W, Zouni A, Biesiadka J (2007) Lipids in photosystem II: Interactions with protein and cofactors. *Biochim Biophys Acta* 1767:509–519
Lu B, Xu C, Awai K, Jones AD, Benning C (2007) A small ATPase protein of *Arabidopsis* , TGD3, involved in chloroplast lipid import. *J Biol Chem* 282:35945–35953
Lykidis A (2007) Comparative genomics and evolution of eukaryotic phospholipid biosynthesis. *Prog Lipid Res* 46:171–199
Ma XY, Browse J (2006) Altered rates of protein transport in *Arabidopsis* mutants deficient in chloroplast membrane unsaturation. *Phytochemistry* 67:1629–1636
Mackender R, Leech R (1974) The galactolipid, phospholipid and fatty acid composition of the chloroplast envelope membrane of *Vicia faba.* L. *Plant Physiol* 53:496–502
Mekhedov S, de Ilarduya OM, Ohlrogge J (2000) Toward a functional catalog of the plant genome. *A survey of genes for lipid biosynthesis. Plant Physiol* 122:389–401
Minnikin DE, Abdolrahimzadeh H, Baddiley J (1974) Replacement of acidic phospholipids by acidic glycolipids in *Pseudomonas diminuta. Nature* 249:268–269
Miquel M, Browse J (1992) *Arabidopsis* mutants deficient in polyunsaturated fatty acid synthesis: Biochemical and genetic characterization of a plant oleoylphosphatidylcholine desaturase. *J Biol Chem* 267:1502–1509
Miquel M, Block MA, Joyard J, Dorne AJ, Dubacq JP, Kader JC, Douce R (1988) Protein-mediated transfer of phosphatidylcholine from liposomes to spinach chloroplast envelope membranes. *Biochim Biophys Acta* 937:219–228
Mongrand S, Bessoule JJ, Cassagne C (1997) A re-examination in vivo of the phosphatidylcholine-galactolipid metabolic relationship during plant lipid biosynthesis. *Biochem J* 327:853–858
Mongrand S, Bessoule J-J, Cabantous F, Cassagne C (1998) The C16:3/C18:3 fatty acid balance in photosynthetic tissues from 468 plant species. *Phytochemistry* 49:1049–1064
Mongrand S, Cassagne C, Bessoule JJ (2000) Import of lyso-phosphatidylcholine into chloroplasts likely at the origin of eukaryotic plastidial lipids. *Plant Physiol* 122:845–852
Moreau P, Bessoule JJ, Mongrand S, Testet E, Vincent P, Cassagne C (1998) Lipid trafficking in plant cells. *Prog Lipid Res* 37:371–391
Morré DJ, Morré JT, Morré SR, Sundqvist C, Sandelius AS (1991a) Chloroplast biogenesis – cell-free transfer of envelope monogalactosylglycerides to thylakoids. *Biochim Biophys Acta* 1070:437–445
Morré DJ, Selldén G, Sundqvist C, Sandelius AS (1991b) Stromal low-temperature compartment derived from the inner membrane of the chloroplast envelope. *Plant Physiol* 97:1558–1564
Mueller-Roeber B, Pical C (2002) Inositol phospholipid metabolism in *Arabidopsis. Characterized and putative isoforms of inositol phospholipid kinase and phosphoinositide-specific phospholipase C. Plant Physiol* 130:22–46
Mühlethaler K, Frey-Wyssling A (1959) The development and structure of proplastids. *J Biophys Biochem Cytol* 6:507–512
Mulichak AM, Theisen MJ, Essigmann B, Benning C, Garavito RM (1999) Crystal structure of SQD1, an enzyme involved in the biosynthesis of the plant sulfolipid headgroup donor UDP-sulfoquinovose. *Proc Natl Acad Sci USA* 96:13097–13102
Muller F, Frentzen M (2001) Phosphatidylglycerophosphate synthases from *Arabidopsis thaliana. FEBS Lett* 509:298–302
Müller MO, Meylan-Bettex M, Eckstein F, Martinoia E, Siegenthaler PA, Bovet L (2000) Lipid phosphorylation in chloroplast envelopes – Evidence for galactolipid CTP-dependent kinase activities. *J Biol Chem* 275:19475–19481
Munné-Bosch S, Alegre L (2002) The function of tocopherols and tocotrienols in plants. *Crit Rev Plant Sci* 21:31–57
Murata N, Nishida I (1987) Lipids of blue-green algae (cyanobacteria). In: Stumpf PK (ed) Lipids: Structure and function, vol 4. Academic, New York, pp 315–347

Nakajyo H, Hisamatsu Y, Sekiguchi M, Goto N, Hasegawa K, Shigemori H (2006) Arabidopside F, a new oxylipin from *Arabidopsis thaliana*. *Heterocycles* 69:295–301

Nakamura Y, Awai K, Masuda T, Yoshioka Y, Takamiya K, Ohta H (2005) A novel phosphatidylcholine-hydrolyzing phospholipase C induced by phosphate starvation in *Arabidopsis*. *J Biol Chem* 280:7469–7476

Nakamura Y, Tsuchiya M, Ohta H (2007) Plastidic phosphatidic acid phosphatases identified in a distinct subfamily of lipid phosphate phosphatases with prokaryotic origin. *J Biol Chem* 282:29013–29021

Neufeld EF, Hall CW (1964) Formation of galactolipids by chloroplasts. *Biochem Biophys Res Commun* 14:503–508

Ohashi T, Ito Y, Okada M, Sakagami Y (2005) Isolation and stomatal opening activity of two oxylipins from *Ipomoea tricolor*. *Bioorg Med Chem Lett* 15:263–265

Ohlrogge J, Browse J (1995) Lipid biosynthesis. *Plant Cell* 7:957–970

Okuley J, Lightner J, Feldmann K, Yadav N, Lark E, Browse J (1994) *Arabidopsis FAD2* gene encodes the enzyme that is essential for polyunsaturated lipid-synthesis. *Plant Cell* 6:147–158

Oursel A, Escoffier A, Kader JC, Dubacq JP, Trémolières A (1987) Last step in the cooperative pathway for galactolipid synthesis in spinach leaves: Formation of monogalactosyldiacylglycerol with C_{18} polyunsaturated acyl groups at both carbon atoms of the glycerol. *FEBS Lett* 219:393–399

Park JM, Cho JH, Kang SG, Jang HJ, Pih KT, Piao HL, Cho MJ, Hwang I (1998) A dynamin-like protein in *Arabidopsis thaliana*. is involved in biogenesis of thylakoid membranes. *EMBO J* 17:859–867

Patterson GW, Hugly S, Harrison D (1993) Sterols and phytyl esters of *Arabidopsis thaliana* under normal and chilling temperatures. *Phytochemistry* 33:1381–1383

Pichler H, Gaigg B, Hrastnik C, Achleitner G, Kohlwein SD, Zellnig G, Perktold A, Daum G (2001) A subfraction of the yeast endoplasmic reticulum associates with the plasma membrane and has a high capacity to synthesize lipids. *Eur J Biochem* 268:2351–2361

Poirier Y, Thoma S, Somerville C, Schifelbein J (1991) A mutant of *Arabidopsis* deficient in xylem loading of phosphate. *Plant Physiol* 97:1087–1093

Pollard M, Ohlrogge J (1999) Testing models of fatty acid transfer and lipid synthesis in spinach leaf using in vivo oxygen-18 labelling. *Plant Physiol* 121:1217–1226

Räntfors M, Evertsson I, Kjellberg JM, Sandelius AS (2000) Intraplastidial lipid trafficking: Regulation of galactolipid release from isolated chloroplast envelope. *Physiol Plant* 110:262–270

Rawsthorne S (2002) Carbon flux and fatty acid synthesis in plants. *Prog Lipid Res* 41:182–196

Roughan PG (1970) Turnover of the glycerolipids of pumpkin leaves. The importance of phosphatidylcholine. *Biochem J* 117:1–8

Roughan PG (1985) Phosphatidylglycerol and chilling sensitivity in plants. *Plant Physiol* 77:740–746

Roughan PG, Slack CR (1982) Cellular organization of glycerolipid metabolism. *Annu Rev Plant Physiol* 33:97–132

Routaboul J-M, Fischer SF, Browse J (2000) Trienoic fatty acids are required to maintain chloroplast function at low temperatures. *Plant Physiol* 124:1697–1705

Ryberg H, Ryberg M, Sundqvist C (1993) Plastid ultrastructure and development. In: Ryberg M, Sundqvist C (eds) Pigment–protein complexes in plastids. Academic, London

Sakaki T, Kondo N, Yamada M (1990) Pathway for the synthesis of triacylglycerols from monogalactosyldiacylglycerols in ozone-fumigated spinach leaves. *Plant Physiol* 94:773–780

Sánchez-Fernández R, Emyr Davies TG, Coleman JOD, Rea PA (2001) The *Arabidopsis thaliana* ABC protein superfamily, a complete inventory. *J Biol Chem* 276:30231–30244

Sastry PS, Kates M (1964) Hydrolysis of monogalactosyl and digalactosyl diglycerides by specific enzymes in runner bean leaves. *Plant Physiol* 3:1280–1287

Schlötz F (1975) Vergrößerung der Kontaktfläche zwischen Chloroplasten und ihrer cytoplasmatischen Umgebung durch tubuläre Ausstülpungen der Plastidenhülle. *Planta* 124:277–285

Schmidt H, Heinz E (1990a) Desaturation of oleoyl groups in envelope membranes from spinach chloroplasts. *Proc Natl Acad Sci USA* 87:9477–9480

Schmidt H, Heinz E (1990b) Involvement of ferredoxin in desaturation of lipid-bound oleate in chloroplasts. *Plant Physiol* 94:214–220

Schmidt H, Heinz E (1993) Direct desaturation of intact galactolipids by a desaturase solubilized from spinach (*Spinacia oleracea*) chloroplast envelopes. *Biochem J* 289:777–782

Schnurr JA, Shockey JM, de Boer GJ, Browse JA (2002) Fatty acid export from the chloroplast Molecular characterization of a major plastidial acyl-coenzyme A synthetase from *Arabidopsis*. *Plant Physiol* 129:1700–1709

Selstam E, Sandelius AS (1984) A comparison between prolamellar bodies and prothylakoid membranes of etioplasts of dark-grown wheat (*Triticum aestivum* cultivar StarkeII) concerning lipid and polypeptide composition. *Plant Physiol* 76:1036–1040

Shockey JM, Fulda MS, Browse JA (2002) *Arabidopsis* contains nine long-chain acyl-coenzyme A synthetase genes that participate in fatty acid and glycerolipid metabolism*Plant Physiol* 129:1710–1722

Siegenthaler PA, Muller MO, Bovet L (1997) Evidence for lipid kinase activities in spinach chloroplast envelope membranes. *FEBS Lett* 416:57–60

Slack CR, Roughan PG, Balasingham N (1977) Labelling studies in vivo on the metabolism of the acyl and glycerol moieties of the glycerolipids in the developing maize leaf. *Biochem J* 162:289–296

Sperling P, Linscheid M, Stocker S, Muhlbach HP, Heinz E (1993) In-vivo desaturation of cis-delta-9-monounsaturated to *cis*-delta-9,12-diunsaturated alkenylether glycerolipids. *J Biol Chem* 268:26935–26940

Staehelin LA (1997) The plant ER: A dynamic organelle composed of a large number of discrete functional domains. *Plant J* 11:1151–1165

Stelmach BA, Muller A, Hennig P, Gebhardt S, Schubert-Zsilavecz M, Weiler EW (2001) A novel class of oxylipins, sn1-*O*-(12-oxophytodienoyl)-sn2-*O*-(hexadecatrienoyl)-monogalactosyl diglyceride, from *Arabidopsis thaliana*. *J Biol Chem* 276:12832–12838

Stenzel I, Hause B, Miersch O, Kurz T, Maucher H, Weichert H, Ziegler J, Feussner I, Wasternack C (2003) Jasmonate biosynthesis and the allene oxide cyclase family of *Arabidopsis thaliana*. *Plant Mol Biol* 51:895–911

Szilágyi A, Selstam E, Akerlund HE (2008) Laurdan fluorescence spectroscopy in the thylakoid bilayer: The effect of violaxanthin to zeaxanthin conversion on the galactolipid dominated lipid environment. *Biochim Biophys Acta* 1778:348–355

van Besouw A, Wintermans JF (1978) Galactolipid formation in chloroplast envelopes. I. Evidence for two mechanisms in galactosylation. *Biochim Biophys Acta* 529:44–53

Wang Q, Sullivan RW, Kight A, Henry RL, Huang J, Jones AM, Korth KL (2004) Deletion of the chloroplast-localized thylakoid formation1 gene product in *Arabidopsis* leads to deficient thylakoid formation and variegated leaves. *Plant Physiol* 136:3594–3604

Westphal S, Heins L, Soll J, Vothknecht UC (2001a) Vipp1 deletion mutant of *Synechocystis* : A connection between bacterial phage shock and thylakoid biogenesis? *Proc Natl Acad Sci USA* 98:4243–4248

Westphal S, Soll J, Vothknecht UC (2001b) A vesicle transport system inside chloroplasts. *FEBS Lett* 506:257–261

Whatley JM, McLean B, Juniper BE (1991) Continuity of chloroplast and endoplasmic reticulum membranes in *Phaseolus vulgaris*. *New Phytol* 117:209–217

Vidi PA, Kanwischer M, Baginsky S, Austin JR, Csucs G, Dörmann P, Kessler F, Brehelin C (2006) Tocopherol cyclase (VTE1) localization and vitamin E accumulation in chloroplast plastoglobule lipoprotein particles. *J Biol Chem* 281:11225–11234

Vothknecht UC, Westhoff P (2001) Biogenesis and origin of thylakoid membranes. *Biochim Biophys Acta* 1541:91–101

Williams JP, Imperial V, Khan MU, Hodson JN (2000) The role of phosphatidylcholine in fatty acid exchange and desaturation in *Brassica napus* L leaves. *Biochem J* 349:127–133

Wintermans JFGM, Van Besouw A, Bogemann G (1981) Galactolipid formation in chloroplast envelopes: 2 Isolation-induced changes in galactolipid composition of envelopes. *Biochim Biophys Acta* 663:99–107

Wisniewska A, Subczynski WK (1998) Effects of polar carotenoids on the shape of the hydrophobic barrier of phospholipid bilayers. *Biochim Biophys Acta* 1368:235–246

Wooding FBP, Northcote DH (1965) Association of the endoplasmic reticulum and the plastids in *Acer* and *Pinus. Amer J Bot* 52:526–531

Xu C, Härtel H, Wada H, Hagio M, Yu B, Eakin C, Benning C (2002) The *pgp1* mutant locus of *Arabidopsis* encodes a phosphatidylglycerolphosphate synthase with impaired activity. *Plant Physiol* 129:594–604

Xu C, Fan J, Riekhof W, Froehlich JE, Benning C (2003) A permease-like protein involved in ER to thylakoid lipid transfer in *Arabidopsis. EMBO J* 22:2370–2379

Xu C, Fan J, Froehlich JE, Awai K, Benning C (2005) Mutation of the TGD1 chloroplast envelope protein affects phosphatidate metabolism in *Arabidopsis. Plant Cell* 17:3094–3110

Xu C, Yu B, Cornish AJ, Froehlich JE, Benning C (2006) Phosphatidylglycerol biosynthesis in chloroplasts of *Arabidopsis* mutants deficient in acyl-ACP glycerol-3-phosphate acyltransferase. *Plant J* 47:296–309

Yadav NS, Wierzbicki A, Aegerter M, Caster CS, Perezgrau L, Kinney AJ, Hitz WD, Booth JR, Schweiger B, Stecca KL, Allen SM, Blackwell M, Reiter RS, Carlson TJ, Russell SH, Feldmann KA, Pierce J, Browse J (1993) Cloning of higher-plant omega-3-fatty-acid desaturases. *Plant Physiol* 103:467–476

Yagi T, Benson AA (1962) Plant sulfolipid. V. *Lysosulfolipid formataion. Biochim Biophys Acta* 57:601–603

Yu B, Xu CC, Benning C (2002) *Arabidopsis* disrupted in *SQD2* encoding sulfolipid synthase is impaired in phosphate-limited growth. *Proc Natl Acad Sci USA* 99:5732–5737

The Role of Metabolite Transporters in Integrating Chloroplasts with the Metabolic Network of Plant Cells

A.P.M. Weber (✉) and K. Fischer

Abstract Chloroplasts are central nodes of the metabolic networks in photosynthetic cells. Metabolic pathways in chloroplasts are highly connected with pathways occurring in the surrounding cytosol and in other organelles. This requires massive flux of solutes across the chloroplast envelope membranes and hence a plethora of metabolite transport systems to facilitate this flux. During the evolution of chloroplasts from a cyanobacterial ancestor, most of these chloroplast envelope membrane transporters have been recruited from the preexisting repertoire of host transporters. We review the current state of knowledge on chloroplast metabolite transporters from an evolutionary perspective and we address the possible functions of apicoplast transporter systems in the Apicomplexa.

1 Introduction

Compartmentation is a fundamental property of eukaryotic cells. In many cases, modules of the metabolic network of these cells are partitioned and distributed between different organelles. This entails that organic small molecules (i.e., metabolites) have to be transported across the membranes bounding cellular organelles. Since most of these small molecules are charged or polar and hence not membrane permeable, various metabolite transporters are required to enable the directional transport of metabolic intermediates and end products between cellular compartments. Transporters thus participate in the cellular metabolic network by partitioning metabolites between cellular compartments. The compartmentation of cellular metabolism augments options for its regulation because it permits the simultaneous operation of competing pathways within the same cell, and thus helps prevent futile cycles. Metabolite transporters thus play a crucial role in connecting parallel and interdependent biosynthetic and catabolic pathways. They consequently represent the integrating elements of metabolic networks, similar to interchanges in road networks.

A.P.M. Weber
Institut für Biochemie der Pflanzen, Heinrich-Heine-Universität, Universitätsstraße 1,
40225 Düsseldorf, Germany
e-mail: andreas.weber@uni-duesseldorf.de

A specific feature of plant cells (*sensu* Archaeplastida; Adl et al. 2005), in contrast to nonphotosynthetic eukaryotes, is that they contain plastids, which are semiautonomous organelles of endosymbiotic origin (Reyes-Prieto et al. 2007) that are bounded by two membranes, the inner and outer plastid envelope membranes. Plastids play essential roles in plants since they represent the sites of photosynthesis and a multitude of other essential metabolic routes. Plastid function is tightly interlinked with that of other cellular compartments. For example, the majority of plastidial proteins are encoded on the nuclear genome and posttranslationally imported into plastids. Plastid function thus depends on the presence of a protein import apparatus in the envelope membrane that permits the specific and efficient uptake of plastid-targeted precursor proteins (Gutensohn et al. 2006; Soll and Schleiff 2004; Chap. 3 in this book). Plastidial metabolism is also intertwined with that of the surrounding cytosol, which causes considerable traffic of metabolic precursors, intermediates, and products across the plastid envelope membrane (Weber 2004; Weber and Fischer 2007; Weber et al. 2005). Understanding metabolite transport between plastid and cytosol is thus of crucial importance for understanding plant metabolism (Weber and Fischer 2007; Weber et al. 2004). In this chapter, we will discuss recent progress in understanding the metabolic connections between plastid and cytosol. Since "nothing in biology makes sense except in the light of evolution" (Dobzhansky 1964), particular emphasis will be given to the evolutionary origin of chloroplast metabolite transport systems and its implications for understanding the endosymbiotic origin of chloroplasts.

2 Endosymbiotic Origin of Chloroplasts and Its Relation to Metabolite Transport

The first plant cell (i.e., the protoalga; McFadden and van Dooren 2004) evolved from a primitive eukaryotic cell that most likely already contained mitochondria, a nucleus, and an endomembrane system through the formation of a permanent association with a prokaryotic photosynthetic cyanobacterium that eventually evolved into a novel organelle, the plastid (Bhattacharya et al. 2004; Margulis 1971, 1975; Mereschkowsky 1905; Reyes-Prieto et al. 2007; Schimper 1885; Chap. 1 in this book). This plastid-containing protoalga gave rise to the Archaeplastida (Adl et al. 2005), i.e., the three photosynthetic eukaryotic lineages containing primary plastids: red and green algae (including land plants), and the glaucocystophytes (Bhattacharya et al. 2004). The two envelope membranes bounding modern plastids are evidence for and relics of the evolutionary origin of plastids by endosymbiosis: the inner-envelope membrane has likely evolved from the cyanobacterial plasma membrane, while the outer-envelope membrane has chimeric origin, being remnant of both cyanobacterial outer membrane and a host membrane derived from the endomembrane system (Cavalier-Smith 2000, 2003). The plastids of the Archaeplastida are of monophyletic origin, i.e., they can be traced back to a single endosymbiotic event (Reyes-Prieto et al. 2007). Additional photosynthetic eukaryotes evolved through secondary and tertiary endosymbioses:

either red or green algae belonging to the Archaeplastida (secondary endosymbioses) or organisms containing secondary plastids (tertiary endosymbiosis) were engulfed by other eukaryotic cells and subsequently reduced to secondary or tertiary plastids, respectively, which are enveloped by three to four membranes (Cavalier-Smith 2002; Li et al. 2006; Yoon et al. 2002, 2005).

How is the endosymbiotic origin of chloroplasts related to metabolite transport? The link comes from the fact that the cyanobacterium introduced the capacity for oxygenic photosynthesis into eukaryotic cells, a process that converted a heterotrophic eukaryotic cell into a photoautotrophic organism. This, however, could only come to fruition if the eukaryotic host cell was able to reap benefit from photosynthetic carbon assimilation by the cyanobacterium (Fig. 1). That is, somehow the host cell must have been able to tap into the cyanobacterial carbon pool, in a controlled manner to avoid killing the new organelle. In other words, the merger of two free-living organisms required the integration and coordination of two previously independent metabolic entities, i.e., a photoautotrophic primary producer and a heterotrophic organism (Weber et al. 2006).

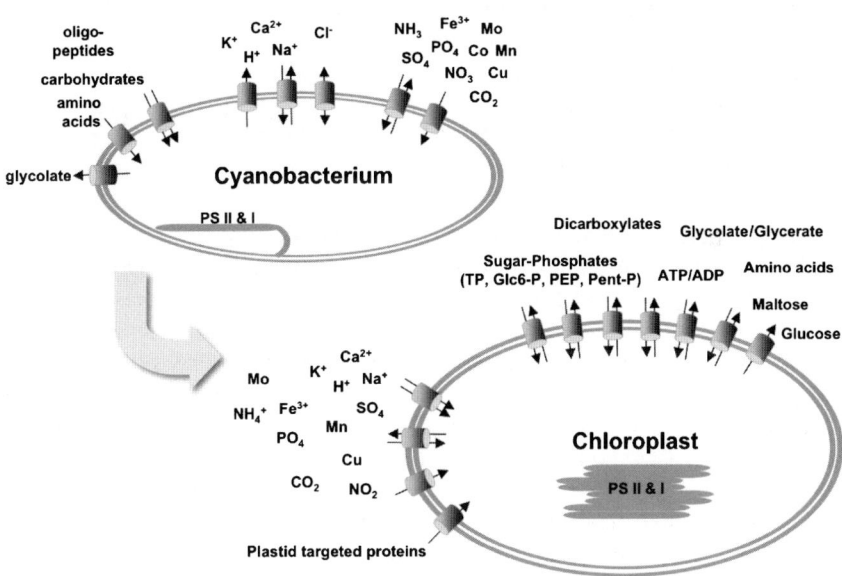

Fig. 1 The evolution of chloroplasts from free-living cyanobacteria required acquisition of novel transport functions in the newly evolving plastid envelope membrane. While the cyanobacterial ancestor of chloroplasts mostly had uptake systems for mineral, carbohydrates, and amino acids, it likely had only a few metabolite export systems. Chloroplasts, however, possess a large variety of metabolite transport systems, for example, those transporting phosphorylated intermediates, carbohydrates, amino acids, dicarboxylic acids, and a large range of other metabolites

2.1 Evolution of the Plastidic Triose Phosphate/Phosphate Translocator Was a Crucial First Step in Establishing Chloroplast–Cytoplasm Metabolic Connection

Intuitively, the export of photosynthate from the cyanobacterium to the host cell seems of key importance in the process of integrating the cyanobacterium into the host cell and for the host to reap benefit from the association. However, establishing a reliable metabolic connection between host and endosymbiont is not as trivial as it may seem at first glance. This is because export of carbon from the plastid to the host must occur in a controlled manner since the continued operation of carbon assimilation and reduction by the Calvin cycle within the plastid can only proceed if withdrawal of cycle intermediates occurs at a rate that does not exceed the rate of net CO_2 assimilation. A feedback mechanism is thus required that prevents uncontrolled export of carbon from the chloroplast. It is also reasonable to posit that the cyanobacterial ancestor of chloroplasts did not possess transporters that permit the export of Calvin cycle intermediates across its plasma membrane, given that the currently known genomes of extant cyanobacteria apparently do not encode such transport systems.

A possible scenario is that a primitive mitochondriate eukaryote was preying on cyanobacteria, probably by phagocytosis. The phagosome membrane surrounding the prey was derived from the endomembrane system and contained protein secretion systems for export of lytic enzymes into the phagosome as well as transporters that permitted the export of metabolites from the phagosome. If we assume that this scenario is reasonable, this implies that the host already possessed a set of metabolite transporters that was targeted to the membrane surrounding the cyanobacterium. Of course, we have to make a sizeable number of additional assumptions to explain the process of transformation of a free-living bacterium into a cellular organelle, such as "survival" in the phagosome, targeting of host-derived transporters beyond the phagosomal membrane (i.e., to the bacterial plasma membrane), transfer of genes to the nucleus, establishment of a protein translocon, coordination of nuclear and organellar genomes, etc. Addressing all these points is beyond the scope of this chapter; we will focus on the establishment of metabolic links between host and endosymbiont.

Insight comes from what is known about metabolite transport in higher plants. It has been known for decades that triose phosphates not required for the regeneration of the CO_2 acceptor ribulose 1,5-bisphosphate represent the first important branch point in the allocation of recently assimilated CO_2 to different metabolic routes. Either they can remain inside the chloroplast to enter plastid-localized metabolic pathways or they can be exported to the cytosol (Tegeder and Weber 2006; Weber 2006). In land plants, the export of the triose phosphates glyceraldehyde 3-phosphate and dihydroxyacetone phosphate to the cytosol is mediated by a triose phosphate/phosphate antiporter (Flügge 1999). This antiporter catalyzes the strict counterexchange of one molecule of triose phosphate with one molecule of orthophosphate (Pi); that is, for each molecule of organically bound phosphate that leaves the chloroplast as triose phosphate, one molecule of inorganic phosphate is returned to the chloroplast, thus avoiding phosphate depletion of the chloroplast stroma (Flügge 1995; Flügge and Heldt 1984; Heldt et al. 1990). The strict stoichiometry of the counterexchange is essential for maintaining phosphate homeostasis in the stroma, since inorganic phosphate is required for the biosynthesis of ATP from ADP and Pi in the light reaction of photosynthesis. Depletion of the plastidial phosphate pool by unbalanced export of organically bound phosphate would rapidly lead to inhibition of photosynthetic electron transport and ultimately to damage of the photosynthetic machinery.

Recent progress in genomics showed that controlled export of reduced carbon from chloroplasts is catalyzed by a triose phosphate/phosphate antiporter not only in green land plants, but in all photosynthetic eukaryotes for which genome or comprehensive expressed sequence tag information is available, including photosynthetic organisms containing secondary plastids, such as diatoms (Weber et al. 2006). In this context it is important to emphasize that phosphate translocators are strict antiporters. Antiporters are perfectly suited to link metabolic routes in neighboring compartments because withdrawal of a metabolite from one compartment is always strictly coupled to the availability of a suitable counterexchange substrate in the other compartment. This permits flexible adaptation to varying metabolic requirements; that is, the direction of transport depends only on substrate concentrations on the *cis* and *trans* sides of the membrane. Directionality of metabolite flux is hence principally governed by metabolic activities on both sides of the membrane, not by regulation of the transporter itself (Weber et al. 2006).

Phylogenomic and phylogenetic analysis of genomic and expressed sequence tag data from a broad range of organisms recently showed that the plastidial triose phosphate/phosphate translocators (TPTs) evolved from transport proteins of the eukaryotic endomembrane system, specifically from sugar nucleotide transporters (NTTs) of the endoplasmic reticulum (ER) and Golgi membranes (Weber et al. 2006). Genes encoding these sugar NTTs are ubiquitously present in eukaryotic genomes but are absent from prokaryotes. They hence likely represent a eukaryote-specific evolutionary innovation. This finding supports the hypothesis that the food vacuole that initially engulfed the cyanobacterium was indeed derived from the host endomembrane system and eventually fused with the bacterial outer membrane to become the outer plastid envelope membrane. A metabolite transporter originally residing in the host endomembrane system was then routed onto the bacterial plasma membrane (i.e., the plastid inner envelope membrane) and thus enabled the host to tap into the photosynthetic carbon pool of the endosymbiont. This hypothetic but nonetheless sensible scenario must have occurred early in plastid evolution because genes encoding members of the plastidic phosphate translocator family (pPT) are found in the genomes of all sequenced photosynthetic eukaryotes, including red algae, green algae, land plants, and photosynthetic organisms containing complex plastids, such as diatoms and dinoflagellates (Weber et al. 2006). The pPTs thus likely evolved at the stage of the protoalga, certainly though before the split of the red and green plant lineages. The pPT family represents one of the best-characterized families of plastid transporters. In the following sections, we will review what is currently known about this important family of plastidial metabolite transporters.

2.2 The Plastid Phosphate Translocator Gene Family

The pPTs are divided into four different subfamilies. The pPTs share inorganic phosphate as a common substrate but have different, though overlapping spectra of counterexchange substrates (Flügge et al. 2003). The pPT subfamilies are named after their main transported substrates, i.e., triose phosphate (Flügge et al. 1989), phosphoenolpyruvate (PEP) (Fischer et al. 1997), glucose 6-phosphate (Glc 6P) (Kammerer et al. 1998), and xylulose 5-phosphate (Eicks et al. 2002). The pPTs connect the primary (and secondary) metabolism of plastids with that of the cytosol (for an overview see Flügge et al. 2003). However, a detailed picture of the physiological function is available only for the TPT, while the physiological function of the other pPTs is not fully understood. The TPT exports triose phosphates to the cytosol, where they are used for the synthesis of several metabolites, such as sucrose, i.e., the TPT delivers the product of photosynthetic carbon fixation to other parts of the cell (Riesmeier et al. 1993; Schneider et al. 2002). The other members of the pPT family import phosphorylated compounds into plastids, where they are fed into different pathways, such as starch synthesis, the oxidative pentose phosphate pathway (OPPP), or the shikimate pathway. Because several excellent reviews have covered many aspects of the pPT family (e.g., Fischer and Weber 2002; Flügge et al. 2003) we are focussing here on three aspects. We will discuss phosphate translocators in red algae, pPTs in apicoplasts, the plastids of Apicomplexa, and finally the role of the Glc 6P/phosphate translocator (GPT) in development and metabolism of seeds.

2.2.1 Plastidic Phosphate Translocators in Red Algae

As outlined already, the pPTs of land plants can be classified into four distinct groups with distinctive substrate specificities (Flügge et al. 2003; Knappe et al. 2003). At least three of these subgroups evolved very early on: using phylogenetic analysis, one can clearly detect genes encoding proteins belonging to the TPT and PEP/phosphate translocator (PPT) families in the genomes of ancient red microalgae, as well as in green plants, indicating that these proteins evolved before the split of the red and green lineages. The clade containing the closest red algal relative of plant GPT and xylulose 5-phosphate/phosphate translocator (XPT) is harder to resolve. Although proteins belonging to this clade are detectable in red algae, it is difficult to decide whether they actually represent functional GPTs or XPTs; experimental analysis of their substrate specificity will be required. Nonetheless, three distinct clades of pPTs are already established in basic red algae.

Functional analysis of the TPT paralog from the red alga *Galdieria sulphuraria* showed that this transporter, in comparison to the corresponding protein from land plants, has narrower substrate specificity. In contrast to the protein from land plants, the red algal TPT only accepts Pi and triose phosphates (i.e., glyceraldehyde 3-phosphate and dihydroxyacetone phosphate) as substrates, whereas the land plant TPT accepts in addition to these substrates also 3-phosphoglyceric acid (3-PGA) (M. Linka and A.P.M. Weber, unpublished results). Thus, the red algal protein is not able to serve as a redox shuttle between plastid stroma and cytoplasm by exchanging triose phosphates for 3-PGA. Since red algae, in contrast to green plants, do not store starch in the chloroplast stroma but in the cytosol, most of the assimilated carbon has to be efficiently exported from the chloroplast in the form of triose phosphates to the cytosol, where they have to be metabolized to recover the Pi required for continued operation of photosynthesis. Since in red algae all of the assimilated carbon needs to leave the chloroplast via TPT at a rate close to the rate of photosynthetic CO_2 assimilation, competition between 3-PGA and triose phosphates for transport across the chloroplast envelope membrane would be counterproductive. Hence, evolution of a narrow-specificity TPT in red algae apparently was advantageous. According to phylogenetic analysis, the phosphate translocators of the chromalveolates are monophyletically derived from the red algal TPTs via secondary endosymbiosis, whereas GPT/XPT and PPT have been lost in the process (Weber et al. 2006). A specific case, the phosphate translocators of the Apicomplexa will be discussed next.

2.2.2 Phosphate Translocators and the Physiological Role of Apicoplasts

In the process of secondary endosymbiosis that led to the evolution of secondary plastids in the chromalveolates, a red alga was captured by a nonphotosynthetic protist and eventually reduced to a complex plastid that is surrounded by four envelope membranes (Li et al. 2006). The two innermost membranes are believed to derive from the inner and outer chloroplast membranes of the red algae, the third layer is a remnant of the red algal plasma membrane, and the outermost envelope membrane is derived from the protist endomembrane system. Interestingly, the gene encoding the red algal TPT was transferred from the degenerating red algal nucleus to the nuclear genome of the host, whereas the genes encoding PPT and XPT/GPT apparently were lost in the process. Once it had arrived at its new location, the TPT gene started to radiate by duplication events and frequently evolved into small gene families (Weber et al. 2006). It is interesting that the recruitment of phosphate translocators for the export of reduced carbon from plastids was recapitulated during the evolution of complex plastids in the chromalveolates, thus emphasizing the importance of a phosphate-balanced, controlled export of triose phosphates from the chloroplast. One group of chromalveolates, the phylum Apicomplexa, comprises several thousand protozoan species which are obligate intracellular parasites, several of them causing important human and animal diseases (Levine 1988). Four *Plasmodium* species are responsible for malaria in humans with *Plasmodium falciparum* being the most lethal one. One of the best-characterized species is *Toxoplasma gondii*, which infects almost all vertebrates, including humans, and any nucleated cell in any tissue within these organisms, although cats (Felidae) are the main host (Wong and Remington 1993).

Apicomplexan parasites differ in some important aspects, for example, with respect to the diseases they cause or their complicated life cycle. In contrast to these differences, their basic biochemistry, genetics, and subcellular architecture are strikingly similar (Roos et al. 2002). Apicomplexan cells contain a single plastid, called the *apicoplast*, which is enclosed by four envelope membranes (Köhler et al. 1997; McFadden et al. 1996; Wilson et al. 1996). In contrast to plant plastids, the physiological function of apicoplasts is not exactly known. However, thorough analyses of apicomplexan genomes and protein localization and inhibitor studies have now provided the first insights into the metabolism of apicoplasts (Fleige et al. 2007; Jomaa et al. 1999; Mazumdar et al. 2006; Ralph et al. 2004). It is now well established that the apicoplast is indispensable because inhibition of its metabolism and replication results in parasite death (Fichera and Roos 1997; He et al. 2001; Jomaa et al. 1999). It has been shown that the apicoplast is involved in the synthesis of at least three different groups of metabolites, namely, fatty acids (via a bacterial type II fatty acid synthase), isopentenyl diphosphate (via the non-mevalonate or 1-deoxy-d-xylulose 5-phosphate, DOXP, pathway, which occurs only in plastids and bacteria), and heme. In addition, apicoplasts contain an almost complete set of glycolytic enzymes, with only enolase and Glc 6P isomerase missing (Fleige et al. 2007). Thus, the metabolism of apicoplasts resembles that of nongreen plastids of higher plants with a few remarkable exceptions. Apicoplasts lack an OPPP; that is, they are apparently not able to metabolize Glc 6P (because they also lack Glc 6P isomerase), which might be attributed to the red algal heritage of apicoplasts (Oesterhelt et al. 2007).

Integration of plastids into the cellular metabolism is achieved by a set of transporters. So far almost nothing is known about transporters of apicoplasts and other organelles in Apicomplexa. Proteins similar to the pPTs are the first apicoplast transporters identified. *P. falciparum* possesses two pPTs (PfiTPT, PfoTPT) which have been shown to be located in the innermost and outermost envelope membranes of the apicoplast but might also be located in the other envelope membranes (Mullin et al. 2006). While PfiTPT possesses a bipartite presequence, which is composed of a signal peptide and an adjacent transit peptide, PfoTPT lacks any presequence. Surprisingly, *T. gondii* and probably other Apicomplexa only possess one pPT which is most similar to PfoTPT and which also lacks a presequence (Fleige et al. 2007). *T. gondii* phosphate translocator has been shown to be localized in different envelope membranes although it was not possible to definitely state that the protein is present in all four membranes (Karnataki et al. 2007). Thus, it is reasonable to assume that the apicomplexan pPTs are responsible for the transport of glycolytic intermediates across some, if not all four envelope membranes. Their physiological functions could be to deliver carbon skeletons, energy, and redox equivalents for the apicoplast fatty acid synthesis and the DOXP pathway, among others (Fig. 2). However, the actual physiological function of these transporters is

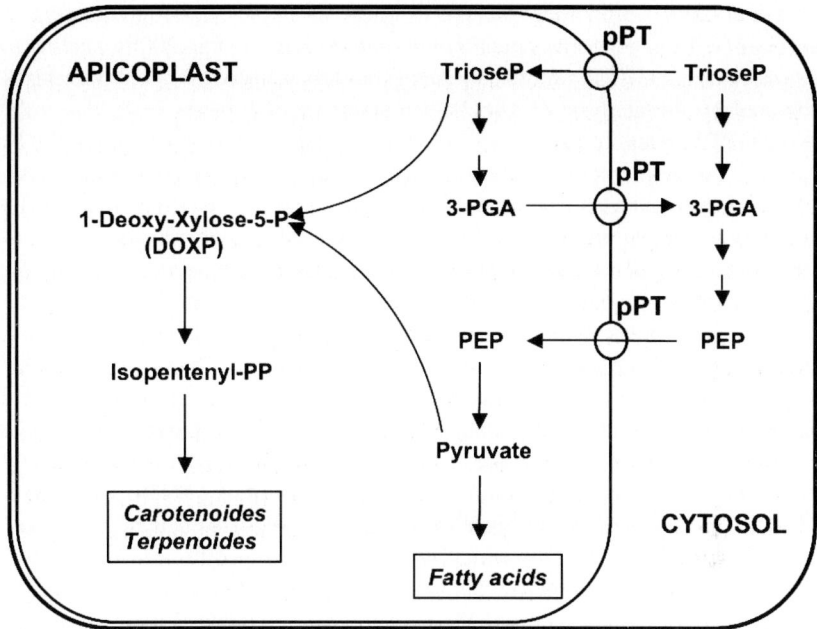

Fig. 2 Putative pathways for the import of carbon into apicoplasts. Triose phosphates are imported into the apicoplast by the plastidic phosphate translocator (*pPT*) and converted to 3-phosphoglycerate (*3-PGA*), which is then exported to the cytosol. There 3-PGA is converted back either to triose phosphates, which results in a reverse triose phosphate/3-PGA shuttle (compared with the situation in plants), or to phosphoenolpyruvate (*PEP*), which is reimported into the apicoplast. The reverse shuttle leads to the indirect import of ATP and redox equivalents which are used, for example, for fatty acid synthesis. PEP serves as a substrate for fatty acid synthesis and the synthesis of terpenoides via the 1-deoxy-d-xylulose 5-phosphate (*DOXP*) pathway

not known so far because their substrate specificities are not known and cannot be deduced from their amino acid sequences because the pPTs from Apicomplexa do not show a particularly higher similarity to one of the well-characterized subfamilies of higher-plant pPTs (Fleige et al. 2007; Knappe et al. 2003).

2.2.3 The GPT Is Important for Gametophyte and Seed Development

Plastids of nongreen tissues of multicellular plants import carbon, mainly in the form of Glc 6P, as a source for biosynthetic pathways and energy. Glc 6P is used for three different purposes. Firstly, it is the substrate for starch synthesis. Secondly, Glc 6P is fed into the OPPP for the production of reducing equivalents (NADPH) and, thirdly, it can deliver carbon skeletons for fatty acid synthesis and other processes via glycolysis. The import of Glc 6P is mediated by the GPT (Kammerer et al. 1998). *Arabidopsis* possesses two different functional GPTs, GPT1 and GPT2

(Knappe et al. 2003). Both are expressed in almost all tissues, although the expression level of GPT1 is generally much higher (Niewiadomski et al. 2005), leading to the assumption that GPT1 is more important for carbon uptake. This hypothesis is corroborated by the analysis of knockout mutants of both genes in *Arabidopsis*. While the GPT2 knockout line shows no obvious phenotype, deletion of the GPT1 gene turned out to be lethal (Niewiadomski et al. 2005). A detailed analysis revealed that the development of the female gametophyte and that of the male gametophyte are impaired. The reason why a defect in Glc 6P transport leads to such severe phenotypes is not clear so far. The existence of several starch-free mutants of *Arabidopsis* that retain full fertility demonstrates that starch accumulation per se is not a prerequisite for gametophyte development (Kofler et al. 2000). Therefore, most probably the interruption of fatty acid synthesis in the GPT1 knockout is responsible for the lethality. It is known that reduction of fatty acid synthesis leads to premature cell death and altered morphology, while a complete loss is lethal (Mou et al. 2000). The disruption of Glc 6P import would not only impair the delivery of carbon skeletons for fatty acid synthesis but also the production of NADPH; however, direct proof of this hypothesis is still lacking.

Besides gametophyte development, GPT1 is also involved in seed maturation. In an attempt to understand the impact of GPT1 on assimilate storage in seeds of *Vicia narbonensis* the expression of the GPT1 gene was reduced by an antisense approach using a seed-specific promoter (Rolletschek et al. 2007). Embryos of these transgenic plants showed a reduction in GPT activity of about 50%, resulting in a decrease of the starch content of the seeds of about 25% and a decrease of the total lipid content of 45%. In contrast, the protein content of seeds increased by 30%. Thus, the reduction of GPT activity leads to a shift in assimilate partitioning from starch/lipids into storage proteins, indicating that the GPT exerts a significant control on seed filling and maturation.

3 Integration of Chloroplast Metabolism with Host Metabolism Was a Host-Driven Process

As outlined already, one of the major chloroplast envelope transporter families (i.e., the pPTs) has evolved from sugar NTTs of the host endomembrane system. Naturally, this discovery prompted the question about the evolutionary origin of chloroplast transporters in extant photosynthetic eukaryotes. Comprehensive phylogenomic and phylogenetic analysis of plastid envelope-localized transporters of *Arabidopsis thaliana* showed that at least 50% of the envelope "permeome" evolved from preexisting host proteins, whereas only a small portion was contributed by the endosymbiont (Tyra et al. 2007). That is, integration of plastid and host metabolism was predominantly a host-driven process. Transport proteins encoded by the host genome acquired targeting signals for routing to the chloroplast and were inserted into the chloroplast envelope membrane. A surprisingly large share of plastid envelope membrane transporters, such as the adenine NTTs, the dicarboxylate translocators (DiTs), and some metal-transporting ATPases have their evolutionary origin in prokaryotic intracellular energy parasites (i.e., *Chlamydia* and/or *Rickettsia*). Most of these genes have already been introduced into the genome of the protoalga by horizontal transfer and they have been maintained throughout plant evolution, emphasizing their importance for connecting the metabolism of plastid and cytosol (Tyra et al. 2007).

3.1 Plastidic Adenine Nucleotide Transporters of Prokaryotic Origin

ATP represents the universal energy currency of all living cells, and is mainly synthesized in mitochondria and, in plant cells, also in chloroplasts. Mitochondria export ATP generated by oxidative phosphorylation in strict counterexchange with ADP via ADP/ATP carriers (AAC). These carriers belong to the large mitochondrial carrier family (MCF), which in *Arabidopsis* consists of 58 proteins (Haferkamp 2007; Picault et al. 2004). An ATP transport activity has been shown to be present also in the envelope membranes of chloroplasts (Heldt 1969) and nongreen plastids (Schünemann et al. 1993) and, recently, in the thylakoid membranes of chloroplasts (Spetea et al. 2004). Within the past decade, the proteins responsible for the plastid ATP transport activities have been identified and their physiological function partly characterized. It turned out that these proteins belong to two different unrelated transporter families, namely, the MCF and the plastid NTTs, the first having eukaryotic, the latter prokaryotic origin (Weber and Fischer 2007).

NTTs can be found in all higher plants and algae. In *Arabidopsis*, the major share of ATP transport activity across the plastid inner envelope membrane is mediated by AtNTT2, one of the two NTTs (Kampfenkel et al. 1995; Reiser et al. 2004). Both NTTs exhibit very similar transport properties but show different expression patterns, with *AtNTT2* being almost ubiquitously expressed, while *AtNTT1* represents a sugar-induced gene (Reiser et al. 2004). Extensive analysis of *NTT* antisense, RNA interference, and knockout plants revealed different physiological functions of the NTTs in different types of plastids. In nongreen plastids the function of the NTTs, especially NTT2, is to supply ATP for several anabolic pathways. These transporters exert significant control on starch synthesis in amyloplasts of potato tubers and on lipid biosynthesis in developing *Arabidopsis* seeds (Geigenberger et al. 2001; Tjaden et al. 1998). However, starch and lipid accumulation in knockout plants is not totally abolished, which indicates that some additional ATP supply mechanisms exist, e.g., ATP regeneration by substrate-level phosphorylation during conversion of PEP to pyruvate (Tegeder and Weber 2006; Voll et al. 2003) and/or a second ATP transport system (Thuswaldner et al. 2007; see below).

In contrast to nongreen plastids, the function of ATP transport in chloroplasts remained elusive. Chloroplasts show a high level of ATP synthesis by photophosphorylation but the ATP transport activity in these organelles is very low (1% of the activity of the TPT). Therefore, it has been proposed that the function of ATP transport is to energize unknown processes in the stroma during the night rather than to transport ATP during photosynthesis (Heldt 1969). Only very recently, this hypothesis could be validated through the analysis of *Arabidopsis* lines lacking both NTTs (Reinhold et al. 2007). These mutants were indistinguishable from wild-type plants when grown under long-day conditions or under short-day conditions at high light intensities but showed a dwarf phenotype with leaf necrotic lesions under short-day conditions at low light intensities. This phenotype is caused by photooxidation due to the accumulation of protoporphyrin IX (Proto IX), which induces the production of reactive oxygen species. The increased Proto IX levels are most likely due to a reduced activity of magnesium chelatase, which requires ATP for catalysis and also for stabilization of the holoenzyme (Reid and Hunter 2004). Thus, one important process that has to be energized during the night by the import of ATP is the magnesium chelatase assembly under conditions when glycolytic ATP production in the stroma in the dark is restricted owing to limited photosynthesis and starch synthesis in the light.

3.2 Plastid-Targeted Members of the Mitochondrial Carrier Family

Some plastid-localized transporters belong to the MCF, including the folates transporter FOLT1 and the *S*-adenosylmethionine transporter SAMT1 (Bouvier et al. 2006; Tyra et al. 2007; Weber and Fischer 2007). In both cases, proteins that in nonphotosynthetic eukaryotes serve as mitochondrial transporters have been recruited to the chloroplast. In addition to the aforementioned NTTs, plastids also harbor ATP transporters that belong to the MCF. These will be discussed in the following paragraphs.

Thylakoid membranes possess, besides proteins involved in photosynthesis, also several transport systems for the transport of inorganic ions (e.g., cation channels; Pottosin and Schönknecht 1996). So far, only one transport activity for organic metabolites has been identified, namely, an ATP transport activity (Spetea et al. 2004). The transport of ATP in strict counterexchange with ADP is catalyzed by the thylakoid ATP carrier (TAAC), which is encoded by the *Arabidopsis* gene At5g01500 (Thuswaldner et al. 2007). The protein belongs to the MCF and shows the highest sequence similarity to the mitochondrial AACs. In addition, the substrate specificities and affinities are very similar to those of the mitochondrial AAC. The TAAC is mainly located in thylakoid membranes but to a lesser extent also in the envelope membranes of chloroplasts and nongreen plastids. This dual localization indicates that TAAC has different physiological functions. In the envelope membranes, it might be involved in the import of ATP in nongreen plastids and chloroplasts in the dark (see Sect. 3.1). In thylakoid membranes, TAAC represents a link between ATP synthesis on the stromal side and nucleotide-dependent reactions in the luminal space (Thuswaldner et al. 2007). These reactions might include the import, folding, and degradation of proteins and their phosphorylation. Phosphoproteins have been recently detected in the thylakoid lumen through proteomic approaches (Rinalducci et al. 2006; Wagner et al. 2006). A phosphate group could also be transferred from ATP to other nucleotides such as GDP by a luminal nucleoside diphosphate kinase (Spetea et al. 2004). The resulting GTP stimulates the turnover of the reaction center D1 protein. Thus, the TAAC may have both direct and indirect roles in the synthesis and degradation of the thylakoid membranes. This view is corroborated by the analysis of knockout mutants lacking the TAAC. These plants have lower amounts of thylakoid membranes (expressed as milligrams of chlorophyll per gram of leaf), resulting in a pale-green phenotype. However, the ATP transport activity in the thylakoid membranes is reduced only to about 50%, pointing to a second yet unidentified transporter.

In plants, another member of the MCF is located in the plastid envelope membranes and transports adenine nucleotides. These proteins are named Brittle1 (BT1) after the *Brittle-1* maize mutant, which has kernels with a collapsed angular appearance (Wentz 1926), a phenotype that is caused by a severely reduced starch content in the endosperm. The *lys5* mutant of barley shows a similar phenotype (Patron et al. 2004). Because the endosperm shows a drastically higher level of ADP-glucose (ADP-Glc) and the incorporation of externally added ADP-Glc into starch in isolated amyloplasts is significantly lower than in wild-type plants it has been concluded that BT1 is an ADP-Glc transporter (Shannon et al. 1998). Direct proof of this hypothesis was very recently achieved by expression of the BT1 gene from maize in *Escherichia coli* and measurement of the transport activity of the protein (Kirchberger et al. 2007). It has been shown that BT1 indeed transports ADP-Glc in strict counterexchange with ADP. Thus, the physiological function of BT1 in cereal endosperm is to deliver the substrate for starch biosynthesis in plastids.

Maize possesses a second BT1 gene (BT1-2), while there are three BT1 genes in rice (Kirchberger et al. 2007). However, BT1-like genes are found in all mono- and dicotyledonous plants. A phylogenetic analysis of the BT1 homologs indicated that the group could be divided into two subgroups. One comprises proteins like BT1 from maize that are found only in cereals. These likely represent ADP-Glc transporters. The other consists of BT1-2 from maize and proteins found in all plants. Because in noncereal plants ADP-Glc synthesis is exclusively localized in plastids these proteins must have a different physiological function (Beckles et al. 2001). Indeed, in these plants the BT1-related protein does not transport ADP-Glc but serves as a uniporter providing the cytosol and other compartments with adenine nucleotides (AMP, ADP, ATP) synthesized in plastids (Leroch et al. 2005). Thus, these closely related transporters not only have different substrate specificities but also show different transport mechanisms (uniport vs. antiport).

4 Nitrogen Assimilation Requires Tight Interaction of Chloroplast and Cytosolic Metabolism

In most land plants, the pathway for the assimilation of inorganic nitrogen into organic compounds such as glutamate is predominantly localized in plastids. The principal form of nitrogen that is converted into organic nitrogen-containing compounds is ammonia. Ammonia yields either from direct uptake from the soil or, in some cases, is provided to the plant by symbiotic, nitrogen-fixing microorganisms. In addition, it can be generated by reduction of nitrate and it results from metabolic reactions, such as photorespiration or phenylalanine ammonia lyase. The reduction of nitrate to nitrite is exclusively located in the cytosol and is catalyzed by assimilatory NADH-dependent nitrate reductase. Nitrite is then imported into the plastid stroma, where it is reduced to ammonia by nitrite reductase. Until recently, a chloroplastic nitrite transporter that functions in nitrite uptake from the cytosol has only been reported from the green alga *Chlamydomonas reinhardtii* (Nar1; Rexach et al. 2000). The currently known genomes of land plants, such as *Arabidopsis*, poplar, or rice do not encode transport proteins related to *C. reinhardtii* Nar1. A putative plastidic nitrite transporter from a land plant was recently identified in de-etiolated cucumber seedlings. This protein (CsNitr1-L) belongs to the proton-dependent oligopeptide transporter family and was localized to the chloroplast envelope membrane (Sugiura et al. 2007). When it was expressed in yeast, nitrite uptake into yeast cells was reduced, which was explained by excretion of nitrite from yeast cells by CsNitr1-L. *Arabidopsis* mutant deficient in the CsNitr1-L paralog At1g68570 accumulated fivefold higher foliar nitrite levels than controls. While these results might be indicative of a role of CsNitr1-L as a plastidic nitrite transporter, it must be cautioned that this putative function has not been unequivocally demonstrated by, e.g., functional analysis of reconstituted recombinant transporter protein.

In land plants, ammonia is assimilated into the organic form by the joint action of glutamine synthetase (GS) and ferredoxin or NADH-dependent glutamate synthase (Fd/NADH-GOGAT). This process consumes two electrons and one molecule of ATP. The major pathway for ammonia assimilation is the plastid located GS/GOGAT reaction cycle (Ireland and Lea 1999; Miflin and Lea 1976, 1980), although the main GS activity in some green plants is predominantly cytosolic (Weber and Flügge 2002) and in red algae is exclusively cytosolic (Terashita et al. 2006). GOGAT activity is exclusively localized in the plastid stroma and in some red algae the gene encoding Fd-GOGAT is even encoded on the plastid genome (Glöckner et al. 2000; Ohta et al. 2003).

The reaction can be summarized as follows:

$$2\text{-oxoglutarate} + \text{glutamate} + \text{ATP} + 2\text{Fd}_{red} + \text{NH}_4^+ \to 2 \text{ glutamate}$$
$$+ \text{ADP} + \text{P}_i + 2\text{Fd}_{ox}.$$

While one glutamate can be withdrawn from the reaction cycle to serve as the principal amino group donor in plant metabolism, the second one reenters the cycle to serve as an acceptor for ammonia in the GS-catalyzed reaction. The GS/GOGAT cycle thus produces one molecule of glutamate from one molecule of 2-oxoglutarate (2-OG) and one molecule of ammonia. De novo glutamate biosynthesis by GS and GOGAT requires the precursor 2-OG, which needs to be imported from the cytosol. The uptake of 2-OG into the plastid stroma is catalyzed by a two-translocator system that is located in the inner plastid envelope membrane (Woo et al. 1987). 2-OG is imported in counterexchange with malate by a 2-OG/malate translocator (DiT1). After conversion of 2-OG into glutamate by GS/GOGAT, glutamate is exported to the cytosol in counter-exchange with malate by a glutamate/malate translocator (DiT2). DiT1 was the first component of the two-translocator system that was identified at the molecular level (Weber and Flügge 2002; Weber et al. 1995), and recently also DiT2 from *Arabidopsis* (Taniguchi et al. 2002) and other plant species (Renné et al. 2003; Taniguchi et al. 2004; Weber and Flügge 2002) was reported.

Phylogenomic analysis showed that DiTs have been introduced to the genomes of the green plant lineage via lateral gene transfer from *Chlamydia* or a related organism (Tyra et al. 2007). Red algal genomes do not encode proteins related to DiTs from plants or bacteria, indicating that DiTs were introduced to plants after the split of the red and the green lineages or, alternatively, have been lost from the red lineage (Tyra et al. 2007). Since GS activity in red algae is cytosolic and GOGAT is plastid-encoded, this leaves open the question how 2-OG and Gln are imported into chloroplasts to drive Glu biosynthesis. Metabolic flux analysis will be required to address this question.

In plants, nitrogen assimilation requires the interaction of three cellular compartments: plastid stroma, cytosol, and mitochondria. The conversion of triose phosphates produced in the Calvin cycle to 2-OG involves the glycolytic pathway (triose phosphate to PEP and pyruvate, respectively), carboxylation of PEP to oxaloacetate by PEP carboxylase, and a partial tricarboxylic acid cycle to accomplish the reduction of oxaloacetate to malate and its subsequent conversion to 2-OG (see Weber and Flügge 2002; Weber 2004, 2006; Weber et al. 2004, 2005 for recent reviews). In leaves of C_3-type plants, approximately 90% of the capacity of the ammonia assimilatory machinery is used for reassimilation of ammonia that is released from the photorespiratory carbon cycle (Coruzzi 2003; Hirel and Lea 2001). During photorespiration, ammonia is generated in the mitochondrial matrix by glycine decarboxylase (Douce et al. 2001; Douce and Neuburger 1999), which is reassimilated by the plastidial isozyme of glutamine synthetase (Blackwell et al. 1987; Wallsgrove et al. 1980). The photorespiratory pathway is highly compartmentalized, involving plastids, cytosol, peroxisomes, and mitochondria. Metabolite transporters are critical to maintain the high net fluxes between these organelles (Weber 2006); however, to date only two transporters involved in this pathway, the plastidial glutamate/malate (DiT2) and 2-OG/malate (DiT1) transporters, have been identified at the molecular level (Renné et al. 2003; Schneidereit et al. 2006).

5 Conclusions

Recent progress in genomics has enabled phylogenomics approaches to understanding metabolite transport across the chloroplast envelope membrane. It turned out that connecting chloroplast to cytosolic metabolism was an early and crucial event in chloroplast evolution and that this process was primarily host-driven. Apparently, it was of evolutionary advantage to recruit substrate antiporters for the exchange of metabolites across the plastid envelope membrane. This mode of transport not only seems to be the preferred one for transport of metabolites across the membranes of plastids, but also of mitochondria and possibly also of the Golgi apparatus and the ER. This is clearly distinct from transport across the plasma membrane and the tonoplast of the vacuole, where transport is frequently energized by proton co- or antiport, or, in the case of ATP-binding cassette type transporters, by hydrolysis of ATP. There are always some exceptions to these general rules, though. For example, the transport of pyruvate into chloroplasts was shown to occur by cotransport with protons or sodium, depending on the plant species (Aoki et al. 1992; Flügge et al. 1985; Huber and Edwards 1977). In some cases, uniport systems coexist with antiporters. For example, chloroplasts have in addition to the phosphate translocator family members also a phosphate/proton cotransporter (Versaw and Harrison 2002). The physiological role of this latter transporter is not yet understood, in particular if it serves as a chloroplastic phosphate importer. In fact, a phosphate uniporter in chloroplasts would counteract the metabolic control exerted by the Pi-coupled TPT by generating a Pi short circuit. A phosphate exporter, however, would be required in starch-synthesizing amyloplasts, such as potato tubers: As outlined above, starch biosynthesis in amyloplasts of dicotyledonous plants requires Glc 6P and ATP, which are both imported from the cytosol by the GPT and NTTs (Kammerer et al. 1998; Linke et al. 2002; Neuhaus and Wagner 2000; Tjaden et al. 1998). Whereas the import of Glc 6P is Pi-balanced (for each Glc 6P imported, one Pi is exported), ATP import is not. ATP is exchanged for ADP; consequently, one surplus Pi accumulates in the plastid for each ATP that is hydrolyzed to ADP and Pi. While unidirectional export of Pi across the envelope of isolated cauliflower bud plastids has been demonstrated (Neuhaus and Maass 1996), the corresponding transporter has not yet been identified.

We conclude that (1) many, if not most, metabolite translocators of the plastid envelope membrane are of host origin and (2) metabolite transporters of the plastid envelope membrane frequently work as substrate antiporters; consequently, the rate and direction of transport commonly depends on metabolic reactions and substrate concentrations on both sides of the membranes. In the evolutionary context, recruiting substrate antiporters for connecting plastid and cytosolic metabolism possibly was advantageous because antiporters, in addition to shuffling metabolites across membranes, also provide a means for intercompartmental crosstalk and for coordinating metabolic activities on both sides of a membrane.

Acknowledgments We are thankful to Mark Linka for assistance with preparing Fig. 1. Work in the authors' laboratories was supported by grants from the Deutsche Forschungsgemeinschaft and the US National Science Foundation.

References

Adl SM, Simpson AG, Farmer MA, Andersen RA, Anderson OR, Barta JR, Bowser SS, Brugerolle G, Fensome RA, Fredericq S, James TY, Karpov S, Kugrens P, Krug J, Lane CE, Lewis LA, Lodge J, Lynn DH, Mann DG, McCourt RM, Mendoza L, Moestrup O, Mozley-Standridge SE, Nerad TA, Shearer CA, Smirnov AV, Spiegel FW, Taylor MF (2005) The new higher level classification of eukaryotes with emphasis on the taxonomy of protists. *J Eukaryot Microbiol* 52:399–451

Aoki N, Ohnishi J, Kanai R (1992) 2 different mechanisms for transport of pyruvate into mesophyll chloroplasts of C4 plants – a comparative study. *Plant Cell Physiol* 33:805–809

Beckles DM, Smith AM, ap Rees T (2001) A cytosolic ADP-glucose pyrophosphorylase is a feature of graminaceous endosperms, but not of other starch-storing organs. *Plant Physiol* 125:818–827

Bhattacharya D, Yoon HS, Hackett JD (2004) Photosynthetic eukaryotes unite: endosymbiosis connects the dots. *Bioessays* 26:50–60

Blackwell RD, Murray AJS, Lea PJ (1987) Inhibition of photosynthesis in barley with decreased levels of chloroplastic glutamine synthetase activity. *J Exp Bot* 38:1799–1809

Bouvier F, Linka N, Isner JC, Mutterer J, Weber APM, Camara B (2006) Arabidopsis SAMT1 defines a plastid transporter regulating plastid biogenesis and plant development. *Plant Cell* 18:3088–3105

Cavalier-Smith T (2000) Membrane heredity and early chloroplast evolution. *Trends Plant Sci* 5:174–182

Cavalier-Smith T (2002) Chloroplast evolution: secondary symbiogenesis and multiple losses. *Curr Biol* 12:R62–R64

Cavalier-Smith T (2003) Genomic reduction and evolution of novel genetic membranes and protein-targeting machinery in eukaryote–eukaryote chimaeras (meta-algae). *Philos Trans R Soc Lond B Biol Sci* 358:109–134

Coruzzi GM (2003) Primary N-assimilation into amino acids in Arabidopsis. In: Somerville CR, Meyerowitz EM (eds) The Arabidopsis book. American Society of Plant Biologists, Rockville, MD, pp 1–17

Dobzhansky T (1964) Biology, molecular and organismic. *Am Zool* 4:443–452

Douce R, Neuburger M (1999) Biochemical dissection of photorespiration. *Curr Opin Plant Biol* 2:214–222

Douce R, Bourguignon J, Neuburger M, Rebeille F (2001) The glycine decarboxylase system: a fascinating complex. *Trends Plant Sci* 6:167–176

Eicks M, Maurino V, Knappe S, Flügge UI, Fischer K (2002) The plastidic pentose phosphate translocator represents a link between the cytosolic and the plastidic pentose phosphate pathways in plants. *Plant Physiol* 128:512–522

Fichera ME, Roos DS (1997) A plastid organelle as a drug target in apicomplexan parasites. *Nature* 390:407–409

Fischer K, Weber A (2002) Transport of carbon in non-green plastids. *Trends Plant Sci* 7:345–351

Fischer K, Kammerer N, Gutensohn M, Arbinger B, Weber A, Häusler RE, Flügge UI (1997) A new class of plastidic phosphate translocators: a putative link between primary and secondary metabolism by the phosphoenolpyruvate/phosphate antiporter. *Plant Cell* 9:453–462

Fleige T, Fischer K, Ferguson DJ, Gross U, Bohne W (2007) Carbohydrate metabolism in the *Toxoplasma gondii* apicoplast: localization of three glycolytic isoenzymes, the single pyruvate dehydrogenase complex, and a plastid phosphate translocator. *Eukaryot Cell* 6:984–996

Flügge UI (1995) Phosphate translocation in the regulation of photosynthesis. *J Exp Bot* 46:1317–1323

Flügge UI (1999) Phosphate translocators in plastids. *Annu Rev Plant Physiol Plant Mol Biol* 50:27–45

Flügge UI, Heldt HW (1984) The phosphate-triose phosphate-phosphoglycerate translocator of the chloroplast. *Trends Biochem Sci* 9:530–533

Flügge UI, Stitt M, Heldt HW (1985) Light-driven uptake of pyruvate into mesophyll chloroplasts from maize. *FEBS Lett* 183:335–339

Flügge UI, Fischer K, Gross A, Sebald W, Lottspeich F, Eckerskorn C (1989) The triose phosphate-3-phosphoglycerate-phosphate translocator from spinach chloroplasts: nucleotide sequence of a full-length cDNA clone and import of the in vitro synthesized precursor protein into chloroplasts. *EMBO J* 8:39–46

Flügge UI, Häusler RE, Ludewig F, Fischer K (2003) Functional genomics of phosphate antiport systems of plastids. *Physiol Plantarum* 118:475–482

Geigenberger P, Stamme C, Tjaden J, Schulz A, Quick PW, Betsche T, Kersting HJ, Neuhaus HE (2001) Tuber physiology and properties of starch from tubers of transgenic potato plants with altered plastidic adenylate transporter activity. *Plant Physiol* 125:1667–1678

Glöckner G, Rosenthal A, Valentin K (2000) The structure and gene repertoire of an ancient red algal plastid genome. *J Mol Evol* 51:382–390

Gutensohn M, Fan E, Frielingsdorf S, Hanner P, Hou B, Hust B, Klosgen RB (2006) Toc, Tic, Tat et al.: structure and function of protein transport machineries in chloroplasts. *J Plant Physiol* 163:333–347

Haferkamp I (2007) The diverse members of the mitochondrial carrier family in plants. *FEBS Lett* 581:2375–2379

He CY, Shaw MK, Pletcher CH, Striepen B, Tilney LG, Roos DS (2001) A plastid segregation defect in the protozoan parasite *Toxoplasma gondii*. *EMBO J* 20:330–339

Heldt HW (1969) Adenine nucleotide translocation in spinach chloroplasts. *FEBS Lett* 5:11–14

Heldt HW, Flügge UI, Borchert S, Brueckner G, Ohnishi J (1990) Phosphate translocators in plastids. *Plant Biol* 10:39–54

Hirel B, Lea PJ (2001) Ammonia assimilation. In: Lea PJ, Morot-Gaudry JF (eds) Plant nitrogen. Springer, Berlin Heidelberg New York, pp 79–100

Huber SC, Edwards GE (1977) Transport in C_4 mesophyll chloroplasts: characterization of the pyruvate carrier *Biochim Biophys Acta* 462:583–602

Ireland RJ, Lea PJ (1999) The enzymes of glutamine, glutamate, asparagine, and aspartate metabolism. In: Singh BK (ed) Plant amino acids. Biochemistry and biotechnology. Dekker, New York, pp 49–109

Jomaa H, Wiesner J, Sanderbrand S, Altincicek B, Weidemeyer C, Hintz M, Turbachova I, Eberl M, Zeidler J, Lichtenthaler HK, Soldati D, Beck E (1999) Inhibitors of the nonmevalonate pathway of isoprenoid biosynthesis as antimalarial drugs. *Science* 285:1573–1576

Kammerer B, Fischer K, Hilpert B, Schubert S, Gutensohn M, Weber A, Flügge UI (1998) Molecular characterization of a carbon transporter in plastids from heterotrophic tissues: the glucose 6-phosphate/phosphate antiporter. *Plant Cell* 10:105–117

Kampfenkel KH, Möhlmann T, Batz O, van Montague M, Inzé D, Neuhaus HE (1995) Molecular characterization of an *Arabidopsis thaliana* cDNA encoding a novel putative adenylate translocator of higher plants. *FEBS Lett* 374:351–355

Karnataki A, Derocher A, Coppens I, Nash C, Feagin JE, Parsons M (2007) Cell cycle-regulated vesicular trafficking of Toxoplasma APT1, a protein localized to multiple apicoplast membranes. *Mol Microbiol* 63:1653–1668

Kirchberger S, Leroch M, Huynen MA, Wahl M, Neuhaus HE, Tjaden J (2007) Molecular and biochemical analysis of the plastidic ADP-glucose transporter (ZmBT1) from Zea mays. *J Biol Chem* 282:22481–22491

Knappe S, Flügge UI, Fischer K (2003) Analysis of the plastidic phosphate translocator gene family in Arabidopsis and identification of new phosphate translocator-homologous transporters, classified by their putative substrate-binding site. *Plant Physiol* 131:1178–1190

Kofler H, Häusler RE, Schulz B, Gröner F, Flügge UI, Weber A (2000) Molecular characterization of a new mutant allele of the plastid phosphoglucomutase in *Arabidopsis*. , and complementation of the mutant with the wild-type cDNA *Mol Gen Genet* 263:978–986

Köhler S, Delwiche CF, Denny PW, Tilney LG, Webster P, Wilson RJM, Palmer JD, Roos DS (1997) A plastid of probable green algal origin in apicomplexan parasites. *Science* 275:1485–1489

Leroch M, Kirchberger S, Haferkamp I, Wahl M, Neuhaus HE, Tjaden J (2005) Identification and characterization of a novel plastidic adenine nucleotide uniporter from *Solanum tuberosum*. *J Biol Chem* 280:17992–18000

Levine ND (1988) Progress in taxonomy of the Apicomplexan protozoa. *J Protozool* 35:518–520

Li S, Nosenko T, Hackett JD, Bhattacharya D (2006) Phylogenomic analysis identifies red algal genes of endosymbiotic origin in the chromalveolates. *Mol Biol Evol* 23:663–674

Linke C, Conrath U, Jeblick W, Betsche T, Mahn A, During K, Neuhaus HE (2002) Inhibition of the plastidic ATP/ADP transporter protein primes potato tubers for augmented elicitation of defense responses and enhances their resistance against *Erwinia carotovora*. *Plant Physiol* 129:1607–1615

Margulis L (1971) Symbiosis and evolution. *Sci Am* 225:48–57

Margulis L (1975) Symbiotic theory of the origin of eukaryotic organelles; criteria for proof. Symp Soc Exp Biol 21–38

Mazumdar J, Wilson EH, Masek K, Hunter CA, Striepen B (2006) Apicoplast fatty acid synthesis is essential for organelle biogenesis and parasite survival in *Toxoplasma gondii*. *Proc Natl Acad Sci USA* 103:13192–13197

McFadden GI, van Dooren GG (2004) Evolution: red algal genome affirms a common origin of all plastids. *Curr Biol* 14: R514–R516

McFadden GI, Reith ME, Munholland J, Lang-Unnasch N (1996) Plastid in human parasites. *Nature* 381:482

Mereschkowsky C (1905) Über Natur und Ursprung der Chromatophoren im Pflanzenreiche. *Biol Centralbl* 25:593–604

Miflin BJ, Lea PJ (1976) The pathway of nitrogen assimilation in plants. *Phytochemistry* 15:873–885

Miflin BJ, Lea PJ (1980) Ammonia assimilation. In: Miflin BJ (ed) The biochemistry of plants. Academic, London, pp 169–202

Mou Z, He Y, Dai Y, Liu X, Li J (2000) Deficiency in fatty acid synthase leads to premature cell death and dramatic alterations in plant morphology. *Plant Cell* 12:405–418

Mullin KA, Lim L, Ralph SA, Spurck TP, Handman E, McFadden GI (2006) Membrane transporters in the relict plastid of malaria parasites. *Proc Natl Acad Sci USA* 103:9572–9577

Neuhaus HE, Maass U (1996) Unidirectional transport of orthophosphate across the envelope of isolated cauliflower-bud amyloplasts. *Planta* 198:542–548

Neuhaus HE, Wagner R (2000) Solute pores, ion channels, and metabolite transporters in the outer and inner envelope membranes of higher plant plastids. *Biochim Biophys Acta* 1465:307–323

Niewiadomski P, Knappe S, Geimer S, Fischer K, Schulz B, Unte US, Rosso MG, Ache P, Flügge UI, Schneider A (2005) The Arabidopsis plastidic glucose 6-phosphate/phosphate translocator GPT1 is essential for pollen maturation and embryo sac development. *Plant Cell* 17:760–775

Oesterhelt C, Klocke S, Holtgrefe S, Linke V, Weber AP, Scheibe R (2007) Redox regulation of chloroplast enzymes in *Galdieria sulphuraria*. in view of eukaryotic evolution *Plant Cell Physiol* 48:1359–1373

Ohta N, Matsuzaki M, Misumi O, Miyagishima SY, Nozaki H, Tanaka K, Shin IT, Kohara Y, Kuroiwa T (2003) Complete sequence and analysis of the plastid genome of the unicellular red alga *Cyanidioschyzon merolae*. *DNA Res* 10:67–77

Patron NJ, Greber B, Fahy BF, Laurie DA, Parker ML, Denyer K (2004) The *lys5*. mutations of barley reveal the nature and importance of plastidial ADP-Glc transporters for starch synthesis in cereal endosperm *Plant Physiol* 135:2088–2097

Picault N, Hodges M, Palmieri L, Palmieri F (2004) The growing family of mitochondrial carriers in Arabidopsis. *Trends Plant Sci* 9:138–146

Pottosin II, Schönknecht G (1996) Ion channel permeable for divalent and monovalent cations in native spinach thylakoid membranes. *J Membr Biol* 152:223–233

Ralph SA, van Dooren GG, Waller RF, Crawford MJ, Fraunholz MJ, Foth BJ, Tonkin CJ, Roos DS, McFadden GI (2004) Tropical infectious diseases: metabolic maps and functions of the *Plasmodium falciparum*. apicoplast *Nat Rev Microbiol* 2:203–216

Reid JD, Hunter CN (2004) Magnesium-dependent ATPase activity and cooperativity of magnesium chelatase from *Synechocystis*. sp. PCC6803. *J Biol Chem* 279:26893–26899

Reinhold T, Alawady A, Grimm B, Beran KC, Jahns P, Conrath U, Bauer J, Reiser J, Melzer M, Jeblick W, Neuhaus HE (2007) Limited nocturnal ATP import into Arabidopsis chloroplasts causes photooxidative damage. *Plant J* 50:293–304

Reiser J, Linka N, Lemke L, Jeblick W, Neuhaus HE (2004) Molecular physiological analysis of the two plastidic ATP/ADP transporters from Arabidopsis. *Plant Physiol* 136: 3524–3536

Renné P, Dreßen U, Hebbeker U, Hille D, Flügge UI, Westhoff P, Weber APM (2003) The *Arabidopsis.* mutant *dct* is deficient in the plastidic glutamate/malate translocator DiT2 *Plant J* 35:316–331

Rexach J, Fernández E, Galván A (2000) The *Chlamydomonas reinhardtii.* Nar1 gene encodes a chloroplast membrane protein involved in nitrite transport *Plant Cell* 12:1441–1453

Reyes-Prieto A, Weber APM, Bhattacharya D (2007) The origin and establishment of the plastid in algae and plants. *Annu Rev Genet* 41:147–168

Riesmeier JW, Flügge UI, Schulz B, Heineke D, Heldt HW, Willmitzer L, Frommer WB (1993) Antisense repression of the chloroplast triose phosphate translocator affects carbon partitioning in transgenic potato plants. *Proc Natl Acad Sci USA* 90:6160–6164

Rinalducci S, Larsen MR, Mohammed S, Zolla L (2006) Novel protein phosphorylation site identification in spinach stroma membranes by titanium dioxide microcolumns and tandem mass spectrometry. *J Proteome Res* 5:973–982

Rolletschek H, Nguyen TH, Hausler RE, Rutten T, Gobel C, Feussner I, Radchuk R, Tewes A, Claus B, Klukas C, Linemann U, Weber H, Wobus U, Borisjuk L (2007) Antisense inhibition of the plastidial glucose-6-phosphate/phosphate translocator in *Vicia.* seeds shifts cellular differentiation and promotes protein storage *Plant J* 51:468–484

Roos DS, Crawford MJ, Donald RG, Fraunholz M, Harb OS, He CY, Kissinger JC, Shaw MK, Striepen B (2002) Mining the *Plasmodium* genome database to define organellar function: what does the apicoplast do? *Philos Trans R Soc Lond B Biol Sci* 357:35–46

Schimper AFW (1885) Untersuchungen über die Chlorophyllkörner und die ihnen homologen Gebilde. *Jahrb Wiss Bot* 16:1–247

Schneider A, Häusler RE, Kolukisaoglu U, Kunze R, van der Graaff E, Schwacke R, Catoni E, Desimone M, Flügge UI (2002) An *Arabidopsis thaliana.* knock-out mutant of the chloroplast triose phosphate/phosphate translocator is severely compromised only when starch synthesis, but not starch mobilisation is abolished *Plant J* 32:685–699

Schneidereit J, Häusler RE, Fiene G, Kaiser WM, Weber APM (2006) Antisense repression reveals a crucial role of the plastidic 2-oxoglutarate/malate translocator DiT1 at the interface between carbon and nitrogen metabolism. *Plant J* 45:206–224

Schünemann D, Borchert S, Flugge UI, Heldt HW (1993) ADP/ATP translocator from pea root plastids (comparison with translocators from spinach chloroplasts and pea leaf mitochondria). *Plant Physiol* 103:131–137

Shannon JC, Pien FM, Cao HP, Liu KC (1998) Brittle-1, an adenylate translocator, facilitates transfer of extraplastidial synthesized ADP-glucose into amyloplasts of maize endosperm. *Plant Physiol* 117:1235–1252

Soll J, Schleiff E (2004) Protein import into chloroplasts. *Nat Rev Mol Cell Biol* 5:198–208

Spetea C, Hundal T, Lundin B, Heddad M, Adamska I, Andersson B (2004) Multiple evidence for nucleotide metabolism in the chloroplast thylakoid lumen. *Proc Natl Acad Sci USA* 101:1409–1414

Sugiura M, Georgescu MN, Takahashi M (2007) A nitrite transporter associated with nitrite uptake by higher plant chloroplasts. *Plant Cell Physiol* 48:1022–1035

Taniguchi M, Taniguchi Y, Kawasaki M, Takeda S, Kato T, Sato S, Tabata S, Miyake H, Sugiyama T (2002) Identifying and characterizing plastidic 2-oxoglutarate/malate and dicarboxylate transporters in *Arabidopsis thaliana. Plant Cell Physiol* 43:706–717

Taniguchi Y, Nagasaki J, Kawasaki M, Miyake H, Sugiyama T, Taniguchi M (2004) Differentiation of dicarboxylate transporters in mesophyll and bundle sheath chloroplasts of maize. *Plant Cell Physiol* 45:187–200

Tegeder M, Weber APM (2006) Metabolite transporters in the control of plant primary metabolism. In: Plaxton WC, McManus MT (eds) Control of primary metabolism in plants. Blackwell, Oxford, UK, pp 85–120

Terashita M, Maruyama S, Tanaka K (2006) Cytoplasmic localization of the single glutamine synthetase in a unicellular red alga, *Cyanidioschyzon merolae* 10D. *Biosci Biotechnol Biochem* 70:2313–2315

Thuswaldner S, Lagerstedt JO, Rojas-Stutz M, Bouhidel K, Der C, Leborgne-Castel N, Mishra A, Marty F, Schoefs B, Adamska I, Persson BL, Spetea C (2007) Identification, expression, and functional analyses of a thylakoid ATP/ADP carrier from Arabidopsis. *J Biol Chem* 282:8848–8859

Tjaden J, Möhlmann T, Kampfenkel K, Henrich G, Neuhaus HE (1998) Altered plastidic ATP/ADP-transporter activity influences potato (*Solanum tuberosum.* L.) tuber morphology, yield and composition of tuber starch *Plant J* 16:531–540

Tyra H, Linka M, Weber APM, Bhattacharya D (2007) Host origin of plastid solute transporters in the first photosynthetic eukaryotes. *Genome Biol* 8:R212

Versaw WK, Harrison MJ (2002) A chloroplast phosphate transporter, PHT2;1, influences allocation of phosphate within the plant and phosphate-starvation responses. *Plant Cell* 14:1751–1766

Voll L, Häusler RE, Hecker R, Weber A, Weissenbock G, Fiene G, Waffenschmidt S, Flügge UI (2003) The phenotype of the Arabidopsis *cue1.* mutant is not simply caused by a general restriction of the shikimate pathway *Plant J* 36:301–317

Wagner V, Gessner G, Heiland I, Kaminski M, Hawat S, Scheffler K, Mittag M (2006) Analysis of the phosphoproteome of *Chlamydomonas reinhardtii.* provides new insights into various cellular pathways *Eukaryot Cell* 5:457–468

Wallsgrove RM, Keys AJ, Bird IF, Cornelius MJ, Lea PJ, Miflin BJ (1980) The location of glutamine-synthetase in leaf-cells and its role in the reassimilation of ammonia released in photo-respiration. *J Exp Bot* 31:1005–1017

Weber APM (2004) Solute transporters as connecting elements between cytosol and plastid stroma. *Curr Opin Plant Biol* 7:247–253

Weber APM (2006) Synthesis, export, and partitioning of the end products of photosynthesis. In: Wise RR, Hoober JK (eds) The structure and function of plastids. Springer, Berlin Heidelberg New York, pp 273–292

Weber APM, Fischer K (2007) Making the connections – the crucial role of metabolite transporters at the interface between chloroplast and cytosol. *FEBS Lett* 581:2215–2222

Weber A, Flügge UI (2002) Interaction of cytosolic and plastidic nitrogen metabolism in plants. *J Exp Bot* 53:865–874

Weber A, Menzlaff E, Arbinger B, Gutensohn M, Eckerskorn C, Flügge UI (1995) The 2-oxoglutarate/malate translocator of chloroplast envelope membranes: molecular cloning of a transporter containing a 12-helix motif and expression of the functional protein in yeast cells. *Biochemistry* 34:2621–2627

Weber APM, Schneidereit J, Voll LM (2004) Using mutants to probe the in vivo function of plastid envelope membrane metabolite transporters. *J Exp Bot* 55:1231–1244

Weber APM, Schwacke R, Flügge UI (2005) Solute transporters of the plastid envelope membrane. *Annu Rev Plant Biol* 56:133–164

Weber APM, Linka M, Bhattacharya D (2006) Single, ancient origin of a plastid metabolite translocator family in Plantae from an endomembrane-derived ancestor. *Eukaryot Cell* 5:609–612

Wentz JB (1926) Heritable characters of maize. XXVI. Concave. *J Hered* 17:327–329

Wilson RJ, Denny PW, Preiser PR, Rangachari K, Roberts K, Roy A, Whyte A, Strath M, Moore DJ, Moore PW, Williamson DH (1996) Complete gene map of the plastid-like DNA of the malaria parasite *Plasmodium falciparum. J Mol Biol* 261:155–172

Wong SY, Remington JS (1993) Biology of *Toxoplasma gondii. Aids* 7:299–316

Woo KC, Flügge UI, Heldt HW (1987) A two-translocator model for the transport of 2-oxoglutarate and glutamate in chloroplasts during ammonia assimilation in the light. *Plant Physiol* 84:624–632

Yoon HS, Hackett JD, Pinto G, Bhattacharya D (2002) The single, ancient origin of chromist plastids. *Proc Natl Acad Sci USA* 99:15507–15512

Yoon HS, Hackett JD, Van Dolah FM, Nosenko T, Lidie KL, Bhattacharya D (2005) Tertiary endosymbiosis driven genome evolution in dinoflagellate algae. *Mol Biol Evol* 22:1299–1308

Retrograde Signalling

L. Dietzel, S. Steiner, Y. Schröter, and T. Pfannschmidt(✉)

Abstract Plastids are organelles typical for plant cells. They are a metabolic and genetic compartment that is involved in most aspects of the life of a plant. Plastids were acquired by plants via endosymbiosis of a photosynthetically active prokaryotic ancestor. Establishment of this endosymbiosis required communication between the endosymbiont and the nucleus of the host cell. During evolution a complex network evolved that embedded development and function of the new organelle into that of the cell. Today the nucleus controls most functions of plastids by providing the essential proteins. However, there exists a backward flow of information from the plastid to the nucleus. This "retrograde" signalling represents a feedback control reporting the functional state of the organelle to the nucleus. By this means extensive communication between the two compartments is established. This helps the plant to perceive and respond properly to varying environmental influences and to developmental signals at the cellular level. Recent observations have extended our understanding of retrograde signalling. Models are presented that provide an overview of the different known pathways.

1 Introduction

Plastids are organelles that are specific for plant and algal cells. They represent a distinct and indispensable biochemical and genetic compartment which is involved in many essential metabolic processes (Buchanan et al. 2002). Plastids originated from an endosymbiotic event in which a photosynthetic active cyanobacterium was engulfed by a heterotrophic eukaryotic cell. During the establishment of this endosymbiosis the cyanobacterium was step-by-step integrated into the cellular processes of the host cell. This was mainly achieved by the transfer

---

T. Pfannschmidt
Department for Plant Physiology, Friedrich-Schiller-University Jena,
Dornburger Str. 159, 07743 Jena, Germany
e-mail: Thomas.Pfannschmidt@uni-jena.de

of most but not all of the cyanobacterial genes to the nucleus of the host cell. This gene transfer gave the host cell control over development and function of this novel compartment since the organelle became functionally dependent on the coordinated expression and import of its nuclear encoded protein components. As a result of this evolutionary invention an autotrophic, eukaryotic cell evolved that was able to perform photosynthesis. From this chlorophyta and, finally, plants evolved 450–500 million years ago (Stoebe and Maier 2002). Today higher plants possess plastids that are typically surrounded by two membranes. The outer one originated from the engulfing host cell, the inner one from the cyanobacterial ancestor. Its further morphology and function exhibits a high plasticity and depend mainly on the tissue context of the respective cell. For instance in photosynthetic tissues cells contain green chloroplasts while in fruits or flowers cells contain coloured chromoplasts (Buchanan et al. 2002). Nevertheless, all plastid types contain an identical genome (the so-called plastome) with a size of around 120–200 kb. It encodes a relatively conserved set of 100–130 genes, which code mainly for components of the photosynthetic apparatus and the plastid-own gene expression machinery which controls the expression of the genetic information on the plastome (Sugiura 1992). However, the largest part of the plastid protein complement is encoded in the nucleus. Current estimates of plastid protein number range from 2,500–4,500 different proteins in these organelles depending on the plastid type (Abdallah et al. 2000; van Wijk 2000; Kleffmann et al. 2004).

All prominent multi-subunit protein complexes of plastids are comprised of a patchwork of plastid and nuclear encoded subunits. Therefore, establishment, assembly and maintenance of these complexes require the coordinated expression of genes in the two different genetic compartments. This coordination is established via extensive flow of information from the nucleus to the plastid (anterograde signalling), i.e. the import of nuclear-encoded plastid proteins. However, there exists a feedback control that signals information about developmental and functional state of the plastid toward the nucleus (retrograde signalling), which induces appropriate changes in the expression of the nuclear-encoded plastid proteins. By this means plastids can control their own protein complement and adapt it to the present developmental and functional state. This complex network of anterograde and retrograde signalling is a major component of plant cell signal networks and contributes to a large extent to plant development and environmental acclimation (Bräutigam et al. 2007). Retrograde signalling has attracted much interest in the last decade and a lot of excellent reviews have been written about it (Rodermel 2001; Jarvis 2001; Papenbrock and Grimm 2001; Gray et al. 2003; Strand 2004; Beck 2005; Nott et al. 2006; Pesaresi et al. 2007). Research data from the last years clearly demonstrated that several different plastid signals exist which are active under different developmental or functional conditions. These signals depend on (1) plastid gene expression (Fig. 1), (2) pigment biosynthesis (tetrapyrrole and carotenoid synthesis) (Figs. 2 and 3) and (3) plastid redox state (photosynthetic electron transport and reactive oxygen species (ROS) accumulation) (Fig. 4). Many of these signals are closely related or functionally linked. Furthermore, plant cells contain a third genetic compartment, the mitochondria, which are also of endosymbiotic origin and contain (in plants) a genome with around 40–50 genes (Burger et al. 2003). Like plastids they import the majority of proteins from the cytosol and assemble them into multi-subunit complexes together with the organellar encoded proteins. Mitochondria are energetically coupled to chloroplasts, however, their signalling events to the nucleus and the potential interaction with plastids are rarely investigated. A number of recent reports demonstrated that this organelle contributes significantly to or interacts with the retrograde plastid signals and thus represent an important player in this signalling network (Pesaresi et al. 2007; Rhoads and Subbaiah 2007).

2 Plastid Signals Depending on Plastid Gene Expression

The first proposal of a retrograde influence by plastid protein synthesis on nuclear gene expression was based on studies with the *albostrians* mutants of barley (Bradbeer et al. 1979). The mutant does not form intact ribosomes in plastids of the basal leaf meristem. This blocks synthesis of plastid polypeptides (Fig. 1) and generates white striped or even completely white leaf tissue. Beside this cytoplasmic protein synthesis of nuclear-encoded plastid proteins, for example the small subunit of RubisCO was decreased despite the existence of intact ribosomes in the cytoplasm (Bradbeer et al. 1979). Furthermore, activities of phosphoribulokinase and NADPH-glyceraldehyde-3-phosphate dehydrogenase were found to be down-regulated in white tissues. These data suggested a retrograde control by plastids affecting nuclear gene expression (Bradbeer et al. 1979). Further studies demonstrated that the expression of various nuclear-encoded proteins involved in photosynthesis, photorespiration or nitrogen assimilation was decreased in un-pigmented *albostrians* mutants (Hess et al. 1991, 1994), whereas several genes encoding proteins for chlorophyll biosynthesis and plastid DNA replication exhibited equal or even an enhanced expression level in white vs. green tissues (Hess et al. 1992, 1993).

Similar mutants exist also in the dicot model organism *Arabidopsis thaliana*. The mutant *sco1* (snowy cotyledons) is mutated in the gene for the plastid elongation factor G (EF-G) (Albrecht et al. 2006), which diminishes chloroplast translation resulting in un-pigmented cotyledons and reduced transcript amounts of nuclear-encoded photosynthesis gene. This phenotypic effect, however, is restricted to the first days of seedling development when strong protein synthesis is required. The *apg3* (albino or pale green mutant 3) mutant of *Arabidopsis* is deficient in the chloroplast ribosome release factor 1. The mutant exhibits an albino phenotype, but can be maintained on agar plates with sucrose. Although 21-day-old plants do not posses detectable amounts of D1, RbcL (large subunit of RubisCO) and RbcS proteins, nuclear transcripts of photosynthesis genes were only slightly decreased (Motohashi et al. 2007). This indicates that (1) impairment of plastid translation *per se* does not cause a repression of nuclear gene expression and (2) that the repression might be restricted to early stages of plant development as seen in the *sco1* mutant.

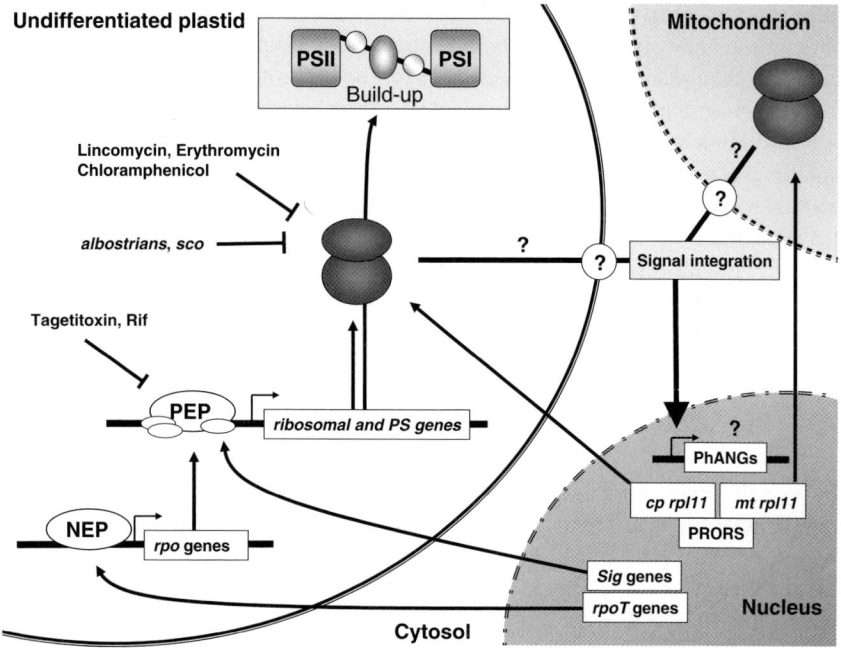

Fig. 1 *Plastid signals depending on organellar gene expression.* The plant cell compartments plastid (in an early undifferentiated form), mitochondrion, cytosol and nucleus are depicted schematically. Organellar ribosomes are given as *dark double ovals*, important enzymes are indicated by *white ovals*. The photosynthetic apparatus which has to be built up during chloroplast development is given schematically in a *separate box*. Genes are shown as *white boxes* labelled with the respective names (for identities compare text). Transcription start sites are indicated by a *small arrow* in front of the genes. Location of the encoded component is indicated by *thin black arrows*. Repressive effects of inhibitors or mutations are given as a *black line with a hammerhead*. Influences of organellar processes on nuclear gene expression are indicated by *thick black arrows*. Transduction of these signals within plastid and mitochondrion, over the respective membrane and through the cytosol (including integration given as a *black dot*) are not known and marked by *question marks*

The role of plastid translation was also extensively studied by application of antibiotics which selectively affect the prokaryotic-type 70S ribosomes (Fig. 1). Chloramphenicol treatment reduced the expression of the nuclear genes *RbcS* and *Lhcb* (encoding the light harvesting proteins of PSII) (Oelmüller and Mohr 1986). Streptomycin treatment reduced *RbcS* transcription in rice (Yoshida et al. 1998), and application of erythromycin and lincomycin resulted in decreased transcript accumulation of *Lhcb* and *RbcS* (Gray et al. 1995; Sullivan and Gray 1999). By this means, the rate of *Lhcb* and *RbcS* transcript accumulation became a kind of molecular marker for the action of the "plastid signal" or "plastid factor" (Oelmüller 1989). During these analyses it became apparent that the antibiotics operated only properly when applied within the first 36–72 h after germination (Oelmüller and Mohr 1986; Gray et al. 1995). This supports the conclusion that the signal originating from plastid gene expression machinery might be restricted to this short time span.

Beside inhibition of translation also inhibition of transcription was found to affect nuclear gene expression (Fig. 1). Treatments with tagetitoxin, nalidixic acid or rifampicin repressed accumulation of *RbcS* and *LhcB* mRNAs, whereas other nuclear genes remained unaffected. This effect was limited to the initial stage of establishment of the plastid transcription machinery since no effect of the drug could be observed in older leaves (Lukens et al. 1987; Rapp and Mullet 1991; Gray et al. 1995; Pfannschmidt and Link 1997). Essential parts of plastid ribosomes (16 and 23 S rRNAs, various ribosomal subunits) and a complete set of tRNAs are encoded on the plastome. Proper assembly and function of plastid ribosomes, therefore, requires a functional transcription. This explains why inhibition of plastid transcription results in the same effects as a treatment with translational inhibitors.

Interestingly, also mitochondrial translation plays a role in plastid-to-nucleus-communication (Fig. 1). Disruption of nuclear genes for ribosomal subunit L11 of both plastids and mitochondria in *Arabidopsis*, resulted in down-regulation of photosynthesis-associated nuclear genes (PhANGs). However, this was observed only when both genes were affected whereas the single mutants revealed no effect (Pesaresi et al. 2006). A similar result could be seen by down-regulation of the essential and dual-targeted prolyl-tRNA-synthetase 1 (PRORS1) (Pesaresi et al. 2006). This enzyme is required for translation in both organelles. While a knock-out of the PRORS1 gene is lethal because of arresting embryo development, leaky mutants exhibiting only down-regulation of the gene are viable. Interestingly, these mutants exhibited a decrease in PhANG expression like that observed in the *rpl11*-mutants. It could be shown that this regulation is light- and photosynthesis-independent and also not caused by oxidative stress. These data suggest a cooperative, synergistic role of translation in both organelles in the modulation of PhANG expression and add an important new facet to this research field.

3 Signals Depending on Pigment Biosynthesis Pathways

Tetrapyrrols and carotenoids are the major pigments of plants that are involved in absorption and quenching of light energy. The expression of the pigment binding proteins such as the Chl a/b binding proteins of the light harvesting complex II (Lhcb) are coupled to the biosynthesis of these pigments. The tetrapyrrole pathway (Fig. 2) plays a crucial role in primary metabolism of a plant and has to be strictly controlled for several reasons. (1) Tetrapyrroles are not only compounds for chlorophyll synthesis but also for heme, phytochrome, and enzymatic cofactors. Therefore, the flux of tetrapyrroles within the different biosynthetic branches has to be controlled. (2) All chlorophylls and its precursors are phototoxic. Once they are produced they have to be rapidly integrated into proteins. (3) All enzymes involved in pigment biosynthesis and light harvesting are plastid-localised while their corresponding genes are encoded in the nucleus. Coordination of pigment biosynthesis and nuclear gene expression, therefore, requires a bi-directional communication between plastids and nucleus (Rüdiger and Grimm 2006).

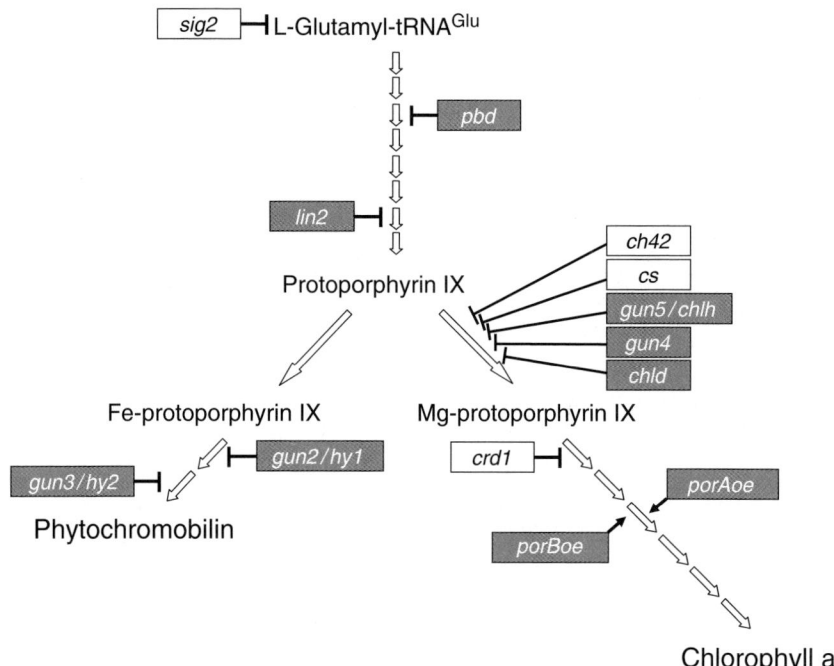

Fig. 2 *Tetrapyrrole biosynthesis pathway*. Only major components of the pathway are given by name. *White arrows* indicate steps of synthesis. Inhibition of single steps by mutational defects is indicated with *black lines with a hammerhead*. The respective mutated genes are given in *boxes* (for identities compare text). *Dark grey boxes* indicate mutants [or overexpressors (oe), *porAoe*, *porBoe* (McCormac and Terry 2004)] which exhibit a *gun* phenotype, *white boxes* mark mutants that do not display such a phenotype. *Lin2* (lesion initiation) encodes coproporphyrinogen oxidase (Ishikawa et al. 2001), *chld* (subunit D of Mg-chelatase) (Strand et al. 2003). The *sig2* mutant has been not tested for a *gun* phenotype

3.1 Tetrapyrrole Biosynthesis

The light-induced expression of Lhc proteins was found to coincide with the greening and maturation process of chloroplasts implying the action of a plastid signal. One potential signal could be attributed to the chlorophyll precursor Mg-Proto-Porphyrin-IX (Mg-Proto-IX). Feeding experiments with the iron chelator dipyridyl led to decreased *Lhcb* mRNA levels in *Chlamydomonas reinhardtii* (Johanningmeier and Howell 1984). The chelation of iron leads to an interruption of the heme feedback inhibition in the tetrapyrrole pathway which in turn causes accumulation of Mg-Proto-IX. This effect could be also observed in higher plants (Kittsteiner et al. 1991). Direct feeding of Mg-Proto-IX to *Chlamydomonas* cell cultures led to induction of nuclear heat-shock genes *HSP70a/b/c* (Kropat et al. 1997) supporting the notion that this chlorophyll precursor could mediate a retrograde signal which affects nuclear gene expression.

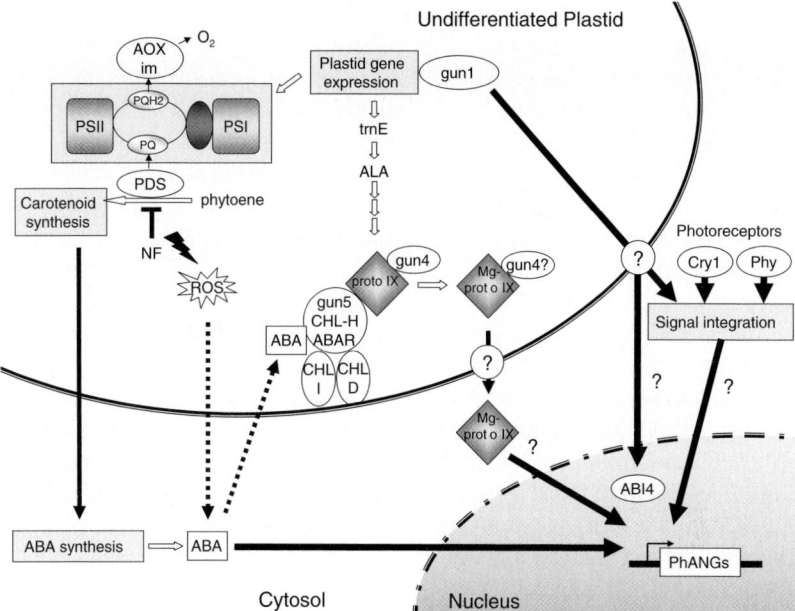

Fig. 3 *Plastid signals depending on tetrapyrrole and carotenoid biosynthesis.* Plastid, cytosol and nucleus are depicted schematically. The photosynthetic apparatus is given schematically in a *separate box*. Genes are shown as *white boxes*, transcription start sites are indicated by a *small arrow* in front of the genes. Protein components are given as *white ovals* (for identity compare text). Tetrapyrrols are represented by *grey squares*. Repressive effects are given as a *black line with a hammerhead*. Influences of organellar processes on nuclear gene expression are indicated by *thick black arrows*, putative diffusion by *dotted arrows*. Transduction of these signals are not known and marked by *question marks*

Another line of evidence for involvement of chlorophyll precursors in retrograde signalling came from studies on carotenoid-deficient plants. Maize seedlings with defects in carotenoid synthesis exhibited a decreased accumulation of *Lhcb* mRNA while other nuclear-encoded transcripts for cytosolic enzymes were not impaired (Mayfield and Taylor 1984). Alternatively, disruption of carotenoid biosynthesis by blocking the phytoene desaturase (PDS) (catalysing an early enzymatic step in this pathway) with the herbicide norflurazon (NF) (Chamovitz et al. 1991) led to comparable effects as the genetic defects. The resulting carotenoid deficiency of plastids led to reduced photosynthetic efficiency followed by photo-oxidative damage of thylakoid membranes due to the loss of non-photochemical de-excitation mechanisms. This photo-oxidative stress within the plastid prevented conversion of proplastids into mature chloroplast and resulted in a decreased expression of nuclear *Lhcb* and *RbcS* genes (Oelmüller and Mohr 1986). Thus, it was concluded that intact plastids are required for expression of nuclear photosynthesis genes and that a "plastid factor" is required for a correct build-up of the photosynthetic machinery (Oelmüller 1989; Taylor 1989).

Using this NF-mediated repression as a tool a genetic approach was performed to get deeper insights into the nature of the plastid signal (Susek et al. 1993). First, a transgenic *Arabidopsis* line carrying a fusion construct consisting of the *Lhcb1.2* (*CAB3*)-promoter (known to be down-regulated by NF treatment, see above) and reporter genes conferring hygromycin resistance and β-glucuronidase activity was created. Then the seed pool of this reporter line was mutagenised with ethyl methane sulphonate (EMS) and the resulting EMS mutant population was grown on plates with an NF-containing medium. By this means the seedling population was screened for individuals exhibiting a genetic defect which interrupted the down-regulation of the *Lhcb* expression under NF and, consequently, conferred hygromycin resistance. These mutants are regarded as defective in plastid signalling and, therefore, were named *genomes uncoupled* (*gun*) mutants. In the second screening step expression of the β-glucuronidase gene was tested and relative activity of the *Lhcb1.2* promoter was estimated. All *gun* mutants exhibited *Lhcb* expression whereas in wild-type plants *Lhcb* transcription was almost abolished. In total six different *gun* mutant lines (*gun1*–*gun6*) were found in this screen (Susek et al. 1993). Since *gun1* is different from *gun2*–*gun5* it is discussed in a separate section (Sect. 4).

The phenotype of the *gun* mutants varies from pale yellowish to undistinguishable from wild-type. The mutants *gun2-gun5* were mapped to the tetrapyrrole synthesis pathway (Fig. 2) (Surpin et al. 2002) and demonstrated reduced accumulation of Mg-Proto-IX under NF treatment which causes down-regulation of *Lhcb* expression in wild-type (Strand et al. 2003). The *gun2* and *gun3* mutant alleles were identified to encode the haem oxygenase and phytochromobiline synthase, respectively. The genetic lesions cause an overproduction of haem which activates a feedback loop that inhibits the *trnE*-reductase (HEMA), the first step of tetrapyrrole biosynthesis. This prevents accumulation of Mg-Proto-IX. Both mutants are allelic with *hy1* and *hy2* (hypocotyl) mutants found in a screen for photomorphogenesis mutant which is consistent with the function of tetrapyrroles as chromophores of phytochromes (Mochizuki et al. 2001). The mutants *gun4* and *gun5* were found to be directly involved in chelation of magnesium into protoporphyrin IX, the step which generates Mg-Proto-IX. *gun4* was found to encode an activator of the Mg-chelatase and *gun5* was affected in CHL-H, a subunit of Mg-chelatase (Fig. 3). GUN4 is a small soluble protein 22-kDa in size that can bind either the substrate proto-IX or the product of the chelation reaction, Mg-Proto-IX. The binding constant of GUN4 and Mg-Proto-IX was found to be lower than that of GUN4 and Proto-IX, however, only the latter couple is able to activate Mg-chelatase. By this means GUN4 could avoid accumulation of phototoxic Mg-Proto-IX and could control the chlorophyll biosynthesis pathway (Mochizuki et al. 2001; Larkin et al. 2003; Strand 2004).

In cyanobacteria GUN4 was shown to modulate enzyme activities of the Mg-chelatase and ferrochelatase that produces haem (Wilde et al. 2004). Thus, GUN4 may function as a global controller of the haem and chlorophyll branches. Since haem or its precursor Proto-IX is exported to mitochondria a control step at this point tetrapyrrole synthesis appears to be ideal for regulation and signalling.

Recently, the presence of Mg-Proto-IX in the cytosol could be visualised by confocal laser scanning technology. The actual low amount of Mg-Proto-IX in the plant cell was increased by circumventing the HEMA feedback inhibition by direct ALA feeding of NF treated seedlings (Ankele et al. 2007). This favours the model that Mg-Proto-IX is directly transported into the cytosol (Strand 2004) over the model which involves a Mg-Proto-IX sensing protein (like GUN4) and a subsequent cytosolic signal transduction cascade (Larkin et al. 2003). How this can be reconciled with the high photo-toxicity of Mg-Proto-IX still has to be resolved.

The *gun5* mutant was found to possess a mutated allele of CHL-H which provides Mg-Proto-IX for the [CHL-I:CHL-D]$_x$ complex in which the Mg insertion into Proto-IX occurs (Willows and Hansson 2003). Mutations in any of the Mg-chelatase subunits resulted in decreased Chl level. Interestingly, mutations in the CHL-I do not result in a *gun* phenotype although these mutants produce even less amounts of Mg-Proto-IX than *gun5* (Mochizuki et al. 2001). This is consistent with the phenotype of a number of other mutants with defects in tetrapyrrole biosynthesis [*ch42* (chlorata), *cs*, *crd1* (copper response defect) (now called chl27)], which all exhibit no *gun* phenotype (Koncz et al. 1990; Tottey et al. 2003). *ch42* and *cs* accumulate less Mg-Proto-IX than wild-type, however, it has not been investigated if this occurs also upon NF treatment. In contrast, the *crd* mutant accumulates more Mg-Proto-IX compared to wild-type. Especially the observations with the *ch42* and *cs* mutants suggest that Mg-Proto-IX levels do not exclusively account for the tetrapyrrole-mediated signal. This is supported by a recent study on *Chlamydomonas* mutants with defects in the Mg-chelatase. These mutants exhibit reduced levels of Mg-tetrapyrroles but increased levels of soluble haem. It was shown that haem can mimic the activating role of Mg-Proto-IX on the induction of HSP70A promoter and other Mg-Proto-IX inducible genes. It was concluded that both tetrapyrroles can act as retrograde signals and that the respective signalling pathways converge at the same *cis*-elements (von Gromoff et al. 2008).

Further support for the idea that a developmental signal descends from the CHL-H (GUN5) subunit came from expression analyses of the nuclear transcripts of *AtSig1-6* genes. They encode sigma factors that are crucial for promoter recognition of the plastid encoded RNA-polymerase (compare Rolland et al. 2008). These factors were found to be repressed after NF-treatment in wild-type but not in the *gun5* background. A similar de-repression in the *gun5* mutant was also found for plastid transcripts depending on PEP activity like *psbA*, *psaA*, *psaC* whereas genes transcribed by the nucleus-encoded RNA-polymerase were not affected. This suggests that GUN5 might act via regulation of nuclear-encoded components of the plastid gene expression machinery in early plastid development (Ankele et al. 2007).

A recent publication reported that CHL-H might be a plastid localised ABA receptor (Fig. 3) (Shen et al. 2006). The authors found that the *Arabidopsis cch* (constitutive chlorina) mutant was deficient in ABA-related responses. The genetic lesion in this mutant was found to be a stronger allele of *gun5*. Therefore, the *cch* mutant displays a "*gun* phenotype". It could be further shown that direct ABA feeding to wild-type plants led to an increase in Mg-Proto-IX levels but to decreased Chl levels. This suggests that a component downstream of the Mg-chelatase plays an additional role in the tetrapyrrole and ABA crosstalk. Whether or not ABA deficiency caused by NF treatment is related to the putative ABA receptor function of the Chl-H subunit has to be studied in the future.

3.2 Integration of Plastid and Cytosolic Signals on Promoter Level

While plastid signal transduction mechanisms still remain elusive, some responsive promoter elements have already been identified. The first studies concluded that light and plastid signals act on the same *cis* elements (Bolle et al. 1996; Kusnetsov et al. 1996) and that these are a complex composition of known transcription factor binding sites (Terzaghi and Cashmore 1995; Puente et al. 1996). Subsequently, it was shown that combination of I- and G-box in a minimal *RbcS* promoter is sufficient to respond to NF triggered plastid signal, sugar, ABA and light (Acevedo-Hernandez et al. 2005). Furthermore, a G-Box and a related sequence motif called CUF (cab upstream factor) element in the *Lhcb1* promoter were shown to be essential for NF-triggered plastid signals (Strand et al. 2003). In addition the *Lhcb1* promoter carries a putative S- (sugar responsive) box which is responsive to ABA. Interestingly, ABI4 (ABA insensitive 4), an AP2-transcription factor, can bind to the respective S-boxes within *Lhcb1* and *RbcS* promoters (Acevedo-Hernandez et al. 2005; Koussevitzky et al. 2007) and the respective ABI4-deficient mutant exhibits a weak *gun* phenotype. This suggests ABA signals and other plastid signals interact at promoter level. This idea is further supported by the recent finding that an ABA (and high light) responsive promoter element within the *Lhcb1* promoter represses the *Lhcb* expression. This promoter element was neither influenced by phytochrome activation nor NF application (Staneloni et al. 2008). A further study in *Chlamydomonas* revealed a distinct *cis*-acting sequence responsive to Mg-Proto-IX and light. The authors concluded that light responsiveness of PhANGs in *Chlamydomonas* is mediated by Mg-Proto-IX (von Gromoff et al. 2006). However, again the light and plastid responsive *cis*-acting elements could not be separated. Further evidence for such an interaction of plastid and cytosolic light-signalling networks came from recent data that the cytosolic blue light photoreceptor *cry1* gene represents a weak *gun* allele (Ruckle et al. 2007) (Sect. 5).

3.3 Carotenoid and ABA Biosynthesis

Carotenoid and chlorophyll biosynthesis as well as *Lhc* gene expression are closely related (Anderson et al. 1995). Phytoene desaturation catalysed by PDS and the subsequent zeta-carotene desaturation are key steps in coordination of photoprotection, chloroplast development and nuclear gene expression. The PDS oxidises phytoene and requires plastoquinone (PQ) as an electron acceptor (Fig. 3). The excess electrons are transferred to oxygen via a plastid terminal oxidase (PTOX). Mutants that lack PTOX (called *immutans*) accumulate phytoene due to the high reduction state of PQ. These mutants show a variegated phenotype indicating that an early step in plastid development is blocked. Interestingly, a complete inhibition of plastid development occurs when PDS is blocked by NF. A similar effect could be found when the subsequent enzyme in the pathway, zeta-carotene desaturase (ZDS), is mutated. A full knock-out of the *SPC1* gene encoding ZDS arrested chloroplast development in the mutant whereas a weaker allele caused only reduction of chlorophyll synthesis due to a down-regulation of genes for components involved in Chl biosynthesis like *PorB* and *CAO* (Dong et al. 2007). Furthermore, the mutation led to carotenoid or ABA deficiency. The ABA-insensitive phenotype of the mutant could be partially restored by exogenously applied ABA.

Although enzymes of the carotenoid and ABA pathway are initially expressed in a light-dependent manner certain enzymes of the xanthophyll cycle pathway are also regulated by the redox state of the photosynthetic electron-transport chain (PET). The application of PET inhibitors DCMU and DBMIB revealed a correlation between the redox state of the PQ pool and the expression of zeaxanthin epoxidase and beta-carotene hydroxylase. Furthermore, a down-regulation of violaxanthin de-epoxidase could be shown after blocking PET (Woitsch and Römer 2003).

Additionally, ABA synthesis was shown to be influenced by lumenal ascorbate availability. Ascorbate is limiting under high light stress conditions and in the *Arabidopsis vtc1* mutant (Pastori et al. 2003; Baier and Dietz 2005). Furthermore, cytosolic events may regulate ABA levels since the last steps of synthesis are cytosolic (Seo and Koshiba 2002). Taken together, ABA synthesis is closely connected to PET, plastid redox state and pigment synthesis. Therefore, ABA levels might be a good indicator for the plastid status during development and under stress. Since ABA is mobile and activates transcription factors it is a reasonable candidate for a retrograde signal.

4 Crosstalk of Signals from Gene Expression and Chlorophyll Biosynthesis

Among the isolated *gun* mutants (Sect. 3) *gun1* is unique since it exhibits de-repression of the *Lhcb* gene not only after NF treatment but also after inhibition of plastid translation with lincomycin or chloramphenicol (Susek et al. 1993). Thus, GUN1 was discussed as a factor potentially involved in both, plastid gene expression and chlorophyll biosynthesis (Fig. 3) (Nott et al. 2006). This was supported by findings that double mutants of *gun1* with either *gun4* or *gun5* exhibit a stronger *gun* phenotype. In addition, earlier microarray data exhibited different expression profiles with only a small overlap in de-regulated genes in *gun1* mutants when compared to *gun2* or *gun5* mutants pointing to two separate but partly redundant signalling pathways (Strand et al. 2003). Recently, a new *gun1* allele was isolated and the GUN1 gene was cloned (Cottage et al. 2007; Koussevitzky et al. 2007). The gene encodes a plastid localised pentatricopeptide repeat (PPR) protein containing a putative DNA binding small mutS related domain. PPR proteins are thought to be involved in interactions with RNA in processing, stability and translation but also with DNA (Saha et al. 2007). For the GUN1 protein, so far, only unspecific DNA binding activity could be demonstrated (Koussevitzky et al. 2007). Interestingly,

GUN1 co-localises with PTAC2, another PPR protein, which is part of the transcriptionally active chromosome of plastids (Pfalz et al. 2006). Until now the precise function of GUN1 is elusive. It was discussed that NF treatment could indirectly affect plastid gene expression (Gray et al. 2003), but treatment with dipyridyl showed that *gun1* mutants retain their *gun* phenotype also in the presence of Mg-Proto-IX suggesting a role of GUN1 downstream of Mg-Proto-IX accumulation and chlorophyll biosynthesis (Koussevitzky et al. 2007).

A novel aspect in retrograde signalling could be inferred from a completely different experimental line. The chlorophyllide a oxygenase (CAO), responsible for the conversion of chlorophyllide a (Chlide) to Chlide b, was found to be involved in the import of *Lhcb* precursors into the plastids (Reinbothe et al. 2006). This offers a potential explanation for transduction of a plastid gene expression derived signal to the nucleus without the need of an export of any signal molecule. Chlorophyll synthesis starts with glutamyl-tRNA which is transformed to δ-aminolevulinic acid (ALA) (Rüdiger and Grimm 2006). Transcription of the plastid encoded gene for glutamyl-tRNA, *trnE*, is exclusively performed by the PEP enzyme in combination with sigma factor 2 (Sig2) (Hanaoka et al. 2003). Thus, any perturbance of plastid transcription or translation by inhibitors or mutations will affect chlorophyll biosynthesis and subsequently signals originating from this pathway. Lack of CAO as in the *chlorina* mutant of *Arabidopsis* prevents the accumulation of Chl b and of the LhcB proteins. A recent study with plastids from the chlorine mutant demonstrated that the major portion of CAO is located at the inner envelope of plastids. It could be cross-linked to Tic40, Tic22 and Tic20 indicating an interaction with the protein import machinery of the plastid inner envelope to form a novel Tic sub-complex distinct from the known Ptsc52 translocon complex (Reinbothe et al. 2006). This complex was found to be responsible for the import of Lhcb1 and Lhcb4 (CP29) proteins but not for the import of a plastocyanin precursor. It was hypothesised that Chlide a binding to CAO and its conversion into Chlide b may prevent the Lhcb precursor from slipping back into the cytosol and supporting its import. On the basis of these data a simple feedback model for plastid signals from Chl biosynthesis was proposed (Bräutigam et al. 2007). Blocking expression of *trnE* either via plastid transcription or translation prevents formation of Chlide a, and thus, Lhcb import is drastically reduced. This would lead to accumulation of Lhcb precursors in the cytosol which could activate a feedback repression of nuclear transcription of PhANGs.

5 Interactions of Plastid and Light-Signalling Networks during Early Plastid Development

Plant photomorphogenesis is regulated by a complex signalling network that activates or represses genes essential for photomorphogenic development (Jiao et al. 2007). These processes include chloroplast biogenesis as well as control of PhANG expression (Fig. 3). Thus, interaction and/or cooperation of retrograde plastid signals with light signals, for example at the level of promoter usage are likely (see above). The pea mutant *lip1* (light-independent photomorphogenesis) exhibits photomorphogenesis in the dark, containing partially developed chloroplasts and elevated levels of PhANG transcripts (Frances et al. 1992). Seedlings of *lip1* treated with inhibitors of plastid translation displayed a reduced accumulation of transcripts of PhANGs in the light. Similar effects were also shown for the *Arabidopsis* photomorphogenetic mutant *cop1-4*. Interestingly, even in the dark plastid translation is necessary for transcript accumulation of PhANGs in *lip1*. Hence, light is not an obligatory requirement for the plastid signal (Sullivan and Gray 1999). This supports the view that early plastid signals have a developmental origin which is distinct from light-induced signals. On the other hand observations exist that plastid signals may affect light-signalling networks. A screen for *cue* (cab under-expressed) mutants from *Arabidopsis* identified plants with defects in de-repression of *cab* (now *Lhcb*) gene expression in response to phytochrome activation. Thus, plastid-derived signals are closely linked to phytochrome control of PhANG expression (Lopez-Juez et al. 1998). In a recent study focussing on weak *gun* alleles a number of new *cryptochrome1* mutants were isolated suggesting that plastid signals may remodel light-signalling networks (Ruckle et al. 2007). Genetic experiments including *cop1-4*, *hy5* and *phy* mutants combined with different light quality treatment suggested that plastid signals are able to convert action of light signalling pathways from a positive into a negative manner by affecting HY5, a positive regulator of PhANG expression. Thus, plastid signals may be required to remodel light-signalling networks in order to integrate information about developmental state of plastids and the environmental light situation.

6 Signals Depending on Photosynthesis and Reactive Oxygen Species

Photosynthetic efficiency is highly dependent on environmental cues. Adverse conditions are, therefore, counteracted by so-called acclimation responses, the aim of which is to compensate for the unfavourable parameter (Anderson et al. 1995; Kanervo et al. 2005; Walters 2005). Many acclimation mechanisms include changes in gene expression both in plastids and nucleus which are controlled by a functional feedback loop via the reduction/oxidation state of components of the PET chain or coupled redox-active molecules (Fig. 4). Such changes in the redox poise are caused by the environment. By this means photosynthesis actively adapts processes in plastids, cytosol or nucleus to its own actual function and coordinates the required changes in the three compartments (Pfannschmidt 2003; Baier and Dietz 2005; Buchanan and Balmer 2005). In addition, reactive oxygen species produced either as a by-product of photosynthesis or as a result of distinct stresses play a major role in redox signalling between plastids and nucleus and add an additional level of regulation (Apel and Hirt 2004; Foyer and Noctor 2005).

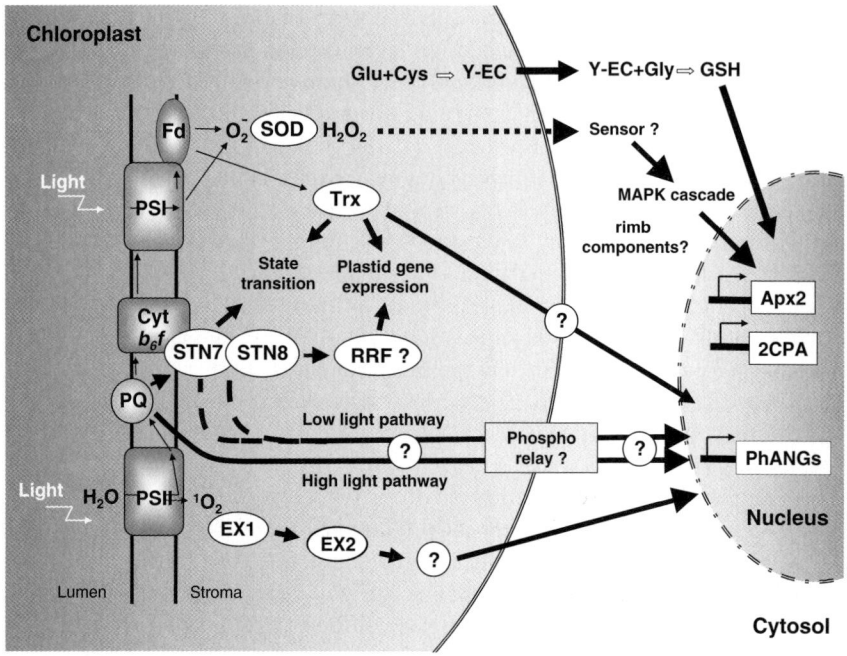

Fig. 4 *Retrograde signals from photosynthetic electron transport and ROS*. The plant cell compartments chloroplast, cytosol and nucleus are depicted schematically. Redox signals generated within the electron transport chain or by generation of ROS and scavenging mechanisms initiate signalling pathways which activate or repress specific target genes in the nucleus (for details see text). Electron flow is given as *very thin black arrows*. Redox signals influencing nuclear gene expression are given by *thick black arrows*. A *dotted arrow* indicates putative diffusion. Unclear or unknown steps or processes are marked by *question marks*. Fd: reduced ferredoxin, SOD: superoxide dismutase, H_2O_2: hydrogen peroxide, GSH: glutathione, $^1O_2$: singlet oxygen, EX1, EX2: Executer 1 and 2, RRF: redox responsive factor. For other abbreviations see text. The figure has been modified from Fig. 2 in Pfannschmidt et al. (2008) Potential regulation of gene expression in photosynthetic cells by redox and energy state – approaches towards better understanding. Annals in Botany (doi: 10.1093/aob/mcn081)

6.1 Signals Originating from Photosynthetic Electron Transport

The first evidence for influences of PET on nuclear gene expression came from experiments with the unicellular algae *Dunaliella tertiolecta* and *Dunaliella salina* (Escoubas et al. 1995; Maxwell et al. 1995). Further studies demonstrated that redox regulation by PET also exist in higher plants (Fig. 4). A study investigating acclimation of *Lemna perpusilla* to varying light intensities indicated that plastoquinone redox state regulates *Lhcb* transcript and LHCII protein accumulation (Yang et al. 2001). In another study potential combinatorial effects of plastid redox state and sugar on *Lhcb* gene expression of *Arabidopsis* were tested (Oswald et al. 2001). DCMU application was able to abolish an increase in *Lhcb* transcript accumulation which usually occurs after sugar depletion. This implies a connection between PET and sugar signalling in *Arabidopsis*. In winter rye *Lhcb* gene expression was found to be regulated by redox signals from PET which were induced by varying light and temperature regimes (Pursiheimo et al. 2001). The authors concluded that the redox state of electron acceptors at the PSI (Photosystem I) acceptor site were a regulating parameter of nuclear gene expression under these conditions. In transgenic tobacco plants carrying a pea *PetE* gene construct it could be demonstrated that the *PetE* construct as well as endogenous *Lhcb1* transcripts decreased upon DCMU treatment. In contrast, nuclear run-on transcription assays indicated up-regulation of the pea *PetE* construct expression suggesting multiple parallel influences at different levels of gene expression (Sullivan and Gray 2002). Redox regulation of post-transcriptional processes was also uncovered. In transgenic tobacco a pea ferredoxin-1 gene construct exhibited light-induced transcript accumulation which could be observed even under the control of a constitutive promoter. Since this response could also be influenced by DCMU it was concluded that PET controlled the transcript accumulation of the transgene. Furthermore, the ribosome loading of the message was affected (Petracek et al. 1997; Petracek et al. 1998). The *Apx2* gene encodes a cytosolic ascorbate peroxidase that detoxifies hydrogen peroxide which accumulates under stress. This gene was found to be induced by high-light treatment, however, DCMU and DBMIB treatments suggested an involvement of the PQ redox state in this regulation at least at an early stage (Karpinski et al. 1997, 1999). This was supported by another study on transgenic tobacco (Yabuta et al. 2004). Other experiments indicated an involvement of leaf transpiration state and abscisic acid in *Arabidopsis Apx2* expression (Fryer et al. 2003). This is consistent with the role of ABA in stress signalling (see above). The *Arabidopsis* mutant *cue-1* (chlorophyll a/b binding protein underexpressing) lacks the phosphoenolpyruvate/phosphate translocator PP1 and exhibits a light intensity dependent under-expression of *Lhc* genes. Measurements of rapid induction kinetics of Chl a fluorescence suggest that a reduced PQ pool size and PET cause the *Lhc* under-expression (Streatfield et al. 1999).

In another experimental approach variations in light quality instead of light quantity were used to manipulate photosynthetic electron transport. Illumination of plants with light sources that preferentially excite either PSI or PSII allow controlled oxidation or reduction of the electron transport chain. This low-light system avoids stress-mediated side effects which may occur under high-light and mimics natural light quality gradients in dense plant populations. The excitation imbalance is counterbalanced in the short-term by state transitions and in the long-term by photosystem stoichiometry adjustment (Dietzel et al. 2008). The latter involves the controlled change of photosynthetic gene expression both in plastids and nucleus. Initial studies with transgenic tobacco together with DCMU and DBMIB treatments indicated PQ redox control of the PC promoter (Pfannschmidt et al. 2001). Further studies with the *nia2* (encoding the cytosolic nitrate reductase) promoter in *Lemna*, *Arabidopsis* and tobacco demonstrated that this control extends also to non-photosynthesis genes (Sherameti et al. 2002). Thus, it can be concluded that PQ redox control of nuclear gene expression occurs both under high- and low-light conditions. Whether this involves two different signal transduction pathways has to be clarified in the future. A recent study with *Chlamydomonas* mutants with different defects in the cytb_6f complex demonstrated that light induction of nuclear genes for tetrapyrrole biosynthesis genes did not depend on PQ redox state but on the integrity of the cytb_6f complex (Shao et al. 2006). This suggests the existence of additional redox signals beside the PQ pool. Whether this regulation mode is also active in higher plants has to be elucidated in the future.

Recent array experiments investigated PET redox effects on nuclear gene expression in a more extended way. Using a macroarray containing genes for proteins with predicted chloroplast localisation (Kurth et al. 2002; Richly et al. 2003) it could be demonstrated that light quality shifts combined with DCMU treatment affect 286 nuclear genes in a redox-dependent manner in *Arabidopsis* (Fey et al. 2005). Identified genes encode proteins not only for photosynthesis but also for gene expression, metabolism and signal transduction indicating broad impact of PET redox signals. Another study used a different array with around 8,000 randomly selected *Arabidopsis* genes and investigated expression profiles in response to different light intensities and light qualities (Piippo et al. 2006). Under these conditions nuclear gene expression responded to redox signals from stromal components such as thioredoxin. Present data are not sufficient to resolve this contradiction arguing for further analyses of these complex regulation events.

6.2 Transduction Pathways from PET Toward the Nucleus

Transduction of redox signals over the plastid envelope and its transmission through the cytosol into the nucleus is not understood yet. So far, the PQ pool is the best-characterised source for redox signals. In a study with *Dunaliella tertiolecta* it could be demonstrated that phosphatase inhibitors were able to reduce the acclimation in response to the high- to low-light shift suggesting that the mediation of the signal might involve a phosphorylation cascade (Escoubas et al. 1995). In a working model a redox-sensitive kinase is proposed to phosphorylate a still unknown plastid protein. After transfer of the signal over the envelope a cytosolic kinase activity might be responsible for phosphorylation of a repressor protein which binds to the *Lhcb* promoter in the nucleus (Durnford and Falkowski 1997). Indeed, several different protein complexes could be observed to interact with the *Lhcb* gene promoter during photoacclimation (Chen et al. 2004). Interestingly this study found that also the trans-thylakoid pH gradient contributes to the *Lhcb* regulation suggesting the existence of several redox signals. Studies with transgenic tobacco lines demonstrated that the promoter for the nuclear gene PsaF (encoding subunit IV of PSI) is regulated by plastid redox signals (Pfannschmidt et al. 2001). Another study investigating the responsiveness of this promoter in more detail demonstrated that it can be activated by a cytosolic kinase activity even when plastid development is arrested by application of norflurazon (Chandok et al. 2001). This supports the idea of a kinase cascade in the transduction of plastid redox signals.

A potential candidate for a redox-sensitive plastid kinase involved in gene regulation is STN7 (Fig. 4). This thylakoid-membrane associated kinase was shown to be required for both state transitions and the long-term response to light-quality gradients (Bellafiore et al. 2005; Bonardi et al. 2005). It changes its activity in dependency on illumination demonstrating all requirements for redox-regulation by the PQ pool, however, its substrate specificity is not known yet. Gene regulation would require that the kinase phosphorylates either an additional protein beside the LHCII or that it initiates an additional signalling branch. Where and whether regulation pathways for state transitions and long-term response to light quality (LTR) form is still unclear. In our model we propose a down-stream redox-responsive factor (RRF) which may mediate the redox signals from the kinase (Fig. 4). Phosphorylation was shown earlier to be a crucial mechanism in controlling plastid gene expression during light-regulated etioplast–chloroplast transition (Tiller and Link 1993). The present picture is still not conclusive. Array analyses with a STN7-deficient mutant indicated only a minor role of STN7 in the regulation of nuclear and plastid gene expression (Bonardi et al. 2005; Tikkanen et al. 2006), but the lack of the orthologue kinase STN8 resulted in clear changes (Bonardi et al. 2005). Some data point to the possibility that the two kinases interact during photoacclimation and gene expression (Fig. 4) (Rochaix 2007; Dietzel et al. 2008), however, more experimental data are necessary to understand the transduction of plastid redox signals toward the nucleus.

6.3 Signals Mediated by Reactive Oxygen Species and Stress-Related Processes

Hydrogen peroxide (H_2O_2) is the principle ROS in plants. It is mainly generated at PSI under conditions when excitation exceeds energy usage by the dark reaction, e.g. under high light or in low temperature. Such conditions lead to over-reduction of the electron transport chain and to electron transfer from ferredoxin to oxygen generating superoxide (Fig. 4). This is detoxified by the superoxide dismutase (SOD) resulting in accumulation of hydrogen peroxide which is reduced to water by antioxidant enzymes such as APX. In this reaction ascorbate is used as electron donor and replenished by reduction via glutathione (Pfannschmidt 2003). Cytosolic APX enzymes are induced by oxidative conditions and therefore represent good markers for cellular stress (Shigeoka et al. 2002). In *Arabidopsis* high-light induction of nuclear genes *apx1* and *apx2* could be correlated to the action of H_2O_2 as a signalling molecule (Karpinski et al. 1997, 1999; Foyer and Noctor 1999). Interestingly *apx2* was also found to be regulated by the PQ pool pointing to a combined action of the two signals. This was confirmed by a recent study which showed that tobacco *apx2* is initially induced by the PQ pool while its later regulation occurred via H_2O_2 (Yabuta et al. 2004). The full impact of H_2O_2 on nuclear gene expression was elucidated by array analyses. 1–2% of the analysed genes exhibited responsiveness to the treatment. Among them many stress-related and defence genes were found (Desikan et al. 2001).

More recent studies support these early observations (Vandenabeele et al. 2004; Davletova et al. 2005). How this regulation is performed, however, is currently being investigated. Working models assume that H_2O_2 pass the chloroplast envelope by diffusion and activate a cytosolic mitogen-activated protein kinase (MAPK) cascade (Kovtun et al. 2000; Apel and Hirt 2004) which links H_2O_2 accumulation and gene expression via a phosphorylation cascade. Studies with *Arabidopsis* suggest that H_2O_2 activates the MAPKK kinases ANP1 or MEKK1 which, in turn, activate other downstream MAPKs (Kovtun et al. 2000). A major problem in our understanding of ROS signalling currently is that H_2O_2 is produced under a number of quite diverse stresses, but that responses to these stresses are very specific. Differences in distribution and local concentrations of H_2O_2 as well as interaction with additional signals are discussed to confer specificity in these processes (Beck 2005).

Another important ROS in stress signalling is singlet oxygen ($^1O_2$), a non-radical ROS. It is continuously produced at PSII by energy transfer from triplet state P680 to oxygen, but its production increases under conditions of over-excitation (Fig. 4). Singlet oxygen possesses a very short half-life (~200 ns) and causes oxidative damage mainly at the site of its generation, i.e. in PSII (op den Camp et al. 2003). Nevertheless, it induces a number of distinct stress responses in the nucleus. Since excess excitation conditions produce several ROS in parallel it is difficult to discern the action of a specific ROS. However, the *Arabidopsis flu* (*fluorescent*) mutant provides a tool to circumvent this problem. The mutant accumulates free protochlorophyllide (Pchlide) when put into darkness and produces enhanced amounts of singlet oxygen upon re-illumination by energy transfer from the Pchlide. This leads to growth inhibition and cell death. However, under continuous illumination when Pchlide is not accumulated the mutant exhibits wild-type like development (Meskauskiene et al. 2001). Transcript profiling with the mutant indicated that around 5% of all genes changed their expression and that 70 genes were specifically activated by singlet oxygen (op den Camp et al. 2003). Surprisingly, the destructive effects of this ROS could be genetically suppressed in a second-site mutant screen of the *flu* mutant indicating that the cell death was not induced by the oxidative damage from singlet oxygen but was initiated by a response programme. In the double mutant *flu/ex1* the singlet oxygen-mediated stress responses were abrogated by the inactivation of a gene called *executer1* (*ex1*) (Wagner et al. 2004). It encodes a plastid-localised protein (EXECUTER1) with a still unknown function. This protein represents a potential sensor and/or mediator of singlet oxygen signals. A more detailed analysis of the cell death response in the *flu* mutant revealed that it is promoted by signalling pathway(s) dependent on ethylene, salicylic and jasmonic acid but that it is blocked by a jasmonic acid precursor (Danon et al. 2005). Recently, a second *executer1*-like gene called *executer2* was identified which is also implicated in singlet oxygen-dependent nuclear gene expression changes. The encoded protein EX2 is also confined to the plastid and appears to interact with EX1. In triple mutants *ex1/ex2/flu* up-regulation of singlet-oxygen regulated genes is almost completely suppressed suggesting that the two proteins are sufficient to confer singlet oxygen-mediated retrograde signalling (Lee et al. 2007). Interestingly, a mutated allele of the haem oyxgenase called *ulf3* can suppress the phenotype of the *flu* mutant (Goslings et al. 2004). *ulf3* is allelic to *gun2* and thus point to a connection between tetrapyrrole and ROS signalling.

In another mutant screen using transgenic *Arabidopsis* carrying a 2-cysteine peroxiredoxin (2CPA)-promoter::luciferase construct several *rimb* (*redox imbalanced*) mutants (Fig. 4) were identified (Heiber et al. 2007). The 2CPA promoter is redox-sensitive and activated by redox signals from components at the PSI acceptor side to increase antioxidant capacity under conditions that could induce photooxidative damage (Baier and Dietz 2005). Thus, these mutants provide a tool for analysing retrograde redox signalling cascades which activate nuclear-encoded antioxidant enzymes. Future work will show whether these signalling pathways are different from that described above.

Reduced glutathione (GSH) is required for re-reduction of the primary ROS scavenger ascorbate. Recent observations indicate that GSH might act as a plastid signal during stress defence programmes (Fig. 4). For instance expression of stress-related genes *apx2* or *pr1* (pathogen related) were correlated with changes in cellular glutathione content (Mullineaux and Rausch 2005). The *Arabidopsis* mutant *rax1-1* (regulator of *Apx2*) is impaired in glutathione synthetase 1 (GSH1) and exhibits a decreased GSH content. As a consequence it exhibits a constitutive high expression of the *Apx2* gene suggesting that low GSH levels activate defence gene expression (Ball et al. 2004). Glutathione is synthesised from glutamate and cysteine forming γ-glutamylcysteine (γ-EC) followed by further addition of glycine. GSH1 catalyses the first step, GSH2 the second. Recent data suggest that in *Arabidopsis* step 1 is confined to the chloroplast while step 2 occurs predominantly in the cytosol (Wachter et al. 2005). This requires that the GSH precursor γ-EC must leave the chloroplast. Its amount, therefore, might well represent a plastid signal reporting potential stress in chloroplasts to the cytosol (Mullineaux and Rausch 2005).

7 Conclusions

In summary, known plastid signals can be classified into two major groups, signals from young, colourless and undifferentiated or damaged plastids and signals from mature and green chloroplasts. Arresting plastid development by inhibitors or mutations typically results in proplastid-like stages that are functionally very restricted (Sullivan and Gray 1999; Nott et al. 2006). The correlating repression of PhANG expression has been interpreted as a hint that the blocked plastid development generates a *negative* signal which represses nuclear gene expression. The observation that in *gun* mutants this repression can be interrupted genetically appears to support this assumption and suggest that tetrapyrrole biosynthesis intermediates might be involved in this negative signalling process. However, different models of negative regulation might also be possible and has been discussed here. A completely different interpretation of the data on inhibiting plastid differentiation would be that the block in plastid development leads to the lack of a *positive* plastid signal resulting in a negative feedback loop. Evolution integrated plastids deeply into the cellular developmental networks. Blocking its biogenesis thus resembles a knock-out of an important hub in the network which subsequently blocks pathways connected to it but which can be disconnected genetically. Since plant development is highly regulated by photoreceptors a connection of plastid developmental signals with light-signalling pathways is a logical consequence and has been reported in several studies. Thus, studies on such undifferentiated plastids help us to learn more about plastid biogenesis and its integration into plant development as a whole. The role of tetrapyrroles as potential signals in green tissues has still to be elucidated. Variegation mutants containing both colourless undifferentiated and mature green plastids as well as transgenic plants with altered tetrapyrrole synthesis provide interesting models for further investigations of this topic (Sakamoto 2003; Alawady and Grimm 2005; Aluru et al. 2006). In contrast, signals from mature chloroplasts do not provide information about biogenesis but about actual function of the plastids. Here, plastids play the role of an active sensor for environmental changes in a mainly photoreceptor-independent manner. Coordination of photosynthetic acclimation and responses to various stresses are the predominant function of these plastid signals and are communicated by pathways to the nucleus which are completely different from that of the developmental signals. Therefore, studying plastid-to-nucleus communication covers two major fields of plant cell biology, understanding of (1) developmental cascades and (2) physiological acclimation to the environment. Detailed analyses of plastid-to-nucleus signalling pathways, therefore, are of greatest interest for many aspects in molecular plant research.

Acknowledgements Our work was supported by grants from the DFG, the "NWP" and "Excellence in Science" programmes of Thuringia to T.P. and to the DFG research group FOR 387.

References

Abdallah F, Salamini F, Leister D (2000) A prediction of the size and evolutionary origin of the proteome of chloroplasts of *Arabidopsis*. Trends Plant Sci 5:141–142

Acevedo-Hernandez GJ, Leon P, Herrera-Estrella LR (2005) Sugar and ABA responsiveness of a minimal RBCS light-responsive unit is mediated by direct binding of ABI4. Plant J 43:506–519

Alawady AE, Grimm B (2005) Tobacco Mg protoporphyrin IX methyltransferase is involved in inverse activation of Mg porphyrin and protoheme synthesis. Plant J 41:282–290

Albrecht V, Ingenfeld A, Apel K (2006) Characterization of the Snowy cotyledon 1 mutant of *Arabidopsis thaliana*: The impact of chloroplast elongation factor G on chloroplast development and plant vitality. Plant Mol Biol 60:507–518

Aluru MR, Yu F, Fu AG, Rodermel S (2006) Arabidopsis variegation mutants: new insights into chloroplast biogenesis. J Exp Bot 57:1871–1881

Anderson JM, Chow WS, Park Y-I (1995) The grand design of photosynthesis: Acclimation of the photosynthetic apparatus to environmental cues. Photosynth Res 46:129–139

Ankele E, Kindgren P, Pesquet E, Strand A (2007) In vivo visualization of Mg-ProtoporphyrinIX, a coordinator of photosynthetic gene expression in the nucleus and the chloroplast. Plant Cell 19:1964–1979

Apel K, Hirt H (2004) Reactive oxygen species: Metabolism, oxidative stress, and signal transduction. Ann Rev Plant Biol 55:373–399.

Baier M, Dietz KJ (2005) Chloroplasts as source and target of cellular redox regulation: a discussion on chloroplast redox signals in the context of plant physiology. J Exp Bot 56:1449–1462

Ball L, Accotto GP, Bechtold U, Creissen G, Funck D, Jimenez A, Kular B, Leyland N, Mejia-Carranza J, Reynolds H, Karpinski S, Mullineaux PM (2004) Evidence for a direct link between glutathione biosynthesis and stress defense gene expression in Arabidopsis. Plant Cell 16:2448–2462

Beck CF (2005) Signaling pathways from the chloroplast to the nucleus. Planta 222:743–756

Bellafiore S, Bameche F, Peltier G, Rochaix JD (2005) State transitions and light adaptation require chloroplast thylakoid protein kinase STN7. Nature 433:892–895

Bolle C, Kusnetsov VV, Herrmann RG, Oelmüller R (1996) The spinach AtpC and AtpD genes contain elements for light-regulated, plastid-dependent and organ-specific expression in the vicinity of the transcription start sites. Plant J 9:21–30

Bonardi V, Pesaresi P, Becker T, Schleiff E, Wagner R, Pfannschmidt T, Jahns P, Leister D (2005) Photosystem II core phosphorylation and photosynthetic acclimation require two different protein kinases. Nature 437:1179–1182

Bradbeer JW, Atkinson YE, Börner T, Hagemann R (1979) Cytoplasmic synthesis of plastid polypeptides may be controlled by plastid synthesized RNA. Nature 279:816–817

Bräutigam K, Dietzel L, Pfannschmidt T (2007) Plastid-nucleus communication: anterograde and retrograde signalling in development and function of plastids. In: Bock R (ed) Cell and molecular biology of plastids. Springer, Berlin, pp 409–455

Buchanan BB, Balmer Y (2005) Redox regulation: a broadening horizon. Ann Rev Plant Biol 56:187–220

Buchanan BB, Gruissem W, Jones RL (2002) Biochemistry and molecular biology of plants. Wiley, Somerset

Burger G, Gray MW, Lang BF (2003) Mitochondrial genomes: anything goes. Trends Genet 19:709–716

Chamovitz D, Pecker I, Hirschberg J (1991) The molecular-basis of resistance to the herbicide norflurazon. Plant Mol Biol 16:967–974

Chandok MR, Sopory SK, Oelmüller R (2001) Cytoplasmic kinase and phosphatase activities can induce PsaF gene expression in the absence of functional plastids: evidence that phosphorylation/dephosphorylation events are involved in interorganellar crosstalk. Mol Gen Genet 264:819–826

Chen YB, Durnford DG, Koblizek M, Falkowski PG (2004) Plastid regulation of Lhcb1 transcription in the chlorophyte alga *Dunaliella tertiolecta*. Plant Physiol 136:3737–3750

Cottage AJ, Mott EK, Wang JH, Sullivan JA, MacLean D, Tran L, Choy MK, Newell CA, Kavanagh TA, Aspinall S, Gray JC (2007) GUN1 (GENOMES UNCOUPLED1) encodes a pentatricopeptide repeat (PPR) protein involved in plastid protein synthesis-responsive retrograde signaling to the nucleus. Photosynth Res 91:276–276

Danon A, Miersch O, Felix G, den Camp RGLO, Apel K (2005) Concurrent activation of cell death-regulating signaling pathways by singlet oxygen in *Arabidopsis thaliana*. Plant J 41:68–80

Davletova S, Rizhsky L, Liang H, Shengqiang Z, Oliver DJ, Coutu J, Shulaev V, Schlauch K, Mittler R (2005) Cytosolic ascorbate peroxidase 1 is a central component of the reactive oxygen gene network of *Arabidopsis*. Plant Cell 17:268–281

Desikan R, Mackerness SAH, Hancock JT, Neill SJ (2001) Regulation of the Arabidopsis transcriptome by oxidative stress. Plant Physiol 127:159–172

Dietzel L, Bräutigam K, Pfannschmidt T (2008) Photosynthetic acclimation: State transitions and adjustment of photosystem stoichiometry – functional relationships between short-term and long-term light quality acclimation in plants. FEBS J 275:1080–1088

Dong HL, Deng Y, Mu JY, Lu QT, Wang YQ, Xu YY, Chu CC, Chong K, Lu CM, Zuo JR (2007) The *Arabidopsis* spontaneous Cell Death1 gene, encoding a zeta-carotene desaturase essential for carotenoid biosynthesis, is involved in chloroplast development, photoprotection and retrograde signalling. Cell Res 17:458–470

Durnford DG, Falkowski PG (1997) Chloroplast redox regulation of nuclear gene transcription during photoacclimation. Photosynth Res 53:229–241

Escoubas JM, Lomas M, Laroche J, Falkowski PG (1995) Light-intensity regulation of cab gene-transcription is signaled by the redox state of the plastoquinone pool. Proc Natl Acad Sci USA 92:10237–10241

Fey V, Wagner R, Brautigam K, Wirtz M, Hell R, Dietzmann A, Leister D, Oelmüller R, Pfannschmidt T (2005) Retrograde plastid redox signals in the expression of nuclear genes for chloroplast proteins of *Arabidopsis thaliana*. J Biol Chem 280:5318–5328

Foyer CH, Noctor G (1999) Plant biology – leaves in the dark see the light. Science 284:599–601

Foyer CH, Noctor G (2005) Redox homeostasis and antioxidant signaling: a metabolic interface between stress perception and physiological responses. Plant Cell 17:1866–1875

Frances S, White MJ, Edgerton MD, Jones AM, Elliott RC, Thompson WF (1992) Initial characterization of a pea mutant with light-independent photomorphogenesis. Plant Cell 4:1519–1530

Fryer MJ, Ball L, Oxborough K, Karpinski S, Mullineaux PM, Baker NR (2003) Control of ascorbate peroxidase 2 expression by hydrogen peroxide and leaf water status during excess light stress reveals a functional organisation of *Arabidopsis* leaves. Plant J 33:691–705

Goslings D, Meskauskiene R, Kim CH, Lee KP, Nater M, Apel K (2004) Concurrent interactions of heme and FLU with Glu tRNA reductase (HEMA1), the target of metabolic feedback inhibition of tetrapyrrole biosynthesis, in dark- and light-grown Arabidopsis plants. Plant J 40:957–967

Gray JC, Sornarajah R, Zabron AA, Duckett CM, Khan MS (1995) Chloroplast control of nuclear gene expression. In: Mathis P (ed) Photosynthesis, from light to biosphere. Kluwer, Dordrecht, pp 543–550

Gray JC, Sullivan JA, Wang JH, Jerome CA, MacLean D (2003) Coordination of plastid and nuclear gene expression. Phil Trans R Soc Lond B 358:135–144

Hanaoka M, Kanamaru K, Takahashi H, Tanaka K (2003) Molecular genetic analysis of chloroplast gene promoters dependent on SIG2, a nucleus-encoded sigma factor for the plastid-encoded RNA polymerase, in *Arabidopsis thaliana*. Nucleic Acids Res 31:7090–7098

Heiber I, Ströher E, Raatz B, Busse I, Kahmann U, Bevan MW, Dietz KJ, Baier M (2007) The redox imbalanced mutants of Arabidopsis differentiate signaling pathways for redox regulation of chloroplast antioxidant enzymes. Plant Physiol 143:1774–1788

Hess WR, Muller A, Nagy F, Borner T (1994) Ribosome-deficient plastids affect transcription of light-induced nuclear genes – genetic-evidence for a plastid-derived signal. Mol Gen Genet 242:305–312

Hess WR, Prombona A, Fieder B, Subramanian AR, Borner T (1993) Chloroplast rps15 and the rpoB/C1/C2 gene-cluster are strongly transcribed in ribosome-deficient plastids – evidence for a functioning non-chloroplast-encoded RNA-polymerase. EMBO J 12:563–571

Hess WR, Schendel R, Borner T, Rudiger W (1991) Reduction of messenger-RNA level for 2 nuclear encoded light regulated genes in the barley mutant *albostrians* is not correlated with phytochrome content and activity. J Plant Physiol 138:292–298

Hess WR, Schendel R, Rudiger W, Fieder B, Borner T (1992) Components of chlorophyll biosynthesis in a barley albino mutant unable to synthesize delta-aminolevulinic-acid by utilizing the transfer-RNA for glutamic-acid. Planta 188:19–27

Ishikawa A, Okamoto H, Iwasaki Y, Asahi T (2001) A deficiency of coproporphyrinogen III oxidase causes lesion formation in *Arabidopsis*. Plant J 27:89–99

Jarvis P (2001) Intracellular signalling: the chloroplast talks! Curr Biol 11:R307–R310

Jiao YL, Lau OS, Deng XW (2007) Light-regulated transcriptional networks in higher plants. Nat Rev Genet 8:217–230

Johanningmeier U, Howell SH (1984) Regulation of light-harvesting chlorophyll-binding protein messenger-RNA accumulation in *Chlamydomonas-reinhardtii* – possible involvement of chlorophyll synthesis precursors. J Biol Chem 259:3541–3549

Kanervo E, Suorsa M, Aro EM (2005) Functional flexibility and acclimation of the thylakoid membrane. Photochem Photobiol Sci 4:1072–1080

Karpinski S, Escobar C, Karpinska B, Creissen G, Mullineaux PM (1997) Photosynthetic electron transport regulates the expression of cytosolic ascorbate peroxidase genes in *Arabidopsis* during excess light stress. Plant Cell 9:627–640

Karpinski S, Reynolds H, Karpinska B, Wingsle G, Creissen G, Mullineaux P (1999) Systemic signaling and acclimation in response to excess excitation energy in Arabidopsis. Science 284:654–657

Kittsteiner U, Brunner H, Rudiger W (1991) The greening process in cress seedlings 2. Complexing agents and 5-aminolevulinate inhibit accumulation of cab-messenger-RNA coding for the light-harvesting chlorophyll a-b protein. Physiol Plant 81:190–196

Kleffmann T, Russenberger D, von Zychlinski A, Christopher W, Sjolander K, Gruissem W, Baginsky S (2004) The *Arabidopsis thaliana* chloroplast proteome reveals pathway abundance and novel protein functions. Curr Biol 14:354–362

Koncz C, Mayerhofer R, Konczkalman Z, Nawrath C, Reiss B, Redei GP, Schell J (1990) Isolation of a gene encoding a novel chloroplast protein by T-DNA tagging in *Arabidopsis-thaliana*. EMBO J 9:1337–1346

Koussevitzky S, Nott A, Mockler TC, Hong F, Sachetto-Martins G, Surpin M, Lim IJ, Mittler R, Chory J (2007) Signals from chloroplasts converge to regulate nuclear gene expression. Science 316:715–719

Kovtun Y, Chiu WL, Tena G, Sheen J (2000) Functional analysis of oxidative stress-activated mitogen-activated protein kinase cascade in plants. Proc Natl Acad Sci USA 97:2940–2945

Kropat J, Oster U, Rudiger W, Beck CF (1997) Chlorophyll precursors are signals of chloroplast origin involved in light induction of nuclear heat-shock genes. Proc Natl Acad Sci USA 94:14168–14172

Kurth J, Varotto C, Pesaresi P, Biehl A, Richly E, Salamini F, Leister D (2002) Gene-sequence-tag expression analyses of 1,800 genes related to chloroplast functions. Planta 215:101–109

Kusnetsov V, Bolle C, Lubberstedt T, Sopory S, Herrmann RG, Oelmüller R (1996) Evidence that the plastid signal and light operate via the same *cis*-acting elements in the promoters of nuclear genes for plastid proteins. Mol Gen Genet 252:631–639

Larkin RM, Alonso JM, Ecker JR, Chory J (2003) GUN4, a regulator of chlorophyll synthesis and intracellular signaling. Science 299:902–906

Lee KP, Kim C, Landgraf F, Apel K (2007) EXECUTER1- and EXECUTER2-dependent transfer of stress-related signals from the plastid to the nucleus of Arabidopsis thaliana. Proc Natl Acad Sci USA 104:10270–10275

Lopez-Juez E, Jarvis RP, Takeuchi A, Page AM, Chory J (1998) New *Arabidopsis* cue mutants suggest a close connection between plastid- and phytochrome regulation of nuclear gene expression. Plant Physiol 118:803–815

Lukens JH, Mathews DE, Durbin RD (1987) Effect of tagetitoxin on the levels of ribulose 1,5-bisphosphate carboxylase, ribosomes, and RNA in plastids of wheat leaves. Plant Physiol 84:808–813

Maxwell DP, Laudenbach DE, Huner NPA (1995) Redox regulation of light-harvesting complex-II and cab messenger-RNA abundance in *Dunaliella-salina*. Plant Physiol 109:787–795

Mayfield SP, Taylor WC (1984) Carotenoid-deficient maize seedlings fail to accumulate light-harvesting chlorophyll a/b binding-protein (LhcP) messenger-RNA. Eur J Biochem 144:79–84

McCormac AC, Terry MJ (2004) The nuclear genes Lhcb and HEMA1 are differentially sensitive to plastid signals and suggest distinct roles for the GUN1 and GUN5 plastid-signalling pathways during de-etiolation. Plant J 40:672–685

Meskauskiene R, Nater M, Goslings D, Kessler F, den Camp RO, Apel K (2001) FLU: A negative regulator of chlorophyll biosynthesis in *Arabidopsis thaliana*. Proc Natl Acad Sci USA 98:12826–12831

Mochizuki N, Brusslan JA, Larkin R, Nagatani A, Chory J (2001) Arabidopsis genomes uncoupled 5 (GUN5) mutant reveals the involvement of Mg-chelatase H subunit in plastid-to-nucleus signal transduction. Proc Natl Acad Sci USA 98:2053–2058

Motohashi R, Yamazaki T, Myouga F, Ito T, Ito K, Satou M, Kobayashi M, Nagata N, Yoshida S, Nagashima A, Tanaka K, Takahashi S, Shinozaki K (2007) Chloroplast ribosome release factor 1 (AtcpRF1) is essential for chloroplast development. Plant Mol Biol 64:481–497

Mullineaux PM, Rausch T (2005) Glutathione, photosynthesis and the redox regulation of stress-responsive gene expression. Photosynth Res 86:459–474

Nott A, Jung HS, Koussevitzky S, Chory J (2006) Plastid-to-nucleus retrograde signaling. Ann Rev Plant Biol 57:739–759

Oelmüller R (1989) Photooxidative destruction of chloroplasts and its effect on nuclear gene-expression and extraplastidic enzyme levels. Photochem Photobiol 49:229–239

Oelmüller R, Mohr H (1986) Photooxidative destruction of chloroplasts and its consequences for expression of nuclear genes. Planta 167:106–113

op den Camp RGL, Przybyla D, Ochsenbein C, Laloi C, Kim CH, Danon A, Wagner D, Hideg E, Gobel C, Feussner I, Nater M, Apel K (2003) Rapid induction of distinct stress responses after the release of singlet oxygen in Arabidopsis. Plant Cell 15:2320–2332

Oswald O, Martin T, Dominy PJ, Graham IA (2001) Plastid redox state and sugars: Interactive regulators of nuclear-encoded photosynthetic gene expression. Proc Natl Acad Sci USA 98:2047–2052

Papenbrock J, Grimm B (2001) Regulatory network of tetrapyrrole biosynthesis – studies of intracellular signalling involved in metabolic and developmental control of plastids. Planta 213:667–681

Pastori GM, Kiddle G, Antoniw J, Bernard S, Veljovic-Jovanovic S, Verrier PJ, Noctor G, Foyer CH (2003) Leaf vitamin C contents modulate plant defense transcripts and regulate genes that control development through hormone signaling. Plant Cell 15:939–951

Pesaresi P, Masiero S, Eubel H, Braun HP, Bhushan S, Glaser E, Salamini F, Leister D (2006) Nuclear photosynthetic gene expression is synergistically modulated by rates of protein synthesis in chloroplasts and mitochondria. Plant Cell 18:970–991

Pesaresi P, Schneider A, Kleine T, Leister D (2007) Interorganellar communication. Curr Opin Plant Biol 10:600–606

Petracek ME, Dickey LF, Huber SC, Thompson WF (1997) Light-regulated changes in abundance and polyribosome association of ferredoxin mRNA are dependent on photosynthesis. Plant Cell 9:2291–2300

Petracek ME, Dickey LF, Nguyen TT, Gatz C, Sowinski DA, Allen GC, Thompson WF (1998) Ferredoxin-1 mRNA is destabilized by changes in photosynthetic electron transport. Proc Natl Acad Sci USA 95:9009–9013

Pfalz J, Liere K, Kandlbinder A, Dietz KJ, Oelmüller R (2006) PTAC2,-6, and-12 are components of the transcriptionally active plastid chromosome that are required for plastid gene expression. Plant Cell 18:176–197

Pfannschmidt T (2003) Chloroplast redox signals: how photosynthesis controls its own genes. Trends Plant Sci 8:33–41

Pfannschmidt T, Link G (1997) The A and B forms of plastid DNA-dependent RNA polymerase from mustard (*Sinapis alba* L.) transcribe the same genes in a different developmental context. Mol Gen Genet 257:35–44

Pfannschmidt T, Schütze K, Brost M, Oelmüller R (2001) A novel mechanism of nuclear photosynthesis gene regulation by redox signals from the chloroplast during photosystem stoichiometry adjustment. J Biol Chem 276:36125–36130

Piippo M, Allahverdiyeva Y, Paakkarinen V, Suoranta UM, Battchikova N, Aro EM (2006) Chloroplast-mediated regulation of nuclear genes in *Arabidopsis thaliana* in the absence of light stress. Physiol Genom 25:142–152

Puente P, Wei N, Deng XW (1996) Combinatorial interplay of promoter elements constitutes the minimal determinants for light and developmental control of gene expression in *Arabidopsis*. EMBO J 15:3732–3743

Pursiheimo S, Mulo P, Rintamäki E, Aro EM (2001) Coregulation of light-harvesting complex II phosphorylation and lhcb mRNA accumulation in winter rye. Plant J 26:317–327

Rapp JC, Mullet JE (1991) Chloroplast transcription is required to express the nuclear genes rbcS and cab plastid DNA copy number is regulated independently. Plant Mol Biol 17:813–823

Reinbothe C, Bartsch S, Eggink LL, Hoober JK, Brusslan J, Andrade-Paz R, Monnet J, Reinbothe S (2006) A role for chlorophyllide a oxygenase in the regulated import and stabilization of light-harvesting chlorophyll a/b proteins. Proc Natl Acad Sci USA 103:4777–4782

Rhoads DM, Subbaiah CC (2007) Mitochondrial retrograde regulation in plants. Mitochondrion 7:177–194

Richly E, Dietzmann A, Biehl A, Kurth J, Laloi C, Apel K, Salamini F, Leister D (2003) Covariations in the nuclear chloroplast transcriptome reveal a regulatory master-switch. EMBO Rep 4:491–498

Rochaix JD (2007) Role of thylakoid protein kinases in photosynthetic acclimation. FEBS Lett 581:2768–2775

Rodermel S (2001) Pathways of plastid-to-nucleus signaling. Trends Plant Sci 6:471–478

Rolland N, Ferro M, Seigneurin-Berny D, Garin J, Block M, Joyard J (2008) The chloroplast envelope proteome and lipidome. Plant Cell Monogr., doi:10.1007/7089_2008_33

Ruckle ME, DeMarco SM, Larkin RM (2007) Plastid signals remodel light signaling networks and are essential for efficient chloroplast biogenesis in Arabidopsis. Plant Cell 19:3944–3960

Rüdiger W, Grimm B (2006) Chlorophyll metabolism, an overview. In: Grimm B, Porra RJ, Rüdiger W, Scheer H (eds) Advances in photosynthesis and respiration. Springer, Dordrecht, pp 133–146

Saha D, Prasad AM, Srinivasan R (2007) Pentatricopeptide repeat proteins and their emerging roles in plants. Plant Physiol Biochem 45:521–534

Sakamoto W (2003) Leaf-variegated mutations and their responsible genes in *Arabidopsis thaliana*. Gen Genet Syst 78:1–9

Seo M, Koshiba T (2002) Complex regulation of ABA biosynthesis in plants. Trends Plant Sci 7:41–48

Shao N, Vallon O, Dent R, Niyogi KK, Beck CF (2006) Defects in the cytochrome b(6)/f complex prevent light-induced expression of nuclear genes involved in chlorophyll biosynthesis. Plant Physiol 141:1128–1137

Shen YY, Wang XF, Wu FQ, Du SY, Cao Z, Shang Y, Wang XL, Peng CC, Yu XC, Zhu SY, Fan RC, Xu YH, Zhang DP (2006) The Mg-chelatase H subunit is an abscisic acid receptor. Nature 443:823–826

Sherameti I, Sopory SK, Trebicka A, Pfannschmidt T, Oelmüller R (2002) Photosynthetic electron transport determines nitrate reductase gene expression and activity in higher plants. J Biol Chem 277:46594–46600

Shigeoka S, Ishikawa T, Tamoi M, Miyagawa Y, Takeda T, Yabuta Y, Yoshimura K (2002) Regulation and function of ascorbate peroxidase isoenzymes. J Exp Bot 53:1305–1319

Staneloni RJ, Rodriguez-Batiller MJ, Casal JJ (2008) Abscisic acid, high-light, and oxidative stress down-regulate a photosynthetic gene via a promoter motif not involved in phytochrome-mediated transcriptional regulation. Mol Plant 1:75–83

Stoebe B, Maier UG (2002) One, two, three: nature's tool box for building plastids. Protoplasma 219:123–130

Strand A (2004) Plastid-to-nucleus signalling. Curr Opin Plant Biol 7:621–625

Strand A, Asami T, Alonso J, Ecker JR, Chory J (2003) Chloroplast to nucleus communication triggered by accumulation of Mg-protoporphyrinIX. Nature 421:79–83

Streatfield SJ, Weber A, Kinsman EA, Hausler RE, Li JM, Post-Beittenmiller D, Kaiser WM, Pyke KA, Flugge UI, Chory J (1999) The phosphoenolpyruvate/phosphate translocator is required for phenolic metabolism, palisade cell development, and plastid-dependent nuclear gene expression. Plant Cell 11:1609–1621

Sugiura M (1992) The chloroplast genome. Plant Mol Biol 19:149–168

Sullivan JA, Gray JC (1999) Plastid translation is required for the expression of nuclear photosynthesis genes in the dark and in roots of the pea lip1 mutant. Plant Cell 11:901–910

Sullivan JA, Gray JC (2002) Multiple plastid signals regulate the expression of the pea plastocyanin gene in pea and transgenic tobacco plants. Plant J 32:763–774

Surpin M, Larkin RM, Chory J (2002) Signal transduction between the chloroplast and the nucleus. Plant Cell 14:S327–S338

Susek RE, Ausubel FM, Chory J (1993) Signal-transduction mutants of *Arabidopsis* uncouple nuclear cab and rbcS gene-expression from chloroplast development. Cell 74:787–799

Taylor WC (1989) Regulatory interactions between nuclear and plastid genomes. Ann Rev Plant Physiol Plant Mol Biol 40:211–233

Terzaghi WB, Cashmore AR (1995) Light-regulated transcription. Ann Rev Plant Physiol Plant Mol Biol 46:445–474

Tikkanen M, Piippo M, Suorsa M, Sirpio S, Mulo P, Vainonen J, Vener AV, Allahverdiyeva Y, Aro EM (2006) State transitions revisited – a buffering system for dynamic low light acclimation of Arabidopsis. Plant Mol Biol 62:779–793

Tiller K, Link G (1993) Phosphorylation and dephosphorylation affect functional characteristics of chloroplast and etioplast transcription systems from mustard (*Sinapis alba* L). EMBO J 12:1745–1753

Tottey S, Block MA, Allen M, Westergren T, Albrieux C, Scheller HV, Merchant S, Jensen PE (2003) *Arabidopsis* CHL27, located in both envelope and thylakoid membranes, is required for the synthesis of protochlorophyllide. Proc Natl Acad Sci USA 100:16119–16124

van Wijk KJ (2000) Proteomics of the chloroplast: experimentation and prediction. Trends Plant Sci 5:420–425

Vandenabeele S, Vanderauwera S, Vuylsteke M, Rombauts S, Langebartels C, Seidlitz HK, Zabeau M, Van Montagu M, Inze D, Van Breusegem F (2004) Catalase deficiency drastically affects gene expression induced by high light in *Arabidopsis thaliana*. Plant J 39:45–58

von Gromoff ED, Schroda M, Oster U, Beck CF (2006) Identification of a plastid response element that acts as an enhancer within the *Chlamydomonas* HSP70A promoter. Nucleic Acids Res 34:4767–4779

von Gromoff ED, Alawady A, Meinecke L, Grimm B, Beck CF (2008) Heme, a plastid-derived regulator of nuclear gene expression in *Chlamydomonas*. Plant Cell 20:552–567

Wachter A, Wolf S, Steininger H, Bogs J, Rausch T (2005) Differential targeting of GSH1 and GSH2 is achieved by multiple transcription initiation: implications for the compartmentation of glutathione biosynthesis in the *Brassicaceae*. Plant J 41:15–30

Wagner D, Przybyla D, Camp ROD, Kim C, Landgraf F, Lee KP, Wursch M, Laloi C, Nater M, Hideg E, Apel K (2004) The genetic basis of singlet oxygen-induced stress responses of *Arabidopsis thaliana*. Science 306:1183–1185

Walters RG (2005) Towards an understanding of photosynthetic acclimation. J Exp Bot 56:435–447

Wilde A, Mikolajczyk S, Alawady A, Lokstein H, Grimm B (2004) The gun4 gene is essential for cyanobacterial porphyrin metabolism. FEBS Lett 571:119–123

Willows R, Hansson M (2003) Mechanism, structure, and regulation of magnesium chelatase. In: Kadish KM, Smith K, Guilard R (eds) The tetrapyrrole handbook II. Academic Press, San Diego, pp 1–48

Woitsch S, Römer S (2003) Expression of xanthophyll biosynthetic genes during light-dependent chloroplast differentiation. Plant Physiol 132:1508–1517

Yabuta Y, Maruta T, Yoshimura K, Ishikawa T, Shigeoka S (2004) Two distinct redox signaling pathways for cytosolic APX induction under photooxidative stress. Plant Cell Physiol 45:1586–1594

Yang DH, Andersson B, Aro EM, Ohad I (2001) The redox state of the plastoquinone pool controls the level of the light-harvesting chlorophyll a/b binding protein complex II (LHC II) during photoacclimation – Cytochrome b(6)f deficient *Lemna perpusilla* plants are locked in a state of high-light acclimation. Photosynth Res 68:163–174

Yoshida R, Sato T, Kanno A, Kameya T (1998) Streptomycin mimics the cool temperature response in rice plants. J Exp Bot 49:221–227

Plastid Division Regulation and Interactions with the Environment

J. Maple, A. Mateo, and S.G. Møller (✉)

Abstract Plastids are vital organelles, fulfilling important metabolic functions that greatly affect plant growth and productivity. As such it is rational that the process of plastid division is carefully controlled, not only to ensure persistence in dividing plant cells and that optimal numbers of plastids are obtained in specialised cell types, but also in response to environmental changes. How this control is exerted by the host nucleus is unclear. Through evolution the plastid division machinery has retained homologues of key proteins of bacterial division apparatus, and also incorporated new proteins of eukaryotic origin. The evolution of this complex process raises intriguing questions as to how the division process has been placed under the control of the host cell. Here we explore the current understanding of the process of plastid division at the molecular and cellular level, with particular respect to the response of plastid division to environmental signals and endogenous cues.

1 Introduction

Plastids are essential organelles that develop from small, colourless, undifferentiated proplastids in dividing meristematic cells. Although research efforts in the plastid field have primarily focused on chloroplasts, a variety of other plastid differentiation pathways occur in specific cell types, including amyloplast formation in tubers, leucoplast formation in petals and chromoplast formation in fruits and flowers (reviewed in Waters and Pyke 2005). Consequently plastids are important, not only for their role in photosynthesis, but also because they house a variety of intermediate metabolic pathways essential for plant cell function and are the site of storage of a wide variety of products (Galili 1995; Ohlrogge and Browse 1995; Tetlow et al. 2005).

S.G. Møller
Centre for Organelle Research, Faculty of Science and Technology, University of Stavanger, 4036, Norway
e-mail: simon.g.moller@uis.no

Plastids are not derived de novo but arise by division of a pre-existing parental organelle (Possingham and Saurer 1969). Whilst very little is known about the cellular and environmental stimuli that trigger plastid division, microscopic observations have demonstrated that plastid division is a tightly regulated process. Of primary importance to plant cell function is the appropriate distribution of plastids during cell division, and in different organisms this creates different challenges. For example unicellular algae, such as *Chlamydomonas* or *C. merolae*, have only one chloroplast per cell and so coordination of cell and chloroplast division is essential to ensure that each daughter cell contains a chloroplast (Kuroiwa 1998; Misumi et al. 2005). In contrast, in higher plants the meristematic cells are estimated to contain between 10 and 20 proplastids and regulation may be less stringent, but proplastid division must still keep pace with cell division so that the daughter cells possess approximately the same number of plastids as the parent cells (Cran and Possingham 1972; Juniper and Clowes 1965; Lyndon and Roberson 1976). Furthermore, in higher plants plastid division is controlled in a cell-type-specific manner to ensure that specialised cell types can carry out their functions optimally. In the best studied example young photosynthetic mesophyll cells of *Arabidopsis* contain only ~14 chloroplasts that must undergo several controlled rounds of division to result in a final complement of ~100 chloroplasts per cell (Marrison et al. 1999).

It is clear that plastid division is intimately linked with the host cell cycle and this raises the intriguing question of how the organelle and its biogenesis have been incorporated into the environment of the cell and placed under the control of the host nucleus. In order to address questions concerning both the mechanism and control of plastid division, it is crucial to have an understanding of the origin and evolution of plastids.

The prevailing endosymbiotic theory holds that plastids arose approximately a billion years ago through symbiosis of a eukaryotic host cell and a free-living cyanobacterial symbiont (Gray 1999; McFadden 1999). During the establishment of a permanent endosymbiosis plastids lost many genes necessary for the free-living state and also relinquished others to the host nucleus. For example, plastid genomes are estimated to encode less than 200 proteins, while perhaps as many as 5,000 gene products are present in the functional organelle (Leister 2003; Martin et al. 2002). A majority of plastid proteins are now translated on cytosolic ribosomes and sophisticated targeting and import machinery has evolved to ensure that plastid proteins are delivered back to their original site of function (reviewed in Bedard and Jarvis 2005; Aronsson and Jarvis 2008). Consequently, all essential plastid processes are dependent on the regulated import of nuclear-encoded proteins. This twist in evolution also means that in addition to controlling plastid function through the controlled supply of nuclear-encoded proteins, the host cell can impart control through the import and subsequent incorporation of novel proteins into existing plastid mechanisms.

Evidently organelle function is dependent upon interactions between nuclear and organelle genetic systems. However, it is too simplistic to consider the plastids to have "lost control" of their biogenesis and whilst most information flows from the nucleus to the chloroplasts, it is also clear that expression of nuclear genes is regulated by signals originating in the chloroplasts (see Sect. 5). Consequently, signals from the environment as well as endogenous cues that influence plastid division can be perceived both directly by the plastid, resulting in a change in plastid state and/or a signal to the nucleus, or by the nucleus/whole cell and relayed to the plastid (Fig. 1).

Questions concerning the evolution of plastids have been a driving force in the study of the plastid division mechanism (Tveitaskog et al. 2007). Reminiscent of their bacterial ancestors chloroplasts divide symmetrically by binary fission, as is evident from image collections that show the chloroplasts to be dumbbell-shaped and progressively centrally constricting during the process of division (Fig. 2). Indeed modern-day plastids have still retained elements of their cyanobacterial ancestor's division machinery. However, many of the proteins required for bacterial cell division have also been lost and during the establishment of the symbiosis the eukaryotic host has also added several new components (Fig. 2). Recent molecular data indicates that overall the mechanisms of bacterial cell and plastid division are remarkably conserved across lineages. The aim of this chapter is to discuss our molecular and cellular knowledge of the plastid division process in higher plants in terms of environmental signals and endogenous cues and to direct the reader towards new research avenues in the field.

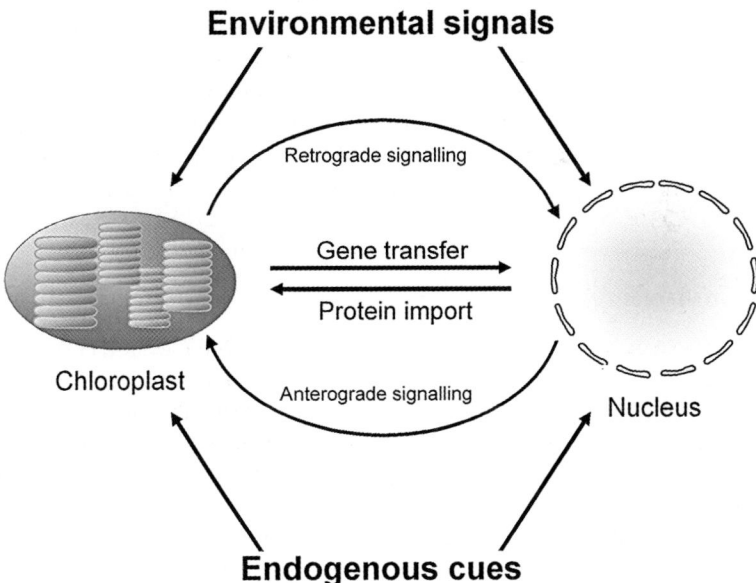

Fig. 1 Plastid and nuclear communication controls plastid division. Environmental and endogenous signals can be perceived directly by the plastid, which ultimately results in signals being transduced to the nucleus (retrograde signalling). In contrast, signals are also perceived directly by the nucleus affecting nuclear gene expression. Control of plastid division may also be imparted by the nucleus through the controlled production of plastid proteins before import or by anterograde signalling

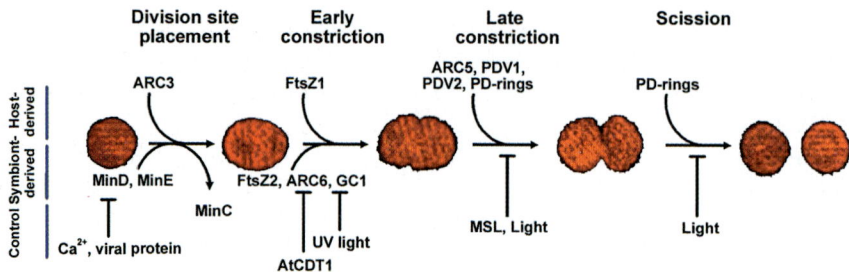

Fig. 2 Model of the evolution of the composition and the regulation of the plastid division machinery. Plastid division can be divided into several stages: division site selection, early and late constriction, and final scission. Extended focus images of chlorophyll autofluorescence from the chloroplast at different stages of division are shown. *Arrows* indicate the inclusion/loss of division components in the plastid division mechanism through evolution. The evolutionary origins of the known division machinery components are indicated as host- (*top*) or symbiont-derived (*middle*). Levels of regulation are indicated (*bottom*)

2 Regulation of the Initiation of Plastid Division

The initiation of plastid division is tightly regulated during plant development with the cell cycle and also in a cell-type-specific manner. There is also evidence that initiation of plastid division can be affected by environmental factors, such as light and pathogen infection (see Sect. 4). How the initiation of plastid division is controlled is not understood but increased knowledge of the evolution, function and regulation of some key components of the machinery now allows us to begin to address this question.

2.1 Controlling Z-Ring Formation and Function

A homologue of the highly conserved, essential bacterial cell-division protein FtsZ was first discovered in the nuclear genome of *Arabidopsis* (Osteryoung and Vierling 1995) and subsequently *FtsZ* genes have been identified in many plant and algae species (reviewed in Gilson and Beech 2001). It is widely accepted that the initiating event in both bacterial cell division and plastid division is the polymerisation of the FtsZ protein at the midcell/midplastid point to form a contractile ring (Z-ring) around the interior surface of the membrane, however the levels at which this might be controlled are less well understood (Errington et al. 2003).

There is evidence that FtsZ is regulated at the transcript level in both tobacco BY-2 cell-suspension cultures and *Chlamydomonas*, where a degree of upregulation is witnessed during plastid division in synchronised cultures (Adams et al. 2008; El-Shami et al. 2002; Suzuki et al. 1994). Furthermore, it is believed that the transcript levels of FtsZ are correlated with chloroplast number and organelle division activity (Gaikwad et al. 2000). The non-synchronous nature of plastid division in *Arabidopsis* has meant that it is not possible to study potential levels of regulation at the transcript level. However, recent molecular data is revealing ways in which the function of FtsZ at the protein level may be regulated to control plastid division.

Whereas the division of *E. coli* and most other bacteria requires only one FtsZ protein, the nuclear genomes of higher plants, mosses and algae have been shown to harbour two FtsZ homologues and most plastidic FtsZ proteins fall into two distinct clades, termed FtsZ1 and FtsZ2 (Osteryoung and McAndrew 2001; Osteryoung et al. 1998; Stokes and Osteryoung 2003). Interestingly, knockout mutants in *Arabidopsis* and *P. patens* reveal an essential role for both FtsZ clades: In both cases the most severely affected mutants harbour just one enlarged chloroplast per cell, instead of approximately 50 or 100 chloroplasts in each cell, respectively (Osteryoung and Pyke 1998; Strepp et al. 1998). The need for two FtsZ proteins was not clear in light of the fact that a single FtsZ protein is sufficient to bring about bacterial cell division. Furthermore, phylogenetic analysis reveals that the two clades of FtsZ protein diverged early in the evolution of photosynthetic eukaryotes (Stokes and Osteryoung 2003) and both retain many characteristics of the bacterial FtsZ proteins, suggesting that the requirement was not mechanistic. It is possible that the addition of an FtsZ was one way in which the host cell could exert control over the division process.

The GTPase activity of seven bacterial FtsZ proteins has been confirmed (de Boer et al. 1992; Lu et al. 1998; Nagahisa et al. 2000; Rajagopalan et al. 2005; Wang and Lutkenhaus 1993), and it is speculated that this polymerisation-induced GTP hydrolysis may provide the force for division. Analysis of the primary structure of plastidic FtsZ proteins reveals that all FtsZ1 and FtsZ2 proteins contain a conserved Rossmann fold which is essential for GTP-hydrolysis and harbours the GTP-binding tubulin signature motif GGGTG(T/S)G (de Boer et al. 1992; RayChaudhuri and Park 1992) associated with the GTPase activity of eukaryotic tubulins and bacterial FtsZ proteins. This suggests that the enzymatic function of the plastidic FtsZ proteins has been conserved, a hypothesis that is further supported by the finding that mutation in the T7 loop in AtFtsZ1 results in a semi-dominant mutation (Yoder et al. 2007) and that the pea FtsZ protein can rescue an ftsZ temperature-sensitive *E. coli* mutant (Gaikwad et al. 2000). Therefore, since both FtsZ clades are predicted to be functional GTPases and have non-redundant functions, GTPase activity alone cannot be sufficient to describe the need for the evolution of two clades of FtsZ proteins. However, in FtsZ1 but not FtsZ2 one highly conserved residue, predicted to contact the guanine nucleotide, is substituted representing an important divergence between these two protein families (Lowe and Amos 1999; Osteryoung and McAndrew 2001; Wang et al. 1997). Experimental confirmation of the GTPase activity of plastidic FtsZs could open important avenues to investigate the function and regulation of the Z-ring in plastids.

A second possible functional distinction is that the defining function of one or both of the FtsZ proteins is as a scaffold protein to bring essential division components to the septum. In *E. coli* the Z-ring is essential for the recruitment of at least ten other proteins (ZipA, FtsA, FtsE, FtsX, FtsK, FtsQ, FtsL, FtsW, FtsI and FtsN) that are required for the progression and completion of cell division (Addinall and Lutkenhaus 1996; Chen et al. 1999; Din et al. 1998; Goehring et al. 2006; Lutkenhaus and Addinall 1997; Ma et al. 1996; Ma and Margolin 1999; Vicente et al. 2006; Wang et al. 1997; Weiss et al. 1999; Yu and Margolin 1997). To date all plastidic FtsZ proteins are published to be imported into chloroplasts and to form Z-rings at the site of plastid division (Fig. 3; Fujiwara and Yoshida 2001; Kuroiwa et al. 2002; McAndrew et al. 2001; Miyagishima et al. 2001b; Vitha et al. 2001). Furthermore, based on structural similarity to the eukaryotic tubulin proteins the C-terminal tails of all FtsZ proteins are predicted to be exposed in the Z-ring and as such able to interact with and recruit accessory proteins (Desai and Mitchison 1997; Lowe and Amos 1998; Nogales et al. 1998a,b). It is highly possible that the evolution of a new FtsZ clade is to recruit novel (potentially eukaryotic) proteins to the ring. Indeed this hypothesis has credibility because of the recent identification of two Z-ring accessory proteins: ARC6, which interacts specifically with FtsZ2 and ARC3, which interacts with FtsZ1 (Fig. 3; Maple et al. 2005, 2007; Vitha et al. 2003).

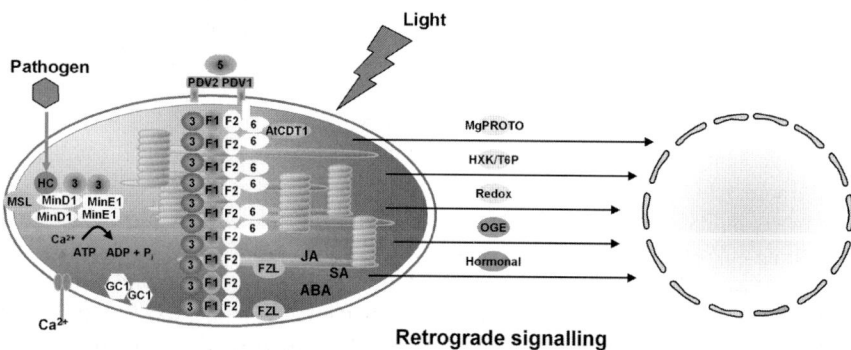

Fig. 3 Simplified working model of plastid division and interorganellar communication. Plastid division is initiated by the assembly of the stromal division proteins AtFtsZ1-1 (F1), AtFtsZ2-1 (F2) at the centre of chloroplasts. ARC6 (6) and ARC3 (3) are then recruited to the Z-ring through specific interactions with AtFtsZ2-1 and AtFtsZ1-1, respectively. AtCDT1 also interacts with ARC6. The placement of the Z-ring requires the combined action of AtMinE1, AtMinD1 and possibly ARC3. GC1 localises to the stromal side of the inner envelope membrane and forms dimers. PDV1 and PDV2 localise to ring-like structures on the cytosolic surface of the outer envelope membrane and recruit ARC5 to form the cytosolic division machinery. Mechanosensitive-like proteins (MSL) are localised to the chloroplast inner membrane as is the FZO-like protein (FZL). The division process could be regulated by the influence of light, calcium concentrations through ion transporters in chloroplast membranes (Ca^{2+}) and pathogens (HC represents HC-pro, an antigen from the Potato virus Y). A fine-tuned coordination between chloroplast and nucleus gene expressions is achieved through a series of so-called retrograde signals: (1) Light-driven changes in the redox state act as signals that regulate the expression of both chloroplasts and nuclear genes (Allen et al. 1995), (2) Sugar signals negatively regulate nuclear gene expression of photosynthetic genes involving hexokinases (HXK) and trealose 6 phosphate (T6P) (Kolbe et al. 2005; Rolland et al. 2006), (3) Mg-protoporphyrin IX (MgPROTO) can modulate the expression of nuclear-encoded genes (Mochizuki et al. 2001; Strand 2004), and (4) Non-photosynthesis-related pathways such as the organelle gene expression (OGE) and possibly hormonal regulation pathway

2.2 Regulation of Division Through the Z-Ring Accessory Proteins

In bacteria the stability and maintenance of the Z-ring is important and this role is performed by ZipA and FtsA (Din et al. 1998; Hale et al. 2000; Liu et al. 1999; Mosyak et al. 2000; Wang et al. 1997; Yan et al. 2000). ZipA and FtsA interact with FtsZ through a highly conserved peptide sequence at the extreme C-terminus known as the CORE domain, a surface-exposed hydrophilic domain, with the highly conserved sequence (D/E-I/V-P-X-F/Y-L) (Ma and Margolin 1999). Interestingly, the CORE domain is present in all FtsZ2 proteins, but not in FtsZ1 proteins and furthermore in *Arabidopsis* this domain is required for the direct interaction of AtFtsZ2 with the essential plastid division protein ARC6 (Fig. 3; Maple et al. 2005). Like ZipA, ARC6 is an integral membrane protein and could anchor the FtsZ proteins to the membrane in plastids (Hale and de Boer 1997; Vitha et al. 2003). Indeed in the *arc6* mutant Z-rings fail to form and plastid division is completely inhibited, as evident by the presence of one or two giant chloroplast per cell in both meristematic and mesophyll cells (Pyke et al. 1994; Robertson et al. 1995; Vitha et al. 2003). Furthermore, using in vivo protein interaction analysis ARC6 has been shown to localise to a discontinuous ring-like structure, indicating that like ZipA and FtsA, ARC6 may not interact with FtsZ2 as a solid ring structure (Maple et al. 2005; Rueda et al. 2003; Vitha et al. 2003).

ARC6 is conserved from cyanobacteria and it is likely that FtsZ2 and ARC6 have been retained through evolution to maintain the integrity of the Z-ring (Koksharova and Wolk 2002; Vitha et al. 2003). However, despite prokaryotic origins of these components, ARC6 has some features distinct from those of FtsA and ZipA, suggesting that ARC6 has evolved to carry out additional functions. Significantly ARC6 has evolved to interact with Cdc10-dependent transcript 1 (AtCDT1), a protein with dual localisation that forms part of the pre-replication complex in the nucleus and which also localises to plastids (Fig. 3; Raynaud et al. 2005). The role of the interaction between AtCDT1 and ARC6 is not clear but it is known that downregulation of AtCDT1 alters both nuclear DNA replication and plastid division. Since changes in endoreduplication have been associated with some stress responses, such as pathogen infection (Lingua et al. 2001), the role of AtCDT1 may be to coordinate events in the nucleus with those in organelles. The identification of AtCDT1 provides the first molecular link between nuclear events and plastid division.

FtsZ1 is more divergent from its cyanobacterial ancestors than FtsZ2, and is unique to plants and green algae. If indeed the Z-ring is stabilised and/or regulated through the conserved action of AtFtsZ2 and ARC6, why have plants evolved to require the FtsZ1 clade of proteins? The C-terminal domain of AtFtsZ1 is not required for the formation or the positioning of the Z-ring at the site of constriction (Maple et al. 2005; Wang et al. 1997), however deletion of this region results in mesophyll cells that contain a reduced number of chloroplasts, indicating that some action of FtsZ1 function is impaired (Yoder et al. 2007). Interestingly, the C-terminal region of AtFtsZ1 is required for the interaction with the stromal plastid division protein ARC3, indicating that this interaction is essential for correct plastid division (Fig. 3; Maple et al. 2007).

ARC3 is composed of an N-terminal FtsZ-like domain and a C-terminal domain containing membrane occupation and recognition nexus (MORN) repeats, linked by a unique middle domain (Shimada et al. 2004). The FtsZ-like domain of ARC3 lacks all predicted catalytic residues suggesting that it is not a functional GTPase protein and the presence of the FtsZ-like domain may have evolved to mediate the interaction of ARC3 with AtFtsZ1. ARC3 was also found to interact with AtMinE1 and AtMinD1, suggesting that ARC3 functions as part of the machinery required to accurately place the Z-ring at the plastid midpoint (see Sect. 2.3). In other proteins MORN domains are known to regulate both enzymatic activity and protein subcellular localisation and dynamics (Im et al. 2007; Ma et al. 2006). Once a more defined function has been assigned to ARC3 it will be interesting to assess whether ARC3 indeed plays a regulatory role.

Homologues of ARC3 have only been identified in *Arabidopsis* and rice and so the evolution of this plastid division component does not explain the requirement for FtsZ1 in plastids of lower plants and algae. Furthermore, in *P. patens* both FtsZ proteins appear to fall into the FtsZ2 family of proteins and in red and brown algae there is a third family, termed FtsZ3 (Stokes and Osteryoung 2003). In *P. patens* it has been suggested that the FtsZ proteins may have additional roles in the cytosol (Kiessling et al. 2004) and clearly there may be further functions associated with each clade of FtsZ proteins and additional Z-ring accessory proteins may have evolved to execute and regulate plastid division at the level of the Z-ring. Identification of the interacting partners of FtsZ proteins in different plastids will be key in understanding the function of multiple FtsZs during plastid division.

2.3 Controlling Plastid Division Through Z-Ring Placement

In bacteria the exact placement of the Z-ring is ensured by the combined effects of the Min family of proteins (de Boer et al. 1989) and homologues of the bacterial MinE and MinD proteins are encoded in the plastid genome of *C. vulgaris* (Wakasugi et al. 1997) and *G. theta* (Douglas and Penny 1999) and in the nuclear genome of *Chlamydomona* and higher plants (Adams et al. 2008; Colletti et al. 2000; Itoh et al. 2001; Kanamaru et al. 2000; Maple et al. 2002).

The plastid Min homologues have retained many characteristics of the bacterial Min system: Like their bacterial homologues the *Arabidopsis* and *Chlamydomonas* Min proteins can form a complex and localise at the plastid poles in *Arabidopsis* in a pattern that is suggested to be reminiscent of the dynamic oscillatory intracellular pattern of prokaryotic MinE (Adams et al. 2008; Maple et al. 2002, 2005). Furthermore, AtMinD1 harbours a conserved deviant Walker A motif required for ATP hydrolysis and AtMinD1 has been experimentally proven to have retained ATPase activity (Aldridge and Møller 2005). The ATPase activity of AtMinD1 is stimulated by AtMinE1 but unlike in *E. coli* this is not dependent on the presence of phospholipids and furthermore AtMinD1 ATPase activity is not stimulated by Mg^{2+}, but by Ca^{2+}, suggesting differences have evolved in the proteins function (Fig. 3; Aldridge and Møller 2005). The ATPase activity of MinD is suggested to drive the oscillatory movement of the Min proteins, to prevent the formation of the Z-ring at sites other than at midcell. If this is the case in plastids then modulation of MinD activity could be sufficient to prevent Z-ring formation at any point in the chloroplasts and thus arrest division before the formation of the Z-ring. The change in the cation dependence of AtMinD1 may represent an evolutionary adaptation to respond to different signals within a plant cell. Indeed, many plant processes are regulated by calcium (Sai and Johnson 2002). Although the ATPase activity of other plastidic MinD proteins is yet to be proven, it will be very interesting to explore whether the change in cation dependence is universal or if the affect of Ca^{2+} is specific to higher plants.

3 Regulation of the Progression of Plastid Division

After formation of the Z-ring it is belived that constriction of the plastid is initiated and that the septum progressively tightens to eventually separate the two new daughter plastids. The molecular mechanisms surrounding these events are largely unknown, however, if environmental or developmental cues change it may be beneficial for a plant cell to slow, speed up or stop plastid division for a period of time. Some of these responses may have been conserved through evolution but it is also likely that the integration of the plastid into the host cell has necessitated the evolution of new levels of control.

3.1 Possible Conserved Mechanisms of Plastid Division Regulation

Bacterial cells regulate the progression of cell division in response to environmental stimuli and whilst the cytosolic environment in which plastids reside is very different to that of free-living bacteria, both are exposed to some common environmental stresses, making it feasible that some regulatory mechanisms may have been conserved throughout evolution.

Of importance to all organisms is the threat of DNA damage resulting from both normal metabolic activities and environmental factors such as UV light and chemical exposure. Many of these lesions cause structural damage to the DNA molecule or can induce potentially harmful mutations in the cell's genome, thus affecting survival. In a subset of bacteria the SulA cell-division inhibitor protein is induced in response to DNA damage as part of the SOS response (Fernandez De Henestrosa et al. 2000; Huisman and D'Ari 1981). SulA interacts with FtsZ and prevents polymerisation, consequently enabling SulA to inhibit cell division until the DNA damage is repaired (Huisman et al. 1984). In *Arabidopsis*, GIANT CHLOROPLAST 1 (GC1, AtSulA; Maple et al. 2004; Raynaud et al. 2004) has weak similarity to SulA proteins and it is possible that GC1 is required to arrest plastid division in response to DNA damage. How this would be possible is not clear: no physical interaction has been detected between GC1 and AtFtsZ1-1 or AtFtsZ2-1. However, there is genetic evidence for an interaction with the FtsZ pathway (Raynaud et al. 2004) and furthermore in Genevestigator*GC1* transcript levels are shown to increase in response to exposure to UVB light (Table 1; Zimmermann et al. 2004). GC1 evolved most likely from the cyanobacterial cell-division protein slr1223 (Raynaud et al. 2004) and investigation of the functions of both slr1223 and GC1 in relation to DNA damage may reveal conserved response mechanisms.

Table 1 Distribution and regulation of known components of the plastid division machinery

Protein	Sy	Cm	Cr	At	Evidence for regulation at protein level	Evidence for regulation at transcript level[d]
MinC	+		−	−	N/A	N/A
MinD	+	+	+	+	Change in cation dependence	+ Cell cycle[e]
MinE	+	+	+	+	Change in domain structure	+ Cell cycle[e] − Isoxaben
ARC3	−	−	−	+	Unique to higher plants. MORN motifs implicated in regulation	None
FtsZ1	−	−[a]	+	+	Evolved via duplication event, unique to higher plants and green algae. Interacts with ARC3	+ Isoxaben, + cell cycle[e], light[f]
FtsZ2	+	−[a]	+	+	Conserved from cyanobacteria, conserved interaction domain motif, interacts with ARC6	+ Cell cycle[e]
ARC6	+[b]	−	+	+	Conserved from cyanobacteria, interacts with FtsZ2 and also PDV1 and AtCDT1	+ Isoxaben − Senescence
GC1	+	+	+	+	Conserved from cyanobacteria. Genetic connection with FtsZ pathway. Possible epimerase function	+ Nematode and light intensity and light quality (UV-A, UV-AB, red, far-red, blue, white) − *A. tumefaciens*
MSL2	+	+	+	+	Conserved from bacteria	None
MSL3	+	+	+	+	Conserved from bacteria. Possible role in response to environmental signals	+ Isoxaben, hormones (Methyl jasmonate and BL/H3BO3) and senescence

(continued)

Table 1 (continued)

Protein	Sy	Cm	Cr	At	Evidence for regulation at protein level	Evidence for regulation at transcript level[d]
PDV1	–	–	–	+	Unique to higher plants. Recruits ARC5. PDV1 interacts with ARC6	None
PDV2	–	–	–	+	Unique to higher plants. Recruits ARC5	+ 6-Benzyladenine and hypoxia
ARC5	–	+[c]	+[c]	+	Unique to higher plants, related to eukaryotic-derived dynamin-family protein	+ Isobaxen and glucose – Senescence
AtCDT1	–	–	–	+	Localises to nucleus and plastids. Interacts with ARC6	+ *A. tumefaciens* and glucose – Senescence and heat
PD rings	–	+	?	+	Of unknown composition, presumed of eukaryotic origin	None

Sy, *Synechocystis*; Cm, *Cyanidioschyzon merolae*; Cr, *Chlamydomonas reinhardtii*; At, *Arabidopsis thaliana*
[a]*C. merolae* harbours two FtsZ3 proteins
[b]The cyanobacterial protein is Ftn2
[c]The *C. merolae* protein is CmDnm2; the *C. reinhardtii* protein is DRP5B
[d]Microarray data from the Genevestigator database for the highest category of upregulation (+) and downregulation (–) in response to different factors (https://www.genevestigator.ethz.ch(Zimmermann et al. 2004)
[e]Adams et al. 2008; El-Shami et al. 2002; Suzuki et al. 1994
[f]Ullanat and Jayabaskaran 2002

A large family of mechanosensitive channels of small conductance-like proteins (MSL) has been identified in the nuclear genome of *Arabidopsis* (Haswell and Meyerowitz 2006). In *E. coli* mechanosensitive ion channels are opened by force, and help to protect cells against osmotic shock (Levina et al. 1999). The *Arabidopsis* MSL2 and MSL3 are stromal plastid proteins and localise to discrete foci, in close proximity to the chloroplast envelope, in an analogous manner to AtMinE1 and AtMinD1 (Fig. 3; Haswell and Meyerowitz 2006; Maple et al. 2002). In *Arabidopsis*, *msl2-1/msl3-1* double loss-of-function mutants harbour large and spherical chloroplasts suggesting increased internal osmotic pressure (Haswell and Meyerowitz 2006). This phenotype is only seen in the double mutant suggesting that MSL2 and MSL3 have redundant functions. It has been suggested that MSL2 and MSL3 are required to release ions from the plastid in response to changes in envelope-membrane tension produced during constriction of the dividing chloroplasts. However, investigation of the response of MSL to different factors reveals that *MSL3* is specifically upregulated in response to some hormones and senescence, whereas *MSL2* is not upregulated (Table 1; Zimmermann et al. 2004). This suggests a model whereby MSL2 is required to regulate envelope-membrane tension during normal plastid division, and MSL3 is called into play in response to environmental factors or during plant cell death.

3.2 Eukaryotic Mechanisms of Plastid Division Regulation

It is widely speculated that the incorporation of new components of eukaryotic origin into the plastid division machinery is not purely to enable the division machinery to carry out the fission of a double membrane organelle, but also to allow the host to exert control over the process.

The first protein of eukaryotic origin to be identified as part of the plastid division machinery was a dynamin-like protein in both *Arabidopsis* (ARC5; Gao et al. 2003) and *C. merolae* (CmDnm2; Miyagishima et al. 2003). The dynamin superfamily of GTPases is well documented to participate in fusion and fission of intracellular membrane structures, mediating events such as endocytosis and mitochondrial division in eukaryotes (Hinshaw 2000). ARC5 and CmDnm2 define a new class of dynamin-like proteins that function specifically in plastid division and have no homologues in prokaryotes, indicating that they evolved from the host cell.

Dynamins are multi-domain proteins and self-assemble into higher-order structures resembling rings. Both ARC5 and CmDnm2 harbour conserved domains and relocate from the cytosol to the surface of the outer chloroplast membrane to form a ring-like structure at the site of constriction (Fig. 3; Gao et al. 2003; Miyagishima et al. 2003). ARC5 and CmDnm2 are closely related and phylogenetic analysis reveals that the role of dynamin proteins in plastid division evolved before the red and green algae diverged. This ancient acquisition of dynamin proteins may have been fuelled by the need to generate the force required to complete the division of the double membrane plastid. Consistent with this idea *arc5* chloroplasts enter division and arrest when they have become centrally constricted, never completing division (Gao et al. 2003; Marrison et al. 1999). More recently, elegant experiments using optical tweezers has shown that dynamin generates the major constrictive force during division in *C. merolae* (Yoshida et al. 2006). It is not yet known whether the role of dynamins is solely to provide a mechanochemical function or whether regulation of late stages of plastid division can be imparted through the dynamins.

ARC5 and CmDnm2 are distantly related to dynamin-like proteins shown to play a role in mitochondrial division in higher plants (ADL2a and ADL2b; Arimura et al. 2004; Arimura and Tsutsumi 2002), yeast (Dnm1p; Bleazard et al. 1999), mammals (Drp1; Smirnova et al. 2001) and red algae (CmDnm1; Nishida et al. 2003). Furthermore, in *Saccharomyces cerevisae* a rhomboid protease has been shown to regulate mitochondrial membrane remodelling through the processing of the dynamin-like protein Mgm1p (McQuibban et al. 2003). Rhomboid proteases are a relatively poorly investigated family but have been shown to be important in signal transduction. There are at least 15 members in the *Arabidopsis* family (Garcia-Lorenzo et al. 2006) and it will be interesting to analyse if any play a role in the regulation of plastid division.

The localisation of ARC5 to the plastid division site requires the concerted action of two proteins: PDV1 and PDV2 (Miyagishima et al. 2006). Although only PDV1 has been fully characterised, based on amino acid similarity and partial functional redundancy, it is predicted that both PDV1 and PDV2 are transmembrane proteins and will localise to a ring-like structure on the outer envelope at the division site, with the N-terminal region facing into the cytosol (Miyagishima et al. 2006). The N-terminal domain harbours a coiled-coil domain suggesting that this domain interacts directly with ARC5 or through other components to recruit ARC5 after constriction has initiated. Deletion of the C-terminal glycine residue of PDV1 (PDV1-c) results in a chloroplast division defect similar to that seen in the arc5 mutant. This is because although PDV1-c is able to integrate into the chloroplast outer envelope, it is unable to recognise the chloroplast division site (Miyagishima et al. 2006). Recent data has shown that this is because PDV1 interacts with the stromal plastid division protein ARC6 (Glynn et al. 2007). There are two consequences of this interaction: Firstly, it identifies a mechanism by which the cytosolic division machinery is targeted to the exact site of constriction after Z-ring formation and secondly it identifies a possible mechanism by which the cytosol communicates with the stromal division machinery to ultimately control plastid division. It will be very interesting to explore whether PDV2 can also interact with ARC6 or whether PDV2 has a unique interacting partner.

Neither PDV1 nor PDV2 have orthologs in the red algae *C. merolae* suggesting that red algae and green plants may use different mechanisms for recruitment of dynamin to the division site and for communication between the cytosolic and stromal division machineries. However, recent data indicates that an analogous connection does exist since a plastid-dividing dynamin FtsZ-ring can be isolated as an intact structure from synchronised *C. merolae* cells and that the linkage between all the rings is continuous and likely free of membrane lipids (Yoshida et al. 2006). The identification of the protein components that mediate the interaction between the stromal and cytosolic machineries in *C. merolae* will be a key finding.

A second dynamin-like protein has been implicated in plastid division, although it is not yet clear as to whether this affect is direct. FZL is related to a mitochondrial dynamin-like protein called Fzo that mediates fusion between the mitochondrial membranes in animals and fungi (reviewed by Mozdy and Shaw 2003). Loss-of-function *Arabidopsis* mutants in *fzl* show abnormalities in both chloroplast and thylakoid morphology and consistent with these phenotypes an FZL-GFO fusion is associated with both the chloroplast inner envelope membrane and the thylakoid membranes (Gao et al. 2006). Although it has been speculated that FZL is not directly involved in plastid division it is possible that FZL will play a role in regulating plastid division with respect to thylakoid integrity.

The formation and constriction of plastids is associated with electron-dense rings termed plastid division (PD) rings. The PD ring structure consists of two rings, one on the outer and one on the inner surface of the chloroplast membrane in numerous plant and algal species, suggesting that the two PD rings represent a universal feature of dividing chloroplasts in all plant cells (Hashimoto 1986). Furthermore, in the unicellular red alga *C. merolae* a third PD ring has been visualised in the intermembrane space and it is possible that it is ubiquitous throughout plant species (Miyagishima et al. 2001a). The timing of assembly and the behaviour of each ring is different and all form after Z-ring formation (Miyagishima et al. 2001b). The apparent broad distribution of the PD rings in plastids makes it possible that universal control mechanisms have evolved with the PD rings, however none of the protein components are known and until these are identified it is not possible to speculate further on possible levels of regulation.

4 Environmental Effects on Plastid Division

Because of their sessile nature higher plants constantly respond to environmental factors by moulding metabolic, redox and developmental processes in accordance to the environmental inputs. The response of a plant to environmental signals can range from slight acclimation, resulting for example in changes in protein phosphorylation and small antioxidants redox state, to more dramatic changes such as modifications of gene expression and ultrastructural alterations. Because plants are often challenged by different factors giving rise to additive, multiplicative or synergetic reactions, a high degree of developmental plasticity in response to environmental signals is required.

Given the importance of photosynthesis to plant survival, light signals are arguably among the most important environmental cues for the development of plants. However, other factors such as temperature, water status, nutrient and CO_2 availability, pollutants and pathogens are also factors that might compromise plant growth and important development processes including plastid division and function (Fig. 1). Although the molecular process of plastid division is starting to become unravelled, limited information exists on environmental signals that affect the process. The following sections describe how environmental signals may influence plastid division in plants.

4.1 Effects of Light on Chloroplast Division

Since light is the predominant source of energy for plant metabolism, growth and development and because the photochemical reactions take place in the chloroplast it is logical to assume that certain aspects of chloroplast division may be regulated by light. Although no evidence is available demonstrating direct regulation by light, several studies underline the fact that light conditions are intimately linked to chloroplast division.

Plants respond to day–night cycles through a highly complex network of events and it is now well established that vital processes such as stomata opening, photosynthesis and carbon fixation are intrinsically regulated by the circadian clock. More specifically, genes encoding chloroplast proteins directly involved in the recollection of light (*LhcA* and *LhcB*) and within the so-called dark reactions (Rubisco small subunit – *RbcS* and Rubisco activase – *Rca*) are regulated by the circadian system (reviewed in Yakir et al. 2007). Recently the *Arabidopsis* mutants *crb1* and *crb2* were described as having altered chloroplast structure and function resulting in poor photosynthetic efficiency. Interestingly, *Crb* encodes a chloroplast RNA binding protein and *Crb*-deficient mutants, as well as other chloroplast retrograde signalling mutants (*stn7* and *gun1*), show distorted circadian rhythms suggesting that chloroplasts and the circadian clock interact to ensure the fine-tuning and close regulation of chloroplast processes (Hassidim et al. 2007). On the basis of such data it is tempting to speculate that either the regulation of chloroplast division genes or protein import into chloroplasts is regulated in a circadian fashion. The bacterial cell division gene *ftsZ* is indeed known to be expressed with a circadian pattern in the cyanobacterium *Synechococcus elongatus* (Mori and Johnson 2001). However, to date no study has shown any photoperiod-related change of expression in chloroplast division genes in plants.

In several plant species there is a clear relationship between the size of the cell and the number of chloroplasts (Leech and Pyke 1988; Pyke 1997). In wheat, chloroplast number in fully expanded mesophyll cells is positively correlated with the so-called cell face area. For example, cells with a relatively small face area have a correspondingly higher density of chloroplasts suggesting a tight regulation of chloroplast number presumably through division control (Ellis and Leech 1985; Jasinski et al. 2003). Interestingly, chloroplast division initiation events increase in rapidly expanding cells (Pyke 1997; Leech and Pyke 1988) and furthermore, rapid chloroplast accumulation has been observed during leaf greening (Lamppa and Bendich 1979). Simple but very informative experiments revealed that plant leaf discs cultured for four days in continuous white light show a chloroplast regeneration time within 24 h. In contrast, this regeneration time is twofold longer when leaf discs are transferred to darkness. It was suggested that the separation of dumbbell-shaped chloroplasts may represent the rate-limiting step in response to light–dark cycles based on the fact that when dark-adapted leaf discs are transferred to light, most dumbbell-shaped chloroplasts undergo complete division in less than 60 min (Fig. 2; Hashimoto and Possingham 1989). Given that the *arc5* mutant of *Arabidopsis* shows a dumbbell-shaped chloroplast phenotype, these observations further suggest that ARC5 activity may be regulated by light but further research is needed to clarify this.

The accumulation of chloroplasts in plants exposed to high light intensities most probably determines the level of future division events. Highly dense chloroplast subpopulations in cells, as a result of high light stress-induced avoidance, would likely stop division in order to prevent further organelle damage. Evidence supporting this idea comes from studies showing that the chloroplast-targeted ion channel proteins MSL2 and MSL3 in *Arabidopsis* might have a role in sensing osmotic pressure and in the sensing of touch (Haswell and Meyerowitz 2006). MSL2 and MSL3 could represent perfect candidates for sensing the chloroplast number/size ratio according to the cell size and/or the compression of chloroplasts in the cell. The *msl2-1/msl3-1* double mutants harbour both larger cells and larger chloroplasts, suggesting an impairment of the sensing of the number of chloroplasts to cell size ratio (Haswell and Meyerowitz 2006).

The effect of the accumulation and replication of chloroplast on photosynthetic performance has been studied in the *arc3*, *arc5* and *arc11* mutants (Austin and Webber 2005). These *arc* mutants were shown to have a reduced leaf area and altered chlorophyll content in both low and high light conditions. Moreover, their photosynthetic efficiency and the light distribution in photochemical and non-photochemical dissipation pathways were altered as shown by the redox state of their plastoquinone (PQ) pool. Oxygen evolution, level of thylakoid stacking as well as starch accumulation was also affected (Austin and Webber 2005). Similar alterations in photosynthetic efficiency were observed in plants with reduced levels of *GC1* (Maple et al. 2004). The corresponding *GC1* gene is interestingly upregulated by changes in light intensities or light qualities (Table 1; Zimmermann et al. 2004). It appears therefore that despite the increase in chloroplast volume the above-mentioned mutants studied did not perform photosynthesis as wild-type and their capacity for acclimation to higher light was seriously hindered. In fact these plants are unable to adjust the composition of the photosynthetic apparatus in response to changes in light intensity. These differences may be due to structural alterations in the *arc* mutants such as the area to volume ratio with consequent impaired import/export of metabolites through the chloroplast envelope. This could further have consequences for sugar signalling and/or on other retrograde signalling pathways such as those based on the redox state of the plastoquinone pool or the carbohydrate level, ultimately affecting (directly or indirectly) processes such as plastid division. Interestingly, microarray experiments performed on the *arc3*, *arc5* and *arc11* mutants does indeed show upregulation of genes related to carbohydrate synthesis and oxidative stress (unpublished data).

In chloroplasts, ion channels participate in pH homeostasis across the inner chloroplast envelope and the thylakoid membranes to ensure appropriate photosynthetic CO_2 fixation. The ion channels appear to be involved in the regulation of ion gradients and the pumping of H^+ from the stroma to the cytosol is thus balanced by an influx of ions through Ca^{2+}, K^+ and Cl^- channels. With respect to this, Ca^{2+} uptake into intact chloroplasts has been shown to be stimulated by illumination as one would expect (Johannes et al. 1991). In contrast, chloroplasts show a large transient spike of free calcium upon transfer from constant light to constant darkness suggesting that calcium may play a role in recognising a lights-off state (Johnson et al. 1995). More recently, a light-induced Ca^{2+} channel, the fast-activating chloroplast cation channel (FACC) was discovered in the inner envelope of chloroplasts (Pottosin et al. 2005). Although a direct correlation between calcium concentrations in the chloroplast and regulation of chloroplast division is still lacking it is interesting to consider the Ca^{2+}-dependent activation of the ATPase activity of AtMinD1 in *Arabidopsis* chloroplasts (Aldridge et al. 2005). It is highly possible that the observed light-dependent fluctuations of free calcium levels within chloroplasts modulate the activity of AtMinD1 and hence chloroplast division in relation to changing light conditions (Figs. 2 and 3).

Very little is known regarding possible pathways regulating the number and size of chloroplasts. However, the high pigment mutant of tomato (*hp-1*) and the *Arabidopsis det1* mutants provide vital clues. When grown in the absence of light, *det1* exhibits light-grown characteristics showing accumulation of light-regulated nuclear and chloroplast transcripts. Interestingly, *det1* root plastids also differentiate into chloroplasts and DET1 is thought to be a negative regulator of light signalling (Chory and Peto 1990). In contrast, the *hp-1* mutant has both an increased chloroplast density and an increased chloroplast size. The *Hp-1* gene has been identified as the UV-damaged DNA binding protein 1 (DDB1) (Lieberman et al. 2004; Liu et al. 2004) and the homologue of DDB1 in *Arabidopsis* has been shown to interact with DET1 (DDB1A, Schroeder et al. 2002). On the basis of studies in several organisms the DDB1/DET1 complex is proposed to interact with chromatin to negatively regulate the expression of nuclear genes, which in turn may affect plastid morphology (Schroeder et al. 2002).

4.2 Pathogen Defence Responses and Plastid Division

Plants have evolved intricate mechanisms to combat infections. Plant disease resistance is frequently correlated with a genetically defined interaction between a plant resistance gene and a corresponding pathogen avirulence gene (Dixon et al. 2000). In tobacco for example, a portion of the tobacco mosaic virus (TMV) replicase is recognised by the tobacco N-gene triggering the hypersensitive response (Erickson et al. 1999). The hypersensitive response is characterised by an oxidative burst, cell death around the area of infection, induction of a systemic acquired resistance (SAR) in non-infected tissues and induction of pathogen-related marker genes (Ryals et al. 1997). Gene-for-gene resistance has been shown to operate during infections with bacterial, insect, fungal and viral pathogens of plants.

There is a clear relationship between viral infections and plastids where changes in the chloroplast inner structures, size and number in infected tissues of several species of plants has recently been reported. These include alterations in thylakoid membranes, degree of stacking, stromal area and starch accumulation, an impairment of the processing and accumulation of plastid rRNA as well as the presence of small vesicles (Hinrichs-Berger et al. 1999; Pompe-Novak et al. 2001, Martínez de Alba et al. 2002; Zechmann et al. 2003; Rodio et al. 2007). Further, studies have reported that the disruption of chloroplast genes leads to resistance to viral agents. For example, by silencing a 33-kDa component of the PSII oxygen-evolving system or inhibiting the photosynthetic electron transport chain with specific inhibitors, virus replication is enhanced (Abbink et al. 2002).

An interesting breakthrough to understand how viral factors may influence chloroplast shape, size and number is from the work of Jin and collaborators (Jin et al. 2007). They first show that infection of potato with Potato virus Y (PVY) leads to altered chloroplast numbers and changes in both chloroplast morphology and photosynthetic efficiency. The viral agent HC-pro, described as a multifunctional protein helper component-proteinase, localises to chloroplasts after infection and has been shown to interact with the chloroplast division protein MinD (Jin et al. 2007). The amino acids responsible for mediating MinD dimerisation also mediate MinD/HC-pro complex formation, which could explain the lower number and altered shape of chloroplasts in cells infected by the virus (Jin et al. 2007). The reason why a viral antigen interacts with a chloroplast division protein remains somewhat unclear. However, it can be postulated that the virus hijacks and inhibits the energy-demanding process of plastid division because it requires the cell machinery for its own purposes (Figs. 2 and 3).

Although further research needs to be conducted to fully understand the interaction between pathogen defence and plastid division it is clear that an interplay between these two seemingly distant processes does occur. Indeed, using Genevestigator we have found that the expression of the *ARC5* plastid division gene is altered in response to the pathogen-associated molecule Syringolin (Table 1; Zimmermann et al. 2004). Moreover, a series of microarray experiments have shown that various plastid division mutants exhibit induction of genes related to defence pathways (unpublished observations).

5 Possible Regulation of Plastid Division

Chloroplasts are genetically semiautonomous organelles that contain their own subset of genes coding for some of the chloroplast proteins, tRNAs and rRNAs. However, in higher plants a great majority of the chloroplast proteins are encoded in the nucleus and must therefore be imported back into the organelle after translation in the cytosol. Because this is the case for all proteins involved in plastid division, retrograde signalling from plastids to the nucleus must occur to ensure temporal regulation of plastid division gene expression. Several signalling pathways by which the chloroplast communicates to the nucleus have recently been described (Gray et al. 2003; Pesaresi et al. 2007) and the following section discusses possible plastid division regulation through retrograde signalling (Fig. 3).

A fine-tuned coordination between photosynthetic electron transport, synthesis of starch and the metabolic demand for reductants is necessary for optimum light use and at the same time to avoid detrimental effects imposed by light. It is thus not surprising that several levels of retrograde regulation seemingly involve reactions related to photosynthesis. Changes in the reduction–oxidation (redox) state of components of the photosynthetic machinery, mirroring excitation imbalances between PSI and PSII, act as signals that regulate the expression of both chloroplasts and nuclear genes (Allen 1995; Pfannschmidt et al. 1999). The use of photosynthetic electron-transport inhibitors suggested the redox state of the PQ pool as a prominent candidate for the origin of the chloroplast redox signals. Changes in photosynthetic efficiency, the PQ pool redox state and the consequent alterations in redox-regulated gene expression have been found in mutants with altered chloroplast morphology (*Prors*, *crb* and *arc* mutants) but so far no specific redox control has been found for chloroplast division genes. However, we observed using the Geninvestigator database that many of the chloroplast division genes are down-regulated when leaves enter senescence (Table 1, Zimmermann et al. 2004). Since one of the first features of senescence is the decay of photosynthesis these observations suggest that photosynthetic efficiency and chloroplast division are at least indirectly linked.

Treatment of organelles with inhibitors of protein synthesis decreases the expression of nuclear photosynthesis genes in a light-independent manner during early stages of plant development (Sullivan and Gray 1999; Gray et al. 2003) showing that impairment in chloroplast gene expression can affect nuclear gene expression. This phenomenon has been called organelle gene expression (OGE) retrograde regulation

(Pesaresi et al. 2007). In tobacco, virus-induced silencing of organellar aminoacyl-tRNA synthetases leads to a severe leaf-yellowing phenotype presumably due to an arrest in chloroplast development and transformed cells harbour a reduced number of chloroplasts with an altogether reduced size. Moreover, these chloroplasts lack most of their thylakoid membranes and appear dumbbell-shaped indicating the impairment of chloroplast division (Kim et al. 2005). The transcription of nuclear genes encoding chloroplast-targeted photosynthetic proteins is also altered in the *Arabidopsis prors* mutants that are mutated in the 5'-untranslated region of the nuclear gene *Prolyl-tRNA Synthetase1* causing defects in both mitochondria and chloroplasts. The mRNA profiling data indicate that the leaky *prors* mutation results in downregulation of genes encoding photosynthetic proteins. This type of regulation is seen in both light- and dark-adapted *prors* plants, indicating that the transcriptional response is independent of light (Pesaresi et al. 2006). Unfortunately, chloroplast ultrastructures are not described in these studies but in the light of the downregulation of the photosynthetic genes described, it is reasonable to assume that both thylakoid membranes and chloroplast morphology are affected.

Plant defence in response to microbial attack is regulated through a complex network of pathways that involve three signalling molecules: salicylic acid (SA), jasmonic acid (JA) and ethylene (Dong 1998). SA is synthesised in response to diverse avirulent pathogens and is responsible for the large-scale transcriptional induction of defence-related genes as well as the establishment of SAR (Ryals et al. 1997; Maleck et al. 2000). The identification of the SA-deficient *Arabidopsis* mutant *sid2*, impaired in a gene encoding a functional isochorismate synthase (ICS), interestingly locates SA biosynthesis in the chloroplast (Wildermuth et al. 2001). Apart from a few studies investigating the role of SA in terms of photosynthetic efficiency, the direct implications of SA being produced in the chloroplast are not known (Mateo et al. 2006). JA is a ubiquitous methyl ester found in several plant species and in a variety of fungi, mosses and ferns. The initial reaction of JA biosynthesis occurs in the stroma of plastids and leads to the 13-lipoxygenase-catalyzed insertion of molecular oxygen into position 13 of α-linolenic acid, most likely released from the plastid envelope. JA is also a key regulator in mechanotransduction after wounding and in response to UV damage in higher plants (Schilmiller and Howe 2005). Abscisic acid (ABA) plays important roles in seed dormancy and in the response of plants to various environmental stresses, among them in response to drought, salt and cold stresses. Of the two possible pathways for the biosynthesis of ABA, the indirect one, in which ABA is derived from farnesyl pyrophosphate and 9-*cis*-violaxanthine is catalyzed by zeaxanthin epoxidase and takes place in the stroma of thylakoids (Marin et al. 1996). In a microarray experiment involving the *arc3*, *arc5* and *arc11* mutants, expression of marker genes for the SA, JA or ABA signalling pathways (PR1; VSP1 and VSP2; COR15 and RD20, respectively) are altered, raising the question as to whether hormonal fluctuations affect plastid division.

Finally, it is worth considering the lipid composition of the chloroplast membranes in relation to chloroplast form and function. Chloroplast membranes are characterised by a high proportion of galactolipids monogalactosyl diacylglycerol (MGD) and digalactosyl diacylglycerol (DGD), a composition reminiscent of that found in cyanobacteria (Siegenthaler 1998). Mutants affected in MGD content show altered chloroplast ultrastructure (Jarvis et al. 2000) whilst mutants impaired in DGD content show impaired chloroplast protein import (Chen and Li 1998). The composition of lipids in the membranes is thought to change in response to environmental conditions and it is tempting to speculate to what extent this affects the plastid division process through inappropriate import of plastid division proteins.

6 Conclusions and Outlook

The mechanism by which the eukaryotic host regulates the timing, frequency and progression of plastid division remains a fundamental question. Although evidence for the regulation of chloroplast number with respect to cell type and the cell cycle has been well documented, the details of how plastid division is regulated with respect to endogenous and environmental cues remain largely unknown. Plant cells coordinately regulate the expression of nuclear and plastid genes in response to different cues. Nuclear genes that regulate chloroplast development and chloroplast gene expression provide part of this control, but information also flows from chloroplasts to the nucleus. The challenge will now be to unravel the interorganellar signalling that controls the initiation and progression of plastid division and ultimately how these signals are perceived by the division machinery. Whilst these questions pose a major challenge it is encouraging that it is only ten years since the first FtsZ protein was shown to be a vital component of the plastid division machinery, and since this time astonishing progress has been made in understanding the plastid division machinery at the molecular and cellular level. Given these foundations and the strength of research in complementary fields spanning single-celled algae to higher plants, it is equally likely that rapid progress will be made in elucidating the regulation of plastid division in connection to environmental cues and other fundamental processes in plants.

Acknowledgements Plastid division research in our laboratory is funded by The Norwegian Research Council/Functional Genomics Program, The Leverhulme Trust, and The European Molecular Biology Organisation.

References

Abbink TE, Peart JR, Mos TN, Baulcombe DC, Bol JF, Linthorst HJ (2002) Silencing of a gene encoding a protein component of the oxygen-evolving complex of photosystem II enhances virus replication in plants. Virology 295:307–319

Adams S, Maple J, Møller SG (2008) Functional conservation of the MIN plastid division homologues of *Chlamydomonas reinhardtii*. Planta. In press. PMID:18270733

Addinall SG, Lutkenhaus J (1996) FtsA is localized to the septum in an FtsZ-dependent manner. J Bacteriol 178:7167–7172

Aldridge C, Møller SG (2005) The plastid division protein AtMinD1 is a Ca^{2+}-ATPase stimulated by AtMinE1. J Biol Chem 280:31673–31678

Aldridge C, Maple J, Møller SG (2005) The molecular biology of plastid division in higher plants. J Exp Bot 56:1061–1077

Allen JF (1995) Thylakoid protein phosphorylation, state1- state 2 transition, and photosystems stoichiometry adjustment: redox control at multiple levels of gene expression. Physiol Plant 93:196–205

Arimura S, Aida GP, Fujimoto M, Nakazono M, Tsutsumi N (2004) *Arabidopsis* dynamin-like protein 2a (ADL2a), like ADL2b, is involved in plant mitochondrial division. Plant Cell Physiol 45:236–242

Arimura S, Tsutsumi N (2002) A dynamin-like protein (ADL2b), rather than FtsZ, is involved in *Arabidopsis* mitochondrial division. Proc Natl Acad Sci USA 99:5727–5731

Aronsson H, Jarvis P (2008) The chloroplast protein import apparatus, its components, and their roles. Plant Cell Monogr., doi:10.1007/7089_2008_40

Austin J, Webber AN (2005) Photosynthesis in *Arabidopsis thaliana* mutants with reduced chloroplast number. Photosynth Res 85:373–384

Bedard J, Jarvis P (2005) Recognition and envelope translocation of chloroplast preproteins. J Exp Bot 56:2287–2320

Bleazard W, McCaffery JM, King EJ, Bale S, Mozdy A, Tieu Q, Nunnari J, Shaw JM (1999) The dynamin-related GTPase Dnm1 regulates mitochondrial fission in yeast. Nat Cell Biol 1:298–304

Chen LJ, Li HM (1998) A mutant deficient in the plastid lipid DGD is defective in protein import into chloroplasts. Plant J 16(1):33–39

Chen JC, Weiss DS, Ghigo JM, Beckwith J (1999) Septal localization of FtsQ, an essential cell division protein in *Escherichia coli*. J Bacteriol 181:521–530

Chory J, Peto CA (1990) Mutations in the *DET1* gene affect cell-type-specific expression of light-regulated genes and chloroplast development in *Arabidopsis*. Proc Natl Acad Sci USA 87:8776–8780

Colletti KS, Tattersall EA, Pyke KA, Froelich JE, Stokes KD, Osteryoung KW (2000) A homologue of the bacterial cell division site-determining factor MinD mediates placement of the chloroplast division apparatus. Curr Biol 10:507–516

Cran DG, Possingham JV (1972) Variation of plastid types in spinach. Protoplasma 74:345–356

de Boer P, Crossley R, Rothfield L (1992) The essential bacterial cell-division protein FtsZ is a GTPase. Nature 359:254–256

de Boer PA, Crossley RE, Rothfield LI (1989) A division inhibitor and a topological specificity factor coded for by the minicell locus determine proper placement of the division septum in *E. coli*. Cell 56:641–649

Desai A, Mitchison TJ (1997) Microtubule polymerization dynamics. Annu Rev Cell Dev Biol 13:83–117

Din N, Quardokus EM, Sackett MJ, Brun YV (1998) Dominant C-terminal deletions of FtsZ that affect its ability to localize in Caulobacter and its interaction with FtsA. Mol Microbiol 27:1051–1063

Dong X (1998) SA, JA, ethylene, and disease resistance in plants. Curr Opin Plant Biol 1:316–323

Douglas SE, Penny SL (1999) The plastid genome of the cryptophyte alga, *Guillardia theta*: complete sequence and conserved synteny groups confirm its common ancestry with red algae. J Mol Evol 48:236–244

Dixon MS, Golstein C, Thomas CM, van Der Biezen EA, Jones JD (2000) Genetic complexity of pathogen perception by plants: the example of *Rcr3*, a tomato gene required specifically by Cf-2. Proc Natl Acad Sci USA 97:8807–8814

Ellis R, Leech RM (1985) Cell size and chloroplast size in relation to chloroplast replication in light-grown wheat leaves. Planta 165:120–125

El-Shami M, El-Kafafi S, Falconet D, Lerbs-Mache S (2002) Cell cycle-dependent modulation of *FtsZ* expression in synchronized tobacco BY2 cells. Mol Genet Genomics 267:254–261

Erickson FL, Dinesh-Kumar SP, Holzberg S, Ustach CV, Dutton M, Handley V, Corr C, Baker BJ (1999) Interactions between tobacco mosaic virus and the tobacco N gene. Philos Trans R Soc Lond B Biol Sci 354:653–658

Errington J, Daniel RA, Scheffers DJ (2003) Cytokinesis in bacteria. Microbiol Mol Biol Rev 67:52–65

Fernandez De Henestrosa AR, Ogi T, Aoyagi S, Chafin D, Hayes JJ, Ohmori H, Woodgate R (2000) Identification of additional genes belonging to the LexA regulon in *Escherichia coli*. Mol Microbiol 35:1560–1572

Fujiwara M, Yoshida S (2001) Chloroplast targeting of chloroplast division FtsZ2 proteins in *Arabidopsis*. Biochem Biophys Res Commun 287:462–467

Gaikwad A, Babbarwal V, Pant V, Mukherjee SK (2000) Pea chloroplast FtsZ can form multimers and correct the thermosensitive defect of an *Escherichia coli* ftsZ mutant. Mol Gen Genet 263:213–221

Galili G (1995) Regulation of lysine and threonine synthesis. *Plant Cell* 7:899–906

Gao H, Kadirjan-Kalbach D, Froehlich JE, Osteryoung KW (2003) ARC5, a cytosolic dynamin-like protein from plants, is part of the chloroplast division machinery. Proc Natl Acad Sci USA 100:4328–4333

Gao H, Sage TL, Osteryoung KW (2006) FZL, an FZO-like protein in plants, is a determinant of thylakoid and chloroplast morphology. Proc Natl Acad Sci USA 103:6759–6764

Garcia-Lorenzo M, Sjodin A, Jansson S, Funk C (2006) Protease gene families in Populus and *Arabidopsis*. BMC Plant Biol 6:30–37

Gilson PR, Beech PL (2001) Cell division protein FtsZ: running rings around bacteria, chloroplasts and mitochondria. Res Microbiol 152:3–10

Glynn JM, Miyagishima SY, Yoder DW, Osteryoung KW, Vitha S (2007) Chloroplast division. Traffic 8:451–461

Goehring NW, Gonzalez MD, Beckwith J (2006) Premature targeting of cell division proteins to midcell reveals hierarchies of protein interactions involved in divisome assembly. Mol Microbiol 61:33–45

Gray MW (1999) Evolution of organellar genomes. Curr Opin Genet Dev 9:678–687

Gray JC, Sullivan JA, Wang JH, Jerome CA, MacLean D (2003) Coordination of plastid and nuclear gene expression. Philos Trans R Soc Lond B Biol Sci 358:135–145

Hale CA, de Boer PA (1997) Direct binding of FtsZ to ZipA, an essential component of the septal ring structure that mediates cell division in *E. coli*. Cell 88:175–185

Hale CA, Rhee AC, de Boer PA (2000) ZipA-induced bundling of FtsZ polymers mediated by an interaction between C-terminal domains. J Bacteriol 182:5153–5166

Hashimoto H (1986) Double ring structure around the constricting neck of the dividing plastids of *Avena sativa*. Protoplasma 135:166–172

Hashimoto H, Possingham JV (1989) Effect of Light on the chloroplast division cycle and DNA synthesis in cultured leaf discs of spinach. Plant Physiol 89:1178–1183

Hassidim M, Yakir E, Fradkin D, Hilman D, Kron I, Keren N, Harir Y, Yerushalmi S, Green RM (2007) Mutations in chloroplast RNA binding provide evidence for the involvement of the chloroplast in the regulation of the circadian clock in *Arabidopsis*. Plant J 51:551–562

Haswell ES, Meyerowitz EM (2006) MscS-like proteins control plastid size and shape in *Arabidopsis thaliana*. Curr Biol 16:1–11

Hinrichs-Berger J, Harfold M, Berger S, Buchenauer H (1999) Cytological responses of susceptible and extremely resistant potato plants to inoculation with potato virus Y. Physiol Mol Plant Pathol 55:143–150

Hinshaw JE (2000) Dynamin and its role in membrane fission. Annu Rev Cell Dev Biol 16:483–519

Huisman O, D'Ari R (1981) An inducible DNA replication-cell division coupling mechanism in *E. coli*. Nature 290:797–799

Huisman O, D'Ari R, Gottesman S (1984) Cell-division control in *Escherichia coli*: specific induction of the SOS function SfiA protein is sufficient to block septation. Proc Natl Acad Sci USA 81:4490–4494

Im YJ, Davis AJ, Perera IY, Johannes E, Allen NS, Boss WF (2007) The N-terminal membrane occupation and recognition nexus domain of *Arabidopsis* phosphatidylinositol phosphate kinase 1 regulates enzyme activity. J Biol Chem 282:5443–5452

Itoh R, Fujiwara M, Nagata N, Yoshida S (2001) A chloroplast protein homologous to the eubacterial topological specificity factor minE plays a role in chloroplast division. Plant Physiol 127:1644–1655

Jarvis P, Doermann P, Peto CA, Lutes J, Benning C, Chory J (2000) Galactolipid deficiency and abnormal chloroplast development in the *Arabidopsis* MGD synthase 1 mutant. Proc Natl Acad Sci USA 97:8175–8179

Jasinski S, Saraiva L, Perennes C, Domenichi S, Stevens R, Raynaud C (2003) NtKIS2: a novel tobacco cyclin-dependent kinase inhibitor is differentially expressed during the cell cycle and plant development. Plant Physiol Biochem 41:503–676

Jin Y, Ma D, Dong J, Li D, Deng C, Jin J, Wang T (2007) The HC-pro protein of potato virus Y interacts with NtMinD of tobacco. Mol Plant Microbe Interact 20(12):1505–1511

Johannes E, Brosnan JM, Sanders (1991) Calcium channels and signals transduction in plant cells. BioEssays 13:331–336

Johnson CH, Knight MR, Kondo T, Masson P, Sedbrook J, Haley A, Trewavas A (1995) Circadian oscillations of cytosolic and chloroplastic free calcium in plants. Science 269(5232):1863–1865

Juniper BE, Clowes FAL (1965) Cytoplasmic organelles and cell growth in root caps. Natur*e* 208:864–865

Kanamaru K, Fujiwara M, Kim M, Nagashima A, Nakazato E, Tanaka K, Takahashi H (2000) Chloroplast targeting, distribution and transcriptional fluctuation of AtMinD1, a Eubacteria-type factor critical for chloroplast division. Plant Cell Physiol 41:1119–1128

Kiessling J, Martin A, Gremillon L, Rensing SA, Nick P, Sarnighausen E, Decker EL, Reski R (2004) Dual targeting of plastid division protein FtsZ to chloroplasts and the cytoplasm. EMBO Rep 5:889–894

Kim YK, Lee JY, Cho HS, Lee SS, Ha HJ, Kim S, Choi D, Pai HS (2005) Inactivation of organellar glutamyl- and seryl-tRNA synthetases leads to developmental arrest of chloroplasts and mitochondria in higher plants. J Biol Chem 280:37098–37106

Koksharova OA, Wolk CP (2002) A novel gene that bears a DnaJ motif influences cyanobacterial cell division. J Bacteriol 184:5524–5528

Kolbe A, Tiessen A, Schluepmann H, Paul M, Ulrich S, Geigenberger P (2005) Trehalose 6- phospahte regulates starch synthesis via posttranslational redox activation of ADP-glucose pyrophosphorylase. Proc Natl Acad Sci USA 102:11118–11123

Kuroiwa T (1998) The primitive red algae *Cyanidium caldarium* and *Cyanidioschyzon merolae*as model system for investigating the dividing apparatus of mitochondria and plastids. BioEssays 20:344–354

Kuroiwa H, Mori T, Takahara M, Miyagishima SY, Kuroiwa T (2002) Chloroplast division machinery as revealed by immunofluorescence and electron microscopy. Planta 215:185–190

Lamppa GK, Bendich AJ (1979) Changes in chloroplast DNA levels during development of Pea (Pisum sativum). Plant Physiol 64:126–130

Leech RM, Pyke KA (1988) Chloroplast division in higher plants with particular reference to wheat. In: Boffey SA, Lloyd D (eds) The division and segregation of organelles. Cambridge University Press, Cambridge, pp 39–62

Leister D (2003) Chloroplast research in the genomic age. Trends Genet 19:47–56

Levina N, Totemeyer S, Stokes NR, Louis P, Jones MA, Booth IR (1999) Protection of *Escherichia coli* cells against extreme turgor by activation of MscS and MscL mechanosensitive channels: identification of genes required for MscS activity. EMBO J 18:1730–1737

Lieberman M, Segev O, Gilboa N, Lalazar A, Levin I (2004) The tomato homolog of the gene encoding UV-damaged DNA binding protein 1 (DDB1) underlined as the gene that causes the high pigment-1 mutant phenotype. Theor Appl Genet 108(8):1574–1581

Lingua G, D'Agostino G, Fusconi A, Berta G (2001) Nuclear changes in pathogen-infected tomato roots. Eur J Histochem 45:21–30

Liu Z, Mukherjee A, Lutkenhaus J (1999) Recruitment of ZipA to the division site by interaction with FtsZ. Mol Microbiol 31:1853–1861

Liu Y, Roof S, Ye Z, Barry C, van Tuinen A, Vrebalov J, Bowler C, Giovannoni J (2004) Manipulation of light signal transduction as a means of modifying fruit nutritional quality in tomato. Proc Natl Acad Sci USA 101(26):9897–9902

Lowe J, Amos LA (1998) Crystal structure of the bacterial cell-division protein FtsZ. Nature 391:203–206

Lowe J, Amos LA (1999) Tubulin-like protofilaments in Ca2+-induced FtsZ sheets. EMBO J 18:2364–2371

Lu C, Stricker J, Erickson HP (1998) FtsZ from *Escherichia coli*, *Azotobacter vinelandii*, and *Thermotoga maritime* – quantitation, GTP hydrolysis, and assembly. Cell Motil Cytoskeleton 40:71–86

Lutkenhaus J, Addinall SG (1997) Bacterial cell division and the Z ring. Annu Rev Biochem 66:93–116

Lyndon RF, Roberson ES (1976) The quantitative ultrastructure of the pea shoot apex in relation to leaf initiation. Protoplasma 87:387–402

Ma H, Lou Y, Lin WH, Xue HW (2006) MORN motifs in plant PIPKs are involved in the regulation of subcellular localization and phospholipid binding. Cell Res 16:466–478

Ma X, Ehrhardt DW, Margolin W (1996) Colocalization of cell division proteins FtsZ and FtsA to cytoskeletal structures in living *Escherichia coli* cells by using green fluorescent protein. Proc Natl Acad Sci USA 93:12998–13003

Ma X, Margolin W (1999) Genetic and functional analyses of the conserved C-terminal core domain of *Escherichia coli* FtsZ. J Bacteriol 181:7531–7544

Maleck K, Levine A, Eulgem T, Morgan A, Schmidt J, Lawton KA, Dangl JL, Dietrich RA (2000) The transcriptome of *Arabidopsis thaliana* during systemic acquired resistance. Nat Genet 26:403–409

Maple J, Aldridge C, Møller SG (2005) Plastid division is mediated by combinatorial assembly of plastid division proteins. Plant J 43:811–823

Maple J, Chua NH, Møller SG (2002) The topological specificity factor AtMinE1 is essential for correct plastid division site placement in *Arabidopsis*. Plant J 31:269–277

Maple J, Fujiwara MT, Kitahata N, Lawson T, Baker NR, Yoshida S, Møller SG (2004) GIANT CHLOROPLAST 1 is essential for correct plastid division in *Arabidopsis*. Curr Biol 14:776–781

Maple J, Vojta L, Soll J, Møller SG (2007) ARC3 is a stromal Z-ring accessory protein essential for plastid division. EMBO Rep 8:293–299

Marin E, Nussaume L, Quesada A, Gonneau M, Sotta B, Hugueney P, Frey A, Marion-Poll A (1996) Molecular identification of zeaxanthin epoxidase of *Nicotiana plumbaginifolia*, a gene involved in abscisic acid biosynthesis and corresponding to the ABA locus of *Arabidopsis thaliana*. EMBO J 15(10):2331–2342

Marrison JL, Rutherford SM, Robertson EJ, Lister C, Dean C, Leech RM (1999) The distinctive roles of five different *ARC* genes in the chloroplast division process in *Arabidopsis*. Plant J 18:651–662

Martin W, Rujan T, Richly E, Hansen A, Cornelsen S, Lins T, Leister D, Stoebe B, Hasegawa M, Penny D (2002) Evolutionary analysis of *Arabidopsis*, cyanobacterial, and chloroplast genomes reveals plastid phylogeny and thousands of cyanobacterial genes in the nucleus. Proc Natl Acad Sci USA 99:12246–12251

Martínez de Alba AE, Flores R, Hernández C (2002) Two chloroplastix viroids induce the accumulation of small RNAs associated with posttranscriptional gene silencing. J Virol 76:13094–13096

Mateo A, Funck D, Mühlenbock P, Kular B, Mullineaux PM, Karpinski S (2006) Controlled levels of salicylic acid are required for optimal photosynthesis and redox homeostasis. J Exp Bot 57:1795–1807

McAndrew RS, Froehlich JE, Vitha S, Stokes KD, Osteryoung KW (2001) Colocalization of plastid division proteins in the chloroplast stromal compartment establishes a new functional relationship between FtsZ1 and FtsZ2 in higher plants. Plant Physiol 127:1656–1666

McFadden GI (1999) Endosymbiosis and evolution of the plant cell. Curr Opin Plant Biol 2:513–519

McQuibban GA, Saurya S, Freeman M (2003) Mitochondrial membrane remodelling regulated by a conserved rhomboid protease. Nature 423:537–541

Misumi O, Matsuzaki M, Nozaki H, Miyagishima S, Mori T, Nishida K, Yagisawa F, Yoshida Y, Kuroiwa H, Kuroiwa T (2005) *Cyanidioschyzon merolae* genome. A tool for facilitating comparable studies on organelle biogenesis in photosynthetic eukaryotes. *Plant Physiol* 137:567–585

Miyagishima S, Kuroiwa H, Kuroiwa T (2001a) The timing and manner of disassembly of the apparatuses for chloroplast and mitochondrial division in the red alga *Cyanidioschyzon merolae*. Planta 212:517–528

Miyagishima S, Takahara M, Mori T, Kuroiwa H, Higashiyama T, Kuroiwa T (2001b) Plastid division is driven by a complex mechanism that involves differential transition of the bacterial and eukaryotic division rings. Plant Cell 13:2257–2268

Miyagishima SY, Froehlich JE, Osteryoung KW (2006) PDV1 and PDV2 mediate recruitment of the dynamin-related protein ARC5 to the plastid division site. Plant Cell 18:2517–2530

Miyagishima SY, Nishida K, Mori T, Matsuzaki M, Higashiyama T, Kuroiwa H, Kuroiwa T (2003) A plant-specific dynamin-related protein forms a ring at the chloroplast division site. Plant Cell 15:655–665

Mochizuki N, Brusslan JA, Larkin R, Nagatani A, Chory J (2001) *Arabidopsis* genomes uncoupled 5 (GUN5) mutants reveals the involvement of Mg-chelatase H subunit in plastid-to-nucleus signal transduction. Proc Natl Acad Sci USA 98:2053–2058

Mori T, Johnson CH (2001) Independence of circadian timing from cell division in cyanobacteria. J Bacteriol 183:2439–2444

Mosyak L, Zhang Y, Glasfeld E, Haney S, Stahl M, Seehra J, Somers WS (2000) The bacterial cell-division protein ZipA and its interaction with an FtsZ fragment revealed by X-ray crystallography. EMBO J 19:3179–3191

Mozdy AD, Shaw JM (2003) A fuzzy mitochondrial fusion apparatus comes into focus. Nat Rev Mol Cell Biol 4:468–478

Nagahisa K, Nakamura T, Fujiwara S, Imanaka T, Takagi M (2000) Characterization of FtsZ homolog from hyperthermophilic archaeon Pyrococcus kodakaraensis KOD1. J Biosci Bioeng 89:181–187

Nishida K, Takahara M, Miyagishima SY, Kuroiwa H, Matsuzaki M, Kuroiwa T (2003) Dynamic recruitment of dynamin for final mitochondrial severance in a primitive red alga. Proc Natl Acad Sci USA 100:2146–2151

Nogales E, Downing KH, Amos LA, Lowe J (1998a) Tubulin and FtsZ form a distinct family of GTPases. Nat Struct Biol 5:451–458

Nogales E, Wolf SG, Downing KH (1998b) Structure of the alpha beta tubulin dimer by electron crystallography. Nature 391:199–203

Ohlrogge J, Browse J (1995) Lipid biosynthesis. Plant Cell 7:957–970

Osteryoung KW, McAndrew RS (2001) The Plastid Division Machine. Annu Rev Plant Physiol Plant Mol Biol 52:315–333

Osteryoung KW, Pyke KA (1998) Plastid division: evidence for a prokaryotically derived mechanism. *Curr Opin Plant Biol* 1:475–479

Osteryoung KW, Stokes KD, Rutherford SM, Percival AL, Lee WY (1998) Chloroplast division in higher plants requires members of two functionally divergent gene families with homology to bacterial ftsZ. *Plant Cell* 10:1991–2004

Osteryoung KW, Vierling E (1995) Conserved cell and organelle division. *Nature* 376:473–474

Pesaresi P, Masiero S, Eubel H, Braun HP, Bhushan S, Glaser E, Salamini F, Leister D (2006) Nuclear photosynthetic gene expression is synergistically modulated by rates of protein synthesis in chloroplasts and mitochondria. *Plant Cell* 18(4):970–991

Pesaresi P, Schneider A, Kleine T, Leister D (2007) Interorganellar communication. Curr Opin Plant Biol 10:600–606

Pfannschmidt T, Nilsson A, Tullberg A, Link G, Allen JF (1999) Direct transcriptional Control of the Chloroplast Genes *psbA* and *psaAB* adjusts photosynthesis to light energy distribution in plants. IUBMB Life 48:271–276

Pompe-Novak M, Wrischer M, Ravnikar M (2001) Ultrastructure of chloroplasts in leaves of potato plants infected by potato virus YNTN. Phyton (Horn, Austria) 41:215–226

Possingham JV, Saurer V (1969) Changes in chloroplast number per cell during leaf development in spinach. Planta 86:186–194

Pottosin J, Muñiz S, Shabala J (2005) Fast-activating Channel Controls Cation Fluxes across the Native Chloroplast Envelope. Membrane Biol 204:145–156

Pyke KA, Rutherford SM, Robertson EJ, Leech RM (1994) arc6, A fertile *Arabidopsis* mutant with only two mesophyll cell chloroplasts. Plant Physiol 106:1169–1177

Pyke KA (1997) The genetic control of plastid division in higher plants. Invited Special Rev Am J Bot 84:1017–1027

Rajagopalan M, Atkinson MA, Lofton H, Chauhan A, Madiraju MV (2005) Mutations in the GTP-binding and synergy loop domains of Mycobacterium tuberculosis ftsZ compromise its function in vitro and in vivo. Biochem Biophys Res Commun 331:1171–1177

RayChaudhuri D, Park JT (1992) *Escherichia coli* cell-division gene ftsZ encodes a novel GTP-binding protein. Nature 359:251–254

Raynaud C, Cassier-Chauvat C, Perennes C, Bergounioux C (2004) An *Arabidopsis* homolog of the bacterial cell division inhibitor SulA is involved in plastid division. Plant Cell 16:1801–1811

Raynaud C, Perennes C, Reuzeau C, Catrice O, Brown S, Bergounioux C (2005) Cell and plastid division are coordinated through the prereplication factor AtCDT1. Proc Natl Acad Sci USA 102:8216–8221

Robertson EJ, Pyke KA, Leech RM (1995) arc6, An extreme chloroplast division mutant of *Arabidopsis* also alters proplastid proliferation and morphology in shoot and root apices. J Cell Sci 108(Pt 9):2937–2944

Rodio ME, Delgado S, De Stradis A, Gómez MD, Flores R, Di Serio F (2007) A viroid RNA with a specific structural motif inhibits chloroplast development. Plant Cell 19(11):3610–3626

Rolland F, Baena-Gonzalez E, Sheen J (2006) Sugar sensing and signalling in plants: conserved and novel mechanisms. Annu Rev Plant Biol 57:675–709

Rueda S, Vicente M, Mingorance J (2003) Concentration and assembly of the division ring proteins FtsZ, FtsA, and ZipA during the *Escherichia coli* cell cycle. J Bacteriol 185:3344–3351

Ryals JA, Neuenschwander UH, Willits MG, Molina A, Steiner HY, Hunt MD (1997) Systemic Acquired Resistance. *Plant Cell* 8:1809–1819

Sai J, Johnson CH (2002) Dark-stimulated calcium ion fluxes in the chloroplast stroma and cytosol. *Plant Cell* 14:1279–1291

Schilmiller AL, Howe GA (2005) Systemic signaling in the wound response. Curr Opin Plant Biol 8:369–377

Schroeder DF, Gahrtz M, Maxwell BB, Cook RK, Kan JM, Alonso JM, Ecker JR, Chory J (2002) De-etiolated 1 and damaged DNA binding protein 1 interact to regulate *Arabidopsis* photomorphogenesis. Curr Biol 12(17):1462–1472

Shimada H, Koizumi M, Kuroki K, Mochizuki M, Fujimoto H, Ohta H, Masuda T, Takamiya K (2004) ARC3, a chloroplast division factor, is a chimera of prokaryotic FtsZ and part of eukaryotic phosphatidylinositol-4-phosphate 5-kinase. Plant Cell Physiol 45:960–967

Siegenthaler P (1998) Molecular organization of acyl lipids in photosynthetic membranes of higher plants. In: Siegenthaler P, Murata N (eds) Lipids in photosynthesis: structure, function and genetics. Kluwer, Dordrecht, pp 119–144

Smirnova E, Griparic L, Shurland DL, van der Bliek AM (2001) Dynamin-related protein Drp1 is required for mitochondrial division in mammalian cells. Mol Biol Cell 12:2245–2256

Stokes KD, Osteryoung KW (2003) Early divergence of the *FtsZ1* and *FtsZ2* plastid division gene families in photosynthetic eukaryotes. Gene 320:97–108

Strand A (2004) Plastid-to-nucleus signalling. Curr Opin Plant Biol 7:621–625

Strepp R, Scholz S, Kruse S, Speth V, Reski R (1998) Plant nuclear gene knockout reveals a role in plastid division for the homolog of the bacterial cell division protein FtsZ, an ancestral tubulin. Proc Natl Acad Sci USA 95:4368–4373

Sullivan JA, Gray JC (1999) Plastid translation is required for the expression of nuclear photosynthesis genes in the dark and in roots of the pea lip1 mutant. Plant Cell 11:901–910

Suzuki K, Ehara T, Osafune T, Kuroiwa H, Kawano S, Kuroiwa T (1994) Behavior of mitochondria, chloroplasts and their nuclei during the mitotic cycle in the ultramicroalga *Cyanidioschyzon merolae*. Eur J Cell Biol 63:280–288

Tetlow IJ, Rawsthorne S, Rines C, Emes MJ (2005) Plastid metabolic pathways. In: Moller SG (ed) Plastids. Blackwell, Oxford, pp 60–109

Tveitaskog AE, Maple J, Møller SG (2007) Plastid division in an evolutionary context. *Biol Chem* 388:937–942

Ullanat R, Jayabaskaran C (2002) Light- and cytokinin-regulated *ftsZ* gene expression in excised cucumber cotyledons (Cucumis sativus). Plant Growth Regul 38:209–218

Vicente M, Rico AI, Martinez-Arteaga R, Mingorance J (2006) Septum enlightenment: assembly of bacterial division proteins. J Bacteriol 188:19–27

Vitha S, Froehlich JE, Koksharova O, Pyke KA, van Erp H, Osteryoung KW (2003) ARC6 is a J-domain plastid division protein and an evolutionary descendant of the cyanobacterial cell division protein Ftn2. Plant Cell 15:1918–1933

Vitha S, McAndrew RS, Osteryoung KW (2001) FtsZ ring formation at the chloroplast division site in plants. J Cell Biol 153:111–120

Wakasugi T, Nagai T, Kapoor M, Sugita M, Ito M, Ito S, Tsudzuki J, Nakashima K, Tsudzuki T, Suzuki Y, Hamada A, Ohta T, Inamura A, Yoshinaga K, Sugiura M (1997) Complete nucleotide sequence of the chloroplast genome from the green alga *Chlorella vulgaris*: the existence of genes possibly involved in chloroplast division. Proc Natl Acad Sci USA 94: 5967–5972

Wang X, Huang J, Mukherjee A, Cao C, Lutkenhaus J (1997) Analysis of the interaction of FtsZ with itself, GTP, and FtsA. J Bacteriol 179:5551–5559

Wang X, Lutkenhaus J (1993) The FtsZ protein of Bacillus subtilis is localized at the division site and has GTPase activity that is dependent upon FtsZ concentration. Mol Microbiol 9:435–442

Waters M, Pyke K (2005) Plastid development and differentiation. In: Moller SG (ed) Plastids. Blackwell, Oxford, pp 30–59

Weiss DS, Chen JC, Ghigo JM, Boyd D, Beckwith J (1999) Localization of FtsI (PBP3) to the septal ring requires its membrane anchor, the Z ring, FtsA, FtsQ, and FtsL. J Bacteriol 181:508–520

Wildermuth MC, Dewdney J, Wu G, Ausubel FM (2001) Isochorismate synthase is required to synthesize salicylic acid for plant defence. *Nature* 417:571–575

Yakir E, Hilman D, Harir Y, Green RM (2007) Regulation of output from the plant circadian clock. *FEBS J* 274:335–345

Yan K, Pearce KH, Payne DJ (2000) A conserved residue at the extreme C-terminus of FtsZ is critical for the FtsA–FtsZ interaction in Staphylococcus aureus. Biochem Biophys Res Commun 270:387–392

Yoder DW, Kadirjan-Kalbach D, Olson BJ, Miyagishima SY, Deblasio SL, Hangarter RP, Osteryoung KW (2007) Effects of mutations in *Arabidopsis* FtsZ1 on plastid division, FtsZ ring formation and positioning, and FtsZ filament morphology in vivo. Plant Cell Physiol 48:775–791

Yoshida Y, Kuroiwa H, Misumi O, Nishida K, Yagisawa F, Fujiwara T, Nanamiya H, Kawamura F, Kuroiwa T (2006) Isolated chloroplast division machinery can actively constrict after stretching. Science 313:1435–1438

Yu XC, Margolin W (1997) Ca^{2+}-mediated GTP-dependent dynamic assembly of bacterial cell division protein FtsZ into asters and polymer networks in vitro. EMBO J 16:5455–5463

Zechmann B, Müller M, Zellnig G (2003) Cytological modifications in zucchini yellow mosaic virus (ZYMV)-infected Styrian pumpkin plants. Arch Virol 148:1119–1133

Zimmermann P, Hirsch-Hoffmann M, Hennig L, Gruissem W (2004) GENIVESTIGATOR. *Arabidopsis* microarray database and analysis toolbox. Plant Physiol 136:2621–2632

Chloroplast Photorelocation Movement

N. Suetsugu and M. Wada(✉)

Abstract Chloroplast photorelocation movement is one of the best-characterized plant organelle movements and is found in various plant species from alga to flowering plants. In general, low-intensity blue light induces a chloroplast accumulation response for efficient light capture and high-intensity blue light induces an avoidance response so that chloroplasts can avoid photodamage. Red light is also effective in some alga, moss, and fern species. It was shown that phototropin (phot) is a blue light receptor in land plants and neochrome (neo), which is a chimera photoreceptor between phytochrome and phototropin, is a red light receptor in ferns and possibly in alga. Although the signal transduction pathways and motility system of chloroplast movement is not clearly understood, several components involved in signaling or motility were identified through molecular genetic research using *Arabidopsis thaliana* and the involvement of actin filaments in the motility system is obvious in most plant species. This chapter summarizes the current progress in research on chloroplast photorelocation movement.

Abbreviations ABD: actinin-type actin binding domain; BDM: 2,3-butanedione monoxime; CBD: chromophore-binding domain; CHUP1: chloroplast unusual positioning1; FMN: flavin mononucleotide; HFR: high-fluence-rate response; HKRD: histidine kinase-related domain; JAC1: J-domain protein required for chloroplast accumulation response 1; LFR: low-fluence-rate response; LOV: light, oxygen and voltage; N: N-terminal hydrophobic region (of CHUP1); NEM: *N*-ethylmaleimide; neo: neochrome; NTE: N-terminal extension (of phytochrome); phot: phototropin; phy: phytochrome; PMI: plastid movement impaired; PRD: PAS-related domain; PRM: proline-rich motif; C: C-terminal conserved region (of CHUP1)

M. Wada
Department of Biology, Faculty of Science, Kyushu University,
Fukuoka, 812-8581, Japan
e-mail: wada@nibb.ac.jb

1 Introduction

Correct positioning and movement of chloroplasts are essential for plant growth and development. According to the cellular environment and functions, chloroplasts relocate in a cell to specific positions. As an adaptive strategy under a fluctuating natural environment, chloroplast relocation movement is induced by various external stimuli; such as touch (Makita and Shihira-Ishikawa 1997; Sato et al. 1999, 2003), submergence (Mommer et al. 2005), water stress (Kondo et al. 2004), chemicals (Tazawa et al. 1991), bacterial elicitors (Boccara et al. 2007), cold treatment (Kodama et al. 2008), etc. Among external stimuli-induced chloroplast movements, light-induced chloroplast movement is universally found from alga to seed plants and is the most intensively studied (Haupt 1999; Haupt and Scheuerlein 1990; Suetsugu and Wada 2007b; Schmidt von Braun and Schleiff 2007; Wada et al. 1993, 2003). Classical comprehensive works by Senn revealed that in various plant species chloroplasts change their position in a cell according to ambient light conditions (Senn 1908). Particularly, chloroplasts move to and relocate at appropriate positions in a cell in response to light intensity (Fig. 1). Weak light induces low-fluence-rate response (LFR) in which chloroplasts move to a position where the highest possible light absorption is ensured. Strong light induces high-fluence-rate response (HFR) where chloroplasts move to a position where the lowest possible light absorption is ensured so that chloroplasts can avoid photodamage. In darkness, the pattern of chloroplast distribution (dark positioning) varies among plant species and among different tissues (Senn 1908). For example, most chloroplasts locate at the bottom of cells in the model flowering plant *Arabidopsis thaliana* (Suetsugu et al. 2005a) and position

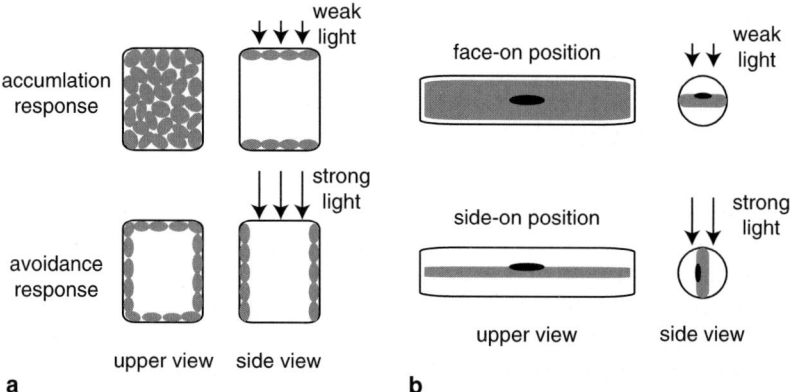

Fig. 1 Chloroplast photorelocation movement in green plant cells. **a** Land plants having multiple small chloroplasts (mosses, ferns, and seed plants). *Gray ovals* indicate chloroplasts. Chloroplasts accumulate under low light conditions to efficiently capture light (accumulation response) and escape from strong light to avoid photodamage (avoidance response). **b** Zygnematales algae, *Mougeotia scalaris* and *Mesotaenium caldariorum*. One giant, flat chloroplast (*gray*) with a nucleus (*black*) in a cell, exposes its entire flat surface to low-intensity light (face or face-on position) and its side to high-intensity light (profile or side-on position)

themselves along cell-cell-bordering anticlinal walls in one-layered prothallial cells of the fern *Adiantum capillus-veneris* (Kagawa and Wada 1993). Different patterning of chloroplast distribution is found among plant species according to the number and size of chloroplasts (Senn 1908; Zurzycki 1980; Haupt and Scheuerlein 1990). In plants having multiple small chloroplasts in a cell, chloroplasts accumulate at an area irradiated with weak light (accumulation response or movement) but escape from strong light (avoidance response or movement) (Fig. 1a). In two Zygnematales algae, *Mougeotia scalaris* and *Mesotaenium caldariorum*, which have one giant and flat chloroplast in a cell, the flat surface faces towards low-intensity light (face or face-on position) while its side faces toward high-intensity light (profile or side-on position) (Fig. 1b). Although LFR and HFR generally mean accumulation response and avoidance response, respectively (for Zygnematales, profile-to-face and face-to-profile orientation, respectively), we do not use these words in this review, since the turning point of light intensity from accumulation response to avoidance response are very different among plant species and between wild-type and mutant lines. In this chapter, we review recent progress in the area of chloroplast photorelocation movement, so for classical detailed analyses readers should refer to the excellent literature on the subject (Senn 1908; Zurzycki 1980; Haupt and Scheuerlein 1990; Haupt 1982, 1999; Wada et al. 1993). The underlying processes of chloroplast photorelocation movement can be separated into three parts: photoperception, signal transduction, and motility system. Recent rapid advance in molecular biology and cell biology have allowed us to discuss the three parts at the molecular level.

2 Photoperception

2.1 General View on Photoreceptor System in Chloroplast Movement

Chloroplast photorelocation movement is found in many green plants from chlorophyta to seed plants (whose chloroplasts originated from primary endosymbiosis of cyanobacteria) and even in some stramenopiles (whose chloroplasts originated from secondary symbiosis). In most plant species, the action spectrum for chloroplast movement revealed that blue light is the most effective in mediating chloroplast movement and flavin-containing proteins were presumed to be photoreceptors (Zurzycki 1980). Since in most cases the blue light effects were not far-red light reversible, it was clear that the blue light receptor other than red/far-red photoreceptor phytochrome (phy) mediates blue light-induced chloroplast movement. Exceptionally, red light-induced and phytochrome-mediated chloroplast movement was found in various cryptogam plants, such as, some Zygnematales (*M. scalaris* and *M. caldariorum*), a moss *Physcomitrella patens* and derived fern species including *A. capillus-veneris* (Suetsugu and Wada 2005, 2007b). A very rare case is green light-induced chloroplast accumulation response in a diatom *Pleurosira laevis* (which is stramenopile alga) (Furukawa et al. 1998). Intensive photobiological analyses of chloroplast movement by polarized light and/or microbeam irradiation suggested that the blue light receptor and phytochrome for chloroplast movement are localized on or close to the plasma membrane (Zurzycki 1980; Haupt and Scheuerlein 1990; Wada et al. 1993). Rapid progress in plant photobiology was brought about by the molecular genetic approach in *A. thaliana*, leading to the identification of genes for blue light receptors (cryptochromes, phototropins, and ZTL/ FKF/LKP proteins) and availability of photoreceptor mutants. Subsequent cloning and characterization of photoreceptor genes in various plants, particularly *A. capillus-veneris*, *P. patens*, and *M. scaralis* revealed that the photoreceptor system for chloroplast movement is highly conserved in Streptophyta.

2.2 Phototropin

2.2.1 Identification of Phototropin as a Blue Light Receptor for Chloroplast Movement

As in most plant species, blue light alone mediates both chloroplast accumulation and avoidance responses in *A. thaliana* (Trojan and Gabry's 1996; Kagawa and Wada 2000). Cryptochrome (cry) is the first blue light photoreceptor identified in plants and two *A. thaliana* cryptochromes, cry1 and cry2, mediate various blue light responses, such as de-etiolation responses, circadian rhythm, and gene expression (Li and Yang 2007). However, normal chloroplast movement was found in the *cry1cry2* double mutant (Kagawa and Wada 2000), indicating that cryptochromes were not the blue light receptors for chloroplast movement. Next, phototropin1 (phot1) was identified as the blue light photoreceptor for phototropism from molecular genetic analyses of *nonphototropic hypocotyl 1* (*nph1*) mutant in *A. thaliana*

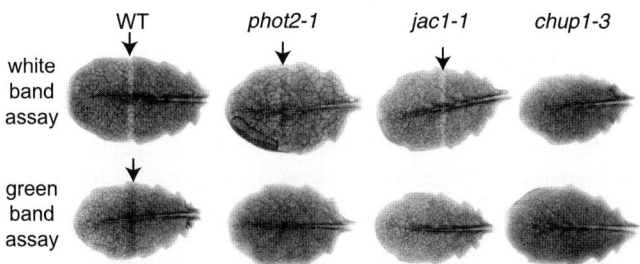

Fig. 2 Band assays for detection of chloroplast photorelocation movement in *A. thaliana*. White band assay is for detection of the chloroplast avoidance response and green band assay is for detection of the chloroplast accumulation response. White or green bands are indicated with *arrows*. With both assays, leaves of wild-type (WT) showed a band whereas *phot2-1* showed a green band with the white band assay. *jac1-1* showed no green band with the green band assay. *chup1-3* did not show any band with both white and green band assay

(Huala et al. 1997). The *phot1* mutant retained both accumulation and avoidance movement but showed slightly lower sensitivity for the accumulation response to weak light (Kagawa and Wada 2000), suggesting that phot1 was not the main photoreceptor for chloroplast movement. A breakthrough resulted from forward genetic analysis of chloroplast avoidance response in *A. thaliana*. We isolated several *cav1* (defective in chloroplast avoidance movement 1) mutants by a novel screening method, white band assay (Kagawa et al. 2001) (Fig. 2). Briefly, detached leaves were irradiated with strong light through an open slit 1 mm in width, resulting in the appearance of a white band because of an increased light transmittance by the avoidance response at the irradiated area. This mutant was defective in the avoidance response and showed an accumulation response even under strong light (Kagawa et al. 2001). Positional cloning of the *CAV1* gene revealed that CAV1 is phot2, a paralog of phot1, and that phot2 is the blue light receptor for the avoidance response. This result was confirmed by a reverse genetic approach (Jarillo et al. 2001). Furthermore, subsequent analysis of the *phot1phot2* double mutant revealed that phot1 and phot2 redundantly mediate the chloroplast accumulation response (Sakai et al. 2001). Therefore, it was shown that phototropins are the long-sought blue light receptor(s) for chloroplast movement. And also phototropins redundantly mediate phototropism (Sakai et al. 2001), stomatal opening (Kinoshita et al. 2001), and leaf expansion (Sakai et al. 2001; Sakamoto and Briggs 2002). Surprisingly, phot2 but not phot1 is also necessary for chloroplast dark positioning (Suetsugu et al. 2005a). Thus, phot2 can regulate all types of chloroplast movement under any light conditions but phot1 mediates only the accumulation response regardless of light intensity (Table 1).

Table 1 Photoreceptors and cytoskeletons for chloroplast movement in various plant species. In *A. thaliana*, red light is ineffective in inducing chloroplast movement. In *P. patens* and *M. scalaris*, dark positioning is not clear

	Photoreceptors					
	Blue light		Red light			
Plant species	Accumulation	Avoidance	Accumulation	Avoidance	Dark positioning	Cytoskeleton
Seed plants (*Arabidopsis thaliana*)	phot1, phot2	phot2			phot2	actin
Ferns (*Adiantum capillus-veneris*)	phot1, phot2?, neo1?	phot2	neo1	neo1	phot1?, phot2	actin
Mosses (*Physcomitrella patens*)	photA1?, A2, B1, B2	photA1, A2, B1, B2	phy1~3?, phy4	phy1~3, phy4?		actin and microtubule
Green algae (*Mougeotia scalaris*)	photA?, photB?	photA?, photB?	neo1?, neo2?	neo1?, neo2?		actin

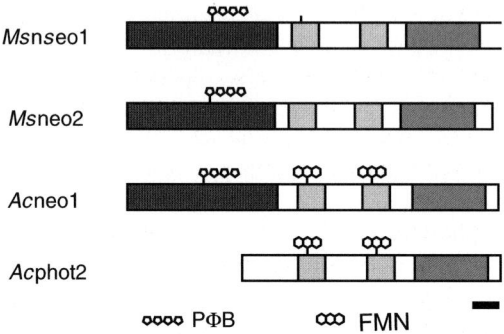

Fig. 3 Phototropins and neochromes. Phototropin (phot) has two LOV domains (*light gray*) binding to flavin mononucleotide (FMN) and C-terminal serine/threonine kinase domain (*gray*). Neochrome (neo) consists of N-terminal phytochrome chromophore (phytochromobilin, PΦB) binding domain and complete phototropin domains. Note that LOV domains of *M. scalaris* neochromes cannot bind to FMN although LOV1 domain of *Ms*NEO1 has conserved cysteine residue (indicated by a *line* without FMN). *Ms*: *Mougeotia scalaris*, *Ac*: *Adiantum capillus-veneris*. *Bar*: 100 amino acids

Phototropin has an N-terminal photosensory domain and a C-terminal serine/threonine kinase domain and shows blue light-activated autophosphorylation activity (Christie 2007) (Fig. 3). The phototropin N-terminal domain consists of two tandem LOV (*l*ight, *o*xygen, and *v*oltage) domains forming a subgroup of the PAS domain superfamily, which is involved in sensory function and/or protein–protein interaction (Taylor and Zhulin 1999). One LOV domain binds one flavin mononucleotide (FMN) and the spectral properties of the LOV domain containing FMN closely match the action spectra for phototropism, chloroplast movement, and stomatal opening (Christie et al. 1999). A highly conserved cysteine residue in LOV domains is essential for the flavin-cysteinyl adduct formation and photocycle (Salomon et al. 2000). The substitution of conserved Cys to Ala of LOV domains revealed that the LOV2 domain but not the LOV1 domain is essential for kinase activation and the regulation of physiological responses (Christie et al. 2002; Matsuoka and Tokutomi 2005). On the basis of the observation of a GFP-fusion protein and cell fractionation analysis, it was found that phototropins localized on the plasma membrane (Sakamoto and Briggs 2002; Harada et al. 2003; Kong et al. 2006; Wan et al. 2008), consistent with the photobiological data that the blue light receptor for chloroplast movement is localized on or close to the plasma membrane (Zurzycki 1980; Haupt and Scheuerlein 1990; Wada et al. 1993).

2.2.2 Prevalence of Phototropin-Mediated Chloroplast Movement in Green Plants

Phototropin was found not only in Streptophyta but also in Chlorophyta, *Ostreococcus* (the Plasinophyceae), *Chlamydomonas reinhardtii* and *Volvox carteri* (the Chlorophyceae) (Suetsugu and Wada 2007b). Since phototropin genes have not yet been found in complete genomes or EST databases of cyanobacteria, red alga, and diatom, phototropin is likely to arise during green plant evolution. Note that the overexpressed *C. reinhardtii PHOT* gene, *CrPHOT*, in *A. thaliana phot-1phot2* can mediate phototropism, chloroplast movement (both the accumulation and avoidance), stomatal opening, and leaf flattening (Onodera et al. 2005), indicating that the basic functions of phototropins have been conserved during green plant evolution.

Involvement of phototropin in chloroplast movement was shown in the fern *A. capillus-veneris* and the moss *P. patens*. *A. capillus-veneris* has two conventional phototropins, *Ac*phot1 and *Ac*phot2. Mutants deficient in the avoidance response but not in the accumulation response carried a nonfunctional *AcPHOT2* gene and transient expression of *AcPHOT2* cDNA but not *AcPHOT1* rescued the defect in the avoidance response, indicating that *Ac*phot2 is the blue light receptor for the chloroplast avoidance response (Kagawa et al. 2004). Importantly, the *Ac*phot2 mutant is impaired also in dark positioning (Tsuboi et al. 2007). Therefore, *Ac*phot2 shares very similar functions to *A. thaliana* phot2 (Table 1). So far, four cloned phototropin genes, *PpPHOTA1, A2, B1,* and *B2*, which are separated into two subgroups (A and B), have been characterized in the moss *P. patens* (Kasahara et al. 2004), although more phototropin genes exist in the near-completely sequenced *P. patens* genome (Rensing et al. 2008). Since gene targeting by homologous recombination can be carried out in this moss (Cove 2005), single (*photA1, photA2, photB1,* and *photB2*), double (*photA1photA2* and *photB1photB2*), and triple knockout lines (*photA2photB1photB2*) of phototropin genes were generated and characterized. Among single *phot* mutants, *photA2* alone showed the defect in chloroplast movement in basal cells of protonemata and lacked the avoidance response. In tip cells, the avoidance response was induced even in *photA2* but not in the *photA1photA2* double knockout. Therefore, these results indicate that *Pp*photAs is essential for the blue light-induced avoidance response in protonemal cells, similar to *A. thaliana* and *A. capillus-veneris* phot2 and suggest that photA1 can be expressed or functional only in tip cells (Kasahara et al. 2004) (Fig. 4a and Table 1). Since *photB1photB2* required several times higher intensity of blue light to induce the avoidance response than wild-type, it is likely that *Pp*photBs are necessary but not essential for the avoidance response or that uncharacterized *Pp*phots may redundantly function with *Pp*photBs (Fig. 4a and Table 1). When *PpPHOTA2* and *PpPHOTB2* were transiently overexpressed, both genes can complement the impairment in avoidance response in the fern *phot2* mutant, showing that both *Pp*photA2 and *Pp*photB2 intrinsically have the ability to induce the avoidance response (Kasahara et al. 2004). Although the accumulation response was induced in all single and double *phot* knockout lines with similar light sensitivity to wild-type, *photA2photB1photB2* showed about 50,000-fold lower sensitivity to induce the blue light-induced accumulation response than wild-type and very weak accumulation movement. Thus, these three phototropins redundantly mediate the accumulation movement and residual accumulation movement in the triple disruptant may be mediated by photA1 or other *Pp*phots (Kasahara et al. 2004) (Fig. 4a and Table 1). In *M. scalaris* whose

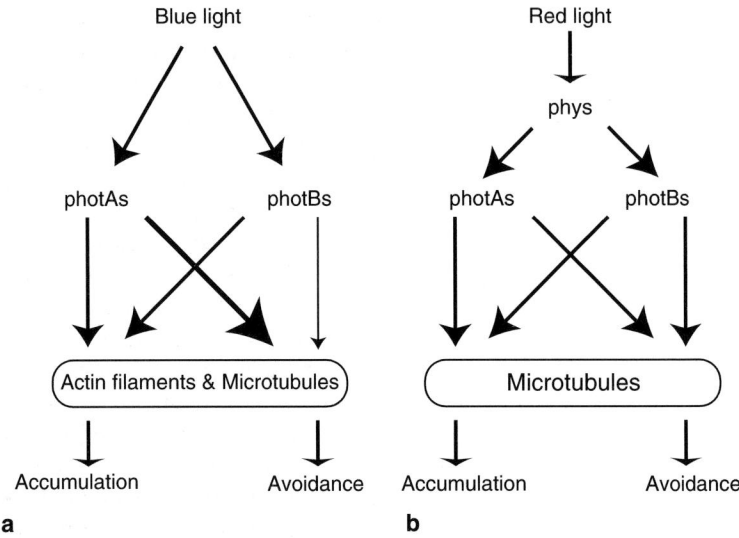

Fig. 4 Photoreceptors and signaling pathway for chloroplast movement in the moss *Physcomitrella patens*. **a** Under the blue light, both photAs and photBs mediate both accumulation and avoidance responses by utilizing both actin filaments and microtubules. photAs are essential for the avoidance response (*thick arrow*). **b** Under red light, four phys absorb red light and utilize the phototropin-signaling pathway to mediate chloroplast movement utilizing microtubules

chloroplast movement is mediated mainly by phytochrome (Haupt 1959), blue light-induced chloroplast movement was not fully cancelled by far red light, indicating the existence of a blue light receptor other than phytochrome for chloroplast movement (Gabry's et al. 1984). Since two phototropin genes, *MsPHOTA* and *MsPHOTB*, were cloned from *M. scalaris* (Suetsugu et al. 2005b) and the photometrical properties of these phototropins were consistent with previous photobiological data (Kagawa and Suetsugu 2007), it is possible that phototropin is also the blue light receptor for chloroplast movement in this alga (Table 1).

2.2.3 Functional Analysis of Phototropin Domains Necessary for Chloroplast Movement

What caused the clear functional difference in chloroplast movement between phot1 and phot2 in *A. thaliana* and *A. capillus-veneris* remained to be determined. One possibility for producing the functional difference is the difference in photochemical properties between phot1 and phot2. Although relative quantum efficiency for photoproduct formation is about tenfold different between the LOV1 and LOV2 domain in phot1 (0.035 and 0.34 for *A. thaliana* phot1; 0.045 and 0.44 for oat phot1; 0.026 and 0.30 for rice phot1, respectively), the difference in phot2 is small (0.13 and 0.27 for *A. thaliana* phot2; 0.33 and 0.29 for rice phot2, respectively) (Salomon et al. 2000; Kasahara et al. 2002b). Importantly, the difference in relative quantum efficiency for photoproduct formation between LOV1 and LOV2 in *Cr*phot, which can induce the avoidance movement in transgenic *A. thaliana* like phot2 (Onodera et al. 2005), is similar to that in phot2 (0.30 and 0.35, respectively) (Kasahara et al. 2002b). Furthermore, dark reversion of LOV2 is faster than that of LOV1 in phot2 and *Cr*phot but not in phot1 of *A. thaliana*, rice and *A. capillus-veneris* (Kasahara et al. 2002b; Kagawa et al. 2004). Moreover, lifetimes of light-activated LOV1 + LOV2 fusion proteins in phot2 are shorter than those in phot1 and that in *Cr*phot is close to that in phot2 (Kasahara et al. 2002b; Kagawa et al. 2004). Interestingly, dark reversion of LOV1 + LOV2 fusion proteins in *Ms*photB is faster than that in *Ms*photA, similar to the relationship between phot1 and phot2 (Kagawa and Suetsugu 2007), suggesting that the functional divergence of duplicated phototropins in photochemical properties may arise several times during plant evolution. Although both phot1 and phot2 of *A. thaliana* are localized mainly on the plasma membrane (Sakamoto and Briggs 2002; Harada et al. 2003; Kong et al. 2006; Wan et al. 2008), the blue light-induced relocalization pattern of GFP-fusion proteins is different between phot1 and phot2 (Sakamoto and Briggs 2002; Kong et al. 2006; Wan et al. 2008). Whereas phot1 re-localizes into cytoplasm (Sakamoto and Briggs 2002; Wan et al. 2008), phot2 is targeted to Golgi (Kong et al. 2006). Note that in both phot1 and phot2 only a small pool of phototropins re-localizes but the remaining large amount of phototropins is still localized on the plasma membrane after blue light irradiation (Sakamoto and Briggs 2002; Kong et al. 2006; Wan et al. 2008). However, in blue light-irradiated samples considerable amounts of phot1 but not phot2 could be detected at the soluble fraction (Sakamoto and Briggs 2002; Knieb et al. 2004; Kong et al. 2006).

Since phototropins have three domains, LOV1, LOV2, and a serine/threonine kinase domain, their importance in mediating chloroplast photorelocation movement was examined by mutational and deletion analysis. Overexpression of mutant phot2 carrying a point mutation Asp720 to Asn (D720N), which lacks kinase activity (Christie et al. 2002), in *phot1phot2* showed no complementation of chloroplast photorelocation movement defect (Kong et al. 2006), and the *phot2-2* allele having Thr767 to Ile in the kinase domain is defective in chloroplast photorelocation movement (Kagawa et al. 2001; Kong et al. 2006). Also the overexpression of *Ac*phot2 carrying Lys709 to Ala mutation in its kinase domain could not rescue the defect in the avoidance response of the *A. capillus-veneris phot2* mutant (Kagawa et al. 2004). Although the expression of the phot2 kinase domain alone could not complement the avoidance response defect in the *phot2* mutant (Kagawa et al. 2004; Kong et al. 2007), it affected chloroplast distribution when expressed in wild-type plants (Kong et al. 2007), where chloroplasts constitutively localized on anticlinal walls similar to the position under the high light condition, even in darkness. This constitutive activity was completely suppressed by the point mutation of the kinase domain (that is D720N) (Kong et al. 2007), indicating that kinase activity is essential for the regulation of chloroplast photorelocation movement.

Cys-to-Ala mutation or deletion of LOV2 but not LOV1 of *Ac*phot2 abolished the ability to mediate the avoidance response, indicating that the photochemical activity of only LOV2 of *Ac*phot2 is necessary for the avoidance response (Kagawa et al. 2004).

Similarly, phot1 LOV2-kinase (lacking the N-terminal domain including LOV1) could mediate the light-induced chloroplast accumulation response in *phot1phot2* (Sullivan et al. 2008), indicating that both the accumulation and the avoidance responses are induced by the photoactivation of the LOV2 domain. Note that the phot1 LOV2-kinase construct used by Sullivan et al. lacks major in vivo phosphorylation sites and thus auto-phosphorylation may not be essential at least for phot1-mediated blue light responses (Sullivan et al. 2008).

The *Ac*phot2 construct with the deletion of the last 40 amino acids of C-terminal extension could not rescue the avoidance response defect in the *phot2* mutant although the deletion of the last 9 or 20 amino acids did not disrupt the *Ac*phot2 function (Kagawa et al. 2004). Phototropin belongs to the AGC kinase family (cAMP-dependent protein kinase *A*, cGMP-dependent protein kinase *G*, and phospholipids-dependent protein kinase *C*) (Bögre et al. 2003) and plant AGC kinases have about 50 amino acids conserved in a C-terminal extension following the kinase domain (Bögre et al. 2003; Kagawa et al. 2004). Therefore, the C-terminal extension may be essential for the function of other AGC kinase as well as phototropin.

2.3 Neochrome

2.3.1 Identification of Neochrome from Derived Fern Species

In some derived fern species, a red light-induced chloroplast accumulation response was found (Yatsuhashi et al. 1985; Yatsuhashi and Kobayashi 1993; Kagawa and Wada 1994). These responses were far-red light reversible and showed the action dichroism, indicating membrane-localized phytochrome involvement (Yatsuhashi et al. 1985; 1987a,b; Yatsuhashi and Kobayashi 1993; Kagawa et al. 1994). To induce the avoidance response in *A. capillus-veneris*, red light at very high fluence rates (more than 230 W m^{-2}) was required and also deduced to be mediated by the membrane-localized phytochrome (Yatsuhashi et al. 1985; Yatsuhashi and Wada 1990). Characteristics of phytochrome for chloroplast movement are very similar to those of deduced phytochrome for red light-induced phototropism in ferns (Wada and Kadota 1989), suggesting that the same phytochrome mediates both phototropism and chloroplast movement in ferns. This hypothesis was supported by two studies on fern photobiology. First, the fern *Pteris vittata*, which belongs to the same family Pteridaceae as *A. capillus-veneris*, lacks red light but not blue light responses both in phototropism and chloroplast movement although in this species phytochrome-dependent germination and cell division occur (Kadota et al. 1989; Tsuboi et al. 2006). More importantly, red light-aphototropic (*rap*) mutants lacked red light-induced chloroplast accumulation and avoidance responses and retained blue light-induced phototropism and chloroplast movement and other phytochrome responses (Kadota and

Wada 1999; Tsuboi et al. 2006). However, the identification of phytochrome for these responses was hampered by the lack of molecular genetic methods for mutant gene cloning.

Genome library screening for *A. capillus-veneris* phytochrome genes made an unexpected discovery, that is the cloning of the *NEOCHROME 1* gene (*NEO1*, formally *PHYTOCHROME3*) (Nozue et al. 1998). NEO1 consists of an N-terminal phytochrome photosensory domain and C-terminal complete phototropin domains (that is two LOV domains and a serine/threonine kinase domain) (Fig. 3). The neo1 phytochrome photosensory domain reconstituted with chromophore phycocyanobilin showed a similar difference spectrum to conventional phytochromes (Nozue et al. 1998) and two LOV domains from neo1 bound FMN with similar spectral properties to phototropins (Christie et al. 1999). Moreover, the serine/threonine kinase domain of neo1 can autophosphorylate in vitro (Kanegae et al. 2006). Therefore, neo1 has characteristics of both phytochrome and phototropin and thus was a strong candidate for red light (and also blue light) receptor for phototropism and chloroplast movement. Subsequently, sequencing of the *NEO1* gene in *rap* mutants and complementation tests revealed that neo1 is the phytochrome for phototropism and chloroplast movement in *A. capillus-veneris* (Kawai et al. 2003) (Table 1). Although basal fern species such as *Osmunda japonica* and *Lygodium japonicum* lack the *NEO1* gene and red light-induced phototropism and chloroplast movement, more derived ferns (Polypod ferns) such as *Dryopteris filix-mas* and *Onoclea sensibilis* have them (Kawai et al. 2003). Since the polypod ferns including most extant ferns diversified in the Cretaceous after angiosperms diversified and formed dense canopy (Schneider et al. 2004), the gain of the *NEO1* gene in polypod ferns may facilitate their rapid diversification under the shade of angiosperms. Synergistic effects between red and blue light were found in chloroplast movement in prothallial cells of *A. capillus-veneris* (Kagawa and Wada 1996) and *rap2* sporophytic leaves drastically impaired phototropism under the weak white light conditions (at least 100-fold hyposensitive) (Kawai et al. 2003). Synergistic effects between red and blue light are found in neo1 expressing *A. thaliana* transgenic plants; simultaneous irradiation with red and blue light synergistically enhanced phototropic response and autophosphorylation activity, indicating that neo1 enhances light sensitivity by synergistically processing red and blue light signaling (Kanegae et al. 2006).

2.3.2 Independent Generation of Neochromes in the Alga *Mougeotia scalaris*

Phytochrome-mediated chloroplast movement in *M. scalaris* is one of the most extensively studied chloroplast movements and it was thought that the membrane-bound phytochrome mediates this response (Haupt 1982, 1999; Haupt and Scheuerlein 1990). Two chimeric photoreceptor genes between phytochrome and phototropin were identified from *M. scalaris*, *MsNEO1* and *MsNEO2* (Suetsugu et al. 2005b). Like fern NEO1, *Ms*NEOs consist of an N-terminal phytochrome photosensory domain and a C-terminal complete phototropin domain (Suetsugu et al.

2005b) (Fig. 3). However, LOV domains from *Ms*neos are highly diverged from those of phototropins and many amino acids necessary for FMN-binding and photocycle in LOV domains were changed (Fig. 3). Actually, recombinant LOV domains (LOV1 or LOV2) from *Ms*neos could not bind FMN (Suetsugu et al. 2005b) and photosensory domains (phytochrome domain plus two LOV domains, PHY + LOV) reconstituted with phytochromobilin showed the same difference spectra as the phytochrome photosensory domains alone (PHY) (Suetsugu et al. 2005b; Kagawa and Suetsugu 2007), indicating that the LOV domains of *Ms*neos do not function as the blue light receptor. Gel filtration analysis revealed that PHY + LOV and the PHY domain of *Ms*neo2 exist as a tetramer and a monomer, respectively, in the solution (Kagawa and Suetsugu 2007), suggesting that the LOV domain region of *Ms*neos is necessary for multimerization as phototropin LOV domains do (Christie 2007). *Ms*neos are strong candidates for red light receptors regulating chloroplast movement in *M. scalaris* (Table 1). First, absorption spectra of photosensory domains from *Ms*neos but not from conventional *M. scalaris* phytochrome 1 closely matched the action spectra for chloroplast movement (Haupt 1959; Suetsugu et al. 2005b; Kagawa and Suetsugu 2007). Second, it is plausible that *Ms*neos are localized on the plasma membrane since they have the complete phototropin domains, consistent with the characteristics of the phytochrome deduced from extensive photobiological experiments (Haupt 1982, 1999). Finally, the transient expression of both *MsNEO1* and *MsNEO2* cDNA rescued the defect in red light-induced chloroplast accumulation response in *A. capillus-veneris rap* (*neo1*) mutants, indicating that *Ms*neos have similar functional properties to fern neo1 (Suetsugu et al. 2005b). Neochrome is found only in polypod ferns and *M. scalaris*. *MsNEO* genes have many introns including two algal phytochrome specific introns (Suetsugu et al. 2005b) whereas fern *NEO* genes have no introns (Nozue et al. 1998). *MsNEO* genes and fern *NEO* genes are phylogenetically distinct (Suetsugu et al. 2005b). Taken together, *NEO* genes arose at least twice during plant evolution.

2.4 Co-Action Between Phototropin and Phytochrome

2.4.1 Red Light-Modulation of Blue Light-Induced Chloroplast Movement

In most plant species except for some cryptogam plants, only blue light induce chloroplast movement. In *A. thaliana*, background red light irradiation enhanced blue light-induced chloroplast movement by activating cytoplasmic motility, although whether this red light effect is phytochrome-dependent remained to be determined (Kagawa and Wada 2000). Photometric analysis of chloroplast movement in *A. thaliana* suggested that phytochromes modulate blue light-induced chloroplast movement; mutants lacking phyA and/or phyB and a phytochrome chromophore-deficient mutant *hy1* showed a greater transmittance change in blue light-induced avoidance response (DeBlasio et al. 2003). Since it is well known that phytochromes regulate gene expression, it is likely that phytochromes modulate blue light-induced chloroplast movement by regulating the expression of genes for components necessary for chloroplast movement.

2.4.2 Role of Phototropin in Phytochrome-Mediated Chloroplast Movement in the Moss *Physcomitrella patens*

Although phototropism and polarotropism are induced by red light in some moss protonemata (Wada and Kadota 1989), red light-induced chloroplast movement has been found so far only in *P. patens* (Kadota et al. 2000; Sato et al. 2001a). Interestingly, blue but not red light-induced chloroplast movement occurs in white light-grown protonemal cells whereas both red and blue light-induced chloroplast movement was found in red light-grown protonemal cells (Kadota et al. 2000). Similar to red light-induced phototropism, red light-induced chloroplast movement in *P. patens* is far-red light-reversible and showed action dichroism (Kadota et al. 2000), indicating that this response is mediated by membrane-localized phytochrome. However, no neochrome-like photoreceptor gene is found in the draft genome sequences of *P. patens* (Rensing et al. 2008), suggesting that a phytochrome(s) distinct from neochrome is the photoreceptor. Four conventional phytochrome genes, *PpPHY1~4*, which are separated into two subgroups (*PHY1* and *PHY3*; *PHY2* and *PHY4*), were so far cloned in *P. patens* (Mittmann et al. 2004), and their involvement in red light-induced chloroplast movement was analyzed by two distinct strategies. First, since this moss is amenable to gene targeting by homologous recombination (Cove 2005), four *PHY* genes were individually mutated by this method (Mittmann et al. 2004). Although four *phy* knockouts showed distinct phenotypes in red light-induced phototropic response, *phy4* knockout lines showed the strongest defects such as lack of positive phototropic response and attenuation of polarotropism (Mittmann et al. 2004). Although four *phy* knockouts showed no defect in nonpolarized red light-induced chloroplast movement, the *phy4* knockout completely lacked a polarized red light-induced chloroplast accumulation response (Mittmann et al. 2004); wild-type chloroplasts tended to accumulate at the cross walls or along the side walls in the center of the protonemal cells when polarized red light with the electrical vector perpendicular or parallel to the protonemal axis, respectively, was irradiated for 24 h (Kadota et al. 2000). These results indicate that four phytochromes redundantly mediate red light-induced chloroplast movement and at least phy4 may function as a membrane-localized phytochrome (Table 1). As the second approach, a transient assay system in the protoplast was developed (Uenaka and Kadota 2007). In this system, very weak chloroplast avoidance but not the accumulation response could be induced by red light only after dark adaptation for more than 18 h although blue light could effectively induce both the accumulation and the avoidance responses (Uenaka and Kadota 2007). However, the transient expression of yellow fluorescent protein (yfp) fusion of phy1~phy3 but not yfp alone potentiated red light-induced avoidance response even after 12 h of dark adaptation (Uenaka and Kadota 2007) (Table 1). Contrary to the previous report (Mittmann et al. 2004), yfp-phy4 was ineffective in potentiating red light-induced avoidance response (Uenaka and Kadota 2007). All of the yfp-phy fusions were localized in the cytoplasm during dark adaptation and red light irradiation could not change cytoplasmic localization of yfp-phy1 and yfp-phy2 (Uenaka and Kadota 2007). The cytoplasmic pool of phy1 and phy2 is essential for the avoidance response because nuclear localization of yfp-phy fusions enforced by the C-terminally fused nuclear localization signal made yfp-phy fusions ineffective (Uenaka and Kadota 2007). Like other plant phytochromes (Rockwell et al. 2006), *Pp*phys consists of four domains, N-terminal extension (NTE), chromophore-binding domain (CBD), PAS-related domain (PRD), and histidine kinase-related domain (HKRD). NTE, CBD, and PRD but not HKRD of phy1 and phy2 are necessary for the induction of the avoidance response and PRD, which can be functionally replaced with GUS by providing the multimerization capacity as is the case with *A. thaliana* phyB (Matsushita et al. 2003), could not be replaced with GUS in both phy1 and phy2 (Uenaka and Kadota 2007), indicating that PRD of phy1 and phy2 has other functions distinct from multimerization.

Although it is clear that conventional phytochromes mediate red light-induced chloroplast movement in *P. patens*, it is very difficult to imagine that cytoplasmic phytochromes show action dichroism and utilize the same mechanism as phototropins do. However, it may be easy to answer this question since phototropins redundantly mediate red light-induced chloroplast movement (Kasahara et al. 2004). Although single (*photA2*) and double knockouts (*photA1photA2* and *photB1photB2*) retained normal fluence rate responses to induce both the accumulation and avoidance responses, the *photA2photB1photB2* triple knockout lacks the avoidance response by red light and showed a very weak accumulation response even at the highest fluence rate of red light (Kasahara et al. 2004). Since phototropins cannot absorb red light (Christie 2007), it is plausible that phototropins function as the downstream components of phytochromes to induce red light-induced chloroplast movement (Fig. 4b). In *P. patens*, phytochrome and phototropin may cooperatively function through direct interaction according to the Rosetta Stone principle; "some pairs of interacting proteins have homologs in another organism fused into a single protein chain" (Marcotte et al. 1999). In this case, the existence of neochromes (fusion between phytochrome and phototropin) in ferns and *M. scalaris* may indicate phytochrome and phototropin can interact with one another in *P. patens*.

3 Signal Transduction

3.1 Nature of the Putative Signal for Chloroplast Photorelocation Movement

The signal transduction pathway from photoreceptor to chloroplast is totally unknown. However, the nature of the signal was deduced by extensive photobiological work in our laboratory (Wada et al. 2003). It is obvious that light-induced gene expression is unnecessary for chloroplast photorelocation movement since both blue and red light-induced chloroplast movement can be induced even in enucleated protonemal cells (Wada 1988). This result is consistent with the facts that phototropins hardly contribute to blue light-induced gene expression in *A. thaliana* (Jiao et al. 2003; Ohgishi et al. 2004). In dark-adapted prothallial cells of *A. capillus-veneris* (Kagawa and Wada 1999) and mesophyll cells of *A. thaliana* (Kagawa and Wada 2000), partial cell irradiation with a microbeam of strong blue light induced chloroplast accumulation towards the irradiated area but chloroplasts could not enter the beam area due to the avoidance response during microbeam irradiation. After irradiation was stopped, chloroplasts entered the beam area for accumulation movement. These results suggest that strong blue light generates the signals for both the accumulation and avoidance response; the signal for the avoidance response can locally override the signal for the accumulation response; the signal for the accumulation response can be transferred to a long distance and is long-lived after light-off; the signal for the avoidance response can function only at the irradiated area and disappears immediately after light-off (Wada et al. 2003). Under the low fluence rate blue light, phot1 and phot2 generate the signal for the accumulation response. Under the high fluence rate blue light, phot2 generates the signal for the avoidance response whereas phot1 generates the signal for the accumulation response, since in *phot2* mutant phot1 induced the accumulation response even under the strong light (Kagawa et al. 2001, 2004; Jarillo et al. 2001). The strength of signal for both the accumulation and avoidance responses is dependent on fluence rate. By the irradiation of two adjacent areas of a protonemata of *A. capillus-veneris* with two different fluence rates of blue or red light (to induce the accumulation response), chloroplasts accumulated towards the higher fluence rate area when the ratio of these fluence rates is higher than 1.5 and this response is independent of the absolute difference (Yatsuhashi 1996; Yatsuhashi et al. 1987a), suggesting that higher fluence light generates more signal for chloroplast accumulation and that chloroplasts can sense the ratio of concentration of the signal (i.e. signal gradient). However, velocities of chloroplast accumulation movement were not dependent on total fluences and on the initial positions of chloroplasts (the average velocity was constant regardless of light quality and fluences) (Kagawa and Wada 1996). In *A. thaliana* mesophyll cells, the signal gradient is likely to be necessary for efficient avoidance movement from the area irradiated with strong light to that with lower fluence rate, since the nearer chloroplasts to the center of the microbeam under the higher fluence rate took longer to start to avoid (Kagawa and Wada 2004), suggesting that chloroplasts may sense the signal gradient between the inside and outside of the beam area and that the increase of the signal by the higher fluence rate of strong light may make the signal gradient obscure. And also velocity of chloroplast avoidance is dependent on fluence rate and on phot2 abundance; the higher fluence rate of blue light and the larger amount of phot2 resulted in the faster avoidance movement of chloroplasts (Kagawa and Wada 2004). When microbeam-irradiated *P. patans* cells in which chloroplasts could not move by the treatment with cytoskeleton-disrupting drugs were washed out with drug-free medium in darkness, chloroplasts moved towards a pre-irradiated area, suggesting that generation and transmission of signal for chloroplast accumulation response occurred even if the cytoskeleton was disrupted and that spatial information of the signal remained at the irradiated area (Sato et al. 2001a).

3.2 Calcium Ion

3.2.1 Phototropin-Mediated Influx of Calcium Ion into Cytosol

Since the roles of the calcium ion (Ca^{2+}) in the signaling pathway is well known in various physiological responses in plants (Hetherington and Brownlee 2004), many studies analyzed the involvement of Ca^{2+} in chloroplast photorelocation movement. The light-regulation of cytosolic Ca^{2+} concentration change was found in several plant species. In *M. scalaris*, phytochrome-mediated uptake of Ca^{2+} (Dreyer and Weisenseel 1979) and UV-A/blue light-induced increase of cytosolic Ca^{2+} concentration (Russ et al. 1991) was found. Blue light but not red light-induced Ca^{2+} transfer into the protonemal cells of *P. patens* was observed and this response was inhibited by a plasma membrane Ca^{2+} channel blocker lanthanum (La^{3+}), suggesting that blue light activates the influx of external Ca^{2+} (Russell et al. 1998). Also in flowering plants (*A. thaliana* and *Nicotiana plumbaginifolia*), blue light but not red light induces a transient increase of cytosolic Ca^{2+} (Baum et al. 1999). Blue light-induced calcium influx in *A. thaliana* was analyzed in detail using different tissues (etiolated or de-etiolated seedlings, rosette leaves, and mesophyll protoplasts) and different techniques (transgenic plants containing aequorin, ion-selective microelectrode ion flux measurements, and the patch-clamp technique) (Baum et al. 1999; Babourina et al. 2002; Stoelzle et al. 2003; Harada et al. 2003). In experiments using aequorin transgenic plants, blue light-induced Ca^{2+} influx was inhibited by a Ca^{2+} chelator O,O'-bis(2-aminoethyl) ethyleneglycol-N,N,N',N'-tetraacetic acid (EGTA) and plasma membrane Ca^{2+} channel blockers (La^{3+}, Co^{2+}, and nifedipine), suggesting that blue light activates the opening of a plasma membrane Ca^{2+} channel (Baum et al. 1999; Harada et al. 2003). Furthermore, the patch-clamp technique revealed that the blue light-activated Ca^{2+} channel, which was also inhibited by Ca^{2+} channel blockers La^{3+} and Gd^{3+}, could be a hyperpolarization-activated, calcium-permeable cation channel (Stoelzle et al. 2003). Although one study suggested that Ca^{2+} release from an internal store might contribute to blue light-induced Ca^{2+} influx due to its partial suppression by phospholipase C inhibitors (neomycin and U-73122) (Harada et al. 2003), ER Ca^{2+} ATPase inhibitor thaspigargin is totally ineffective in blocking the blue light-induced Ca^{2+} influx in another study (Baum et al. 1999). Mutant analyses revealed that phot1 and phot2 redundantly mediate blue light-induced Ca^{2+} influx and activation of a calcium-permeable cation channel (Baum et al. 1999; Babourina et al. 2002; Stoelzle et al. 2003; Harada et al. 2003), although the extent of influence of the *phot1* or *phot2* mutation on this regulation varies among these studies. Blue light-induced Ca^{2+} influx is also phototropin-mediated in the moss *P. patens*, since this response was attenuated in *photA2photB1photB2* triple knockout lines (Tucker et al. 2005).

3.2.2 Controversial Roles of the Calcium Ion During Chloroplast Photorelocation Movement

Ca^{2+} chelator EGTA attenuated chloroplast movement in various plant species (Kadota and Wada 1992a; Tlalka and Gabry's 1993; Tlalka and Fricker 1999; Shihira-Ishikawa et al. 2007), and the supply of external Ca^{2+} or the treatment with Ca^{2+} ionophore A23187 could induce the change of chloroplast distribution (Serlin and Roux 1984; Weidinger and Ruppel 1985; Kadota and Wada 1992a; Tlalka and Gabry's 1993; Tlalka and Fricker 1999). However, plasma membrane Ca^{2+} channel blockers were totally ineffective in inhibiting chloroplast photorelocation movement in *M. scalaris* (Schönbohm et al. 1990), *P. patens* (Sato et al. 2003), and *A. capillus-veneris* (Sato et al. 2001b). In *Lemna trisluca*, inhibitors for Ca^{2+} release from internal stock were more potent to inhibit chloroplast photorelocation movement than plasma membrane Ca^{2+} channel blockers (Tlalka and Gabry's 1993; Tlalka and Fricker 1999). These reports are inconsistent with the results that plasma membrane Ca^{2+} channel blockers effectively inhibited phototropin-mediated influx of external Ca^{2+} (Russell et al. 1998; Baum et al. 1999; Stoelzle et al. 2003; Harada et al. 2003). Interestingly, both mechanical stimulation-induced chloroplast avoidance response in fern protonemal cells and the accumulation response in bryophyte protonemal cells are dependent on external Ca^{2+}, since Ca^{2+} channel blockers (La^{3+} and Gd^{3+}) and the depletion of Ca^{2+} from the culture medium effectively inhibited these responses in *A. capillus-veneris* and *P. patens* (Sato et al. 2001b, 2003). But the same treatments with channel blockers and Ca^{2+} depletion were completely ineffective in inhibiting chloroplast photorelocation movement (Sato et al. 2001b, 2003), strongly suggesting that the influx of external Ca^{2+} is not the signal for chloroplast photorelocation movement at least in ferns and mosses. Furthermore, although phototropins mediate both red and blue light-induced chloroplast movement in *P. patens* (Kasahara et al. 2004), only blue light but not red light induced an increase of cytosolic Ca^{2+} concentration (Russell et al. 1998), suggesting that phototropin may regulate chloroplast movement by a mechanism other than Ca^{2+} influx from external store.

3.3 Putative Signaling Proteins Identified from Mutant Analyses in *Arabidopsis thaliana*

3.3.1 JAC1

To isolate mutants defective in chloroplast accumulation response in *A. thaliana*, the white band assay (Kagawa et al. 2001) was improved so that we could isolate the mutants defective in chloroplast accumulation response as well as in the avoidance response (Suetsugu et al. 2005a). Briefly, a whole area of detached leaves was irradiated with strong light to induce the avoidance response and then with weak light through an open slit 1 mm in width, resulting in the appearance of a green band because of the decrease of light transmittance by the accumulation response at the irradiated area (Fig. 2). With this screening method, *jac1* (*J*-domain protein required for chloroplast *ac*cumulation response *1*) mutant plants were isolated (Fig. 2), which were completely deficient in accumulation response (Suetsugu et al. 2005a). This mutant retained the avoidance response and its chloroplasts tended to take the position at the anticlinal walls irrespective of light conditions, as if the avoidance response was induced. Therefore, in the *jac1* mutant, chloroplasts do not accumulate on the cell bottom in darkness, indicating that *jac1* is also defective in chloroplast dark positioning (Suetsugu et al. 2005a). Few chloroplasts were on the upper cell surface in *jac1* mutant leaves and the avoidance movement was induced by a microbeam irradiation of weak blue light, which induced the accumulation response in wild-type (Suetsugu et al. 2005a). In the *phot2jac1* double mutant, both chloroplast accumulation and avoidance responses were not induced and more chloroplasts resided on the periclinal walls than in the *jac1* mutant, suggesting that phot2 can generate the signal for the avoidance response even under the low light condition and that JAC1 is necessary for the signal transduction pathway specifically in the accumulation response (and in dark positioning) (Suetsugu et al. 2005a). Positional cloning revealed that the *JAC1* gene encodes a C-terminal J-domain protein (Fig. 5). J-domain functions by regulating the activity of 70-kDa heat-shock proteins and has the diagnostic tripeptide HPD (Kelley 1998; Walsh et al. 2004). A J-domain of JAC1 has this HPD sequence and is highly similar to that of auxilin, which is required for the ATP-dependent dissociation of clathrin from clathrin-coated vesicles together with 70-kDa heat shock cognate protein during endocytosis (Eisenberg and Greene 2007). We have only ever isolated five *jac1* alleles from screening for mutants deficient in the accumulation response and dark positioning, and all of the *jac1* alleles carry a mutation which introduces premature stop codon and will have J-domain-truncated JAC1 proteins if expressed (Suetsugu et al. 2005a; Suetsugu and Wada 2007b; Suetsugu and Wada, unpublished results) (Fig. 5), indicating that the J-domain of JAC1 is essential for chloroplast accumulation response and dark positioning. At least in *jac1-1* and *jac1-2*, truncated JAC1 mutant proteins could not be detected in western blot analysis although mRNA was detected (Suetsugu et al. 2005a). *A. thaliana* has seven auxilin-like C-terminal J-domain proteins including JAC1 and it was shown that one among them (At4g12770) has clathrin-uncoating activity (Lam et al. 2001). Although the N-terminus of JAC1 has no homology to other proteins, those of JAC1 and another six *A. thaliana* auxilin-like proteins have a short motif FxDxF (x is any amino acid), which is defined as an adaptor protein-binding motif found in several accessory proteins involved in assembly of clathrin-coated vesicles (Brett et al. 2002). Although the involvement of JAC1 in clathrin-mediated endocytosis has not yet been shown, the lack of apparent defects of growth and development in the *jac1* mutant suggested that JAC1 is not essential for clathrin uncoating and/or that other auxilin-like proteins redundantly mediate clathrin-mediated endocytosis. However, it was shown that ectopic overexpression of the *JAC1* gene in *A. thaliana* roots attenuated endocytosis and aluminum ion uptake in root hair cells (Ezaki et al. 2007), although no *JAC1* gene expression was detected in root by RT-PCR (Suetsugu et al. 2005a). Therefore, JAC1 may be able to regulate clathrin-mediated endocytosis.

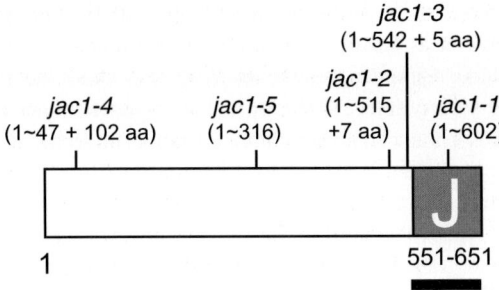

Fig. 5 JAC1 protein structure and the position of mutations in *jac1* mutant alleles. JAC1 has a C-terminal auxilin-like J-domain (amino acid 551-651). Five *jac1* alleles deduced to encode truncated JAC1 protein lacking J-domain. *Bar*: 100 amino acids

3.3.2 PMI1

By measuring changes in red-light transmittance through leaves in response to light irradiation, several *plastid movement impaired (pmi)* mutants were isolated (DeBlasio et al. 2005). Among them, the *pmi1* mutant showed the most severe phenotypes; any chloroplast movement in *pmi1-1* could not be detected in response to weak and strong blue light when analyzed in both red-light transmittance and time-lapse microscopy although the pattern of actin filaments was similar to that of wild-type (DeBlasio et al. 2005). Under dark adaptation performed by DeBlasio et al. (2003), the involvement of PMI1 in chloroplast dark positioning is not clear since chloroplast accumulation on the cell bottom was not clearly detected even in wild-type (DeBlasio et al. 2005). PMI1 is a plant specific protein and has no homology to proteins with known functions (DeBlasio et al. 2005). Two PMI1-like proteins (At5g20610 and At5g26160) exist in *A. thaliana* although these are less similar to PMI1 than PMI1-like proteins from other plant species.

3.3.3 PMI2 and PMI15

pmi2 mutant was isolated through the mutant screening method by which *pmi1* was isolated (DeBlasio et al. 2005). It showed the defect specifically in the avoidance response (Luesse et al. 2006). The *pmi2* mutant was less sensitive to blue light to induce the avoidance response and red-light transmittance after chloroplast avoidance was saturated under 100 μmol m^{-2} s^{-1} blue light that was several times lower than that in wild-type (Luesse et al. 2006). Similar to the *pmi1* mutant (DeBlasio et al. 2005), actin organization was not perturbed in *pmi2* (Luesse et al. 2006). PMI2 is a plant specific coiled-coil protein and has a putative nuclear localization signal (NLS) at the C-terminus (Luesse et al. 2006). However, in transgenic plants expressing GFP-PMI2(509-607) including NLS, GFP fluorescence was found in cytosol but not in the nucleus, suggesting that the NLS-like sequence of PMI2 cannot function as NLS (Luesse et al. 2006). *A. thaliana* has another PMI2-like protein,

PMI15, which is closer to *A. thaliana* PMI2 than to PMI2-like proteins from other plant species (Luesse et al. 2006). The *pmi15* mutant showed a very slight defect in chloroplast avoidance response (Luesse et al. 2006). And the *pmi2pmi15* double mutant had a more severe defect in the avoidance response than *pmi2* single mutant and showed a slightly enhanced accumulation response under weak blue light (Luesse et al. 2006).

4 Motility System

4.1 Actin Filaments

As in other plant organelle movement such as mitochondria, peroxisomes, and Golgi bodies (Wada and Suetsugu 2004), chloroplast movement in most plant species is dependent on actin filaments (Takagi 2003; Wada et al. 2003; Suetsugu and Wada 2007b). Anti-actin drugs (cytochalasin B and D, latrunculin B, and trifluoperazine) but not anti-microtubule drugs (oryzalin and colchicine) effectively inhibited both chloroplast accumulation and avoidance movement and induced aberrant chloroplast positioning in various green plant species (Wagner et al. 1972; Blatt et al. 1980; Izutani et al. 1990; Kadota and Wada 1992a; Tlalka and Gabry's 1993; Liebe and Menzel 1995; Sinclair and Hall 1995; Malec et al. 1996; Gorton et al. 1999; Sato et al. 1999; Augustynowicz et al. 2001; Kandasamy and Meagher 1999) (Table 1).

In various plant species, a possible interaction with chloroplasts and structures around chloroplasts of actin filaments were observed by rhodamine-phalloidin staining, anti-actin immunolabeling, or electron microscopy (Cox et al. 1987; Kadota and Wada 1989, 1992b; Mineyuki et al. 1995; Dong et al. 1996, 1998; Kandasamy and Meagher 1999; Sakai and Takagi 2005; Sakurai et al. 2005; Kumatani et al. 2006; Krzeszowiec et al. 2007). In protonemal cells of *A. capillus-veneris*, both blue and red light induced formation of circular structures made of actin filaments on the plasma membrane side of chloroplasts (Kadota and Wada 1989, 1992b). Detailed analysis revealed that these structures were formed on chloroplasts that stayed in the low light microbeam or within the shaded area adjacent to the edge of the strong light microbeam but were not found on chloroplasts moving towards a low light microbeam or escaping from a strong light microbeam light (Kadota and Wada 1992b), suggesting that these circular actin structures represent the structures for chloroplast anchoring. In epidermal cells of *Vallisuneria gigantea*, red light induced actin structures similar to that of *A. capillus-veneris*, with honeycomb structures (Dong et al. 1996, 1998). Red light-induced formation of the honeycomb structures correlated to decreased chloroplast motility and the resistance of chloroplasts to centrifugal forces (Dong et al. 1996, 1998), suggesting that honeycomb actin structures are also involved in chloroplast anchoring. In the following studies, short and thick actin bundles were found to associate with chloroplasts on the periclinal walls in epidermal cells of *V. gigantea* (Sakai and Takagi 2005; Sakurai et al. 2005). High-intensity blue light induced chloroplast avoidance movement towards anticlinal walls concomitant with disappearance of short actin bundles (Sakurai et al. 2005). Then, straight actin bundles without association with chloroplasts appeared and the short and thick actin bundles were formed around chloroplasts accumulated on the anticlinal walls (Sakai and Takagi 2005; Sakurai et al. 2005). Therefore, the release of anchorage from short actin bundles is likely to be a prerequisite for chloroplast avoidance response in this plant. Similar actin configurations were also found in spinach mesophyll cells in response to strong blue light (Kumatani et al. 2006). In *A. thaliana*, however, blue light-induced changes of actin organization were not detected but red light modulated patterns of actin filaments in the *phot2* mutant but not in wild-type (Krzeszowiec et al. 2007). Long actin bundles were formed under weak red light and subcircular structures appeared under strong red light only in *phot2* (Krzeszowiec et al. 2007). It is very unlikely that these actin patterns of the red light-irradiated *phot2* mutant observed in the study by Krzeszowiec et al. (2007) are involved in chloroplast photorelocation movement since red light is ineffective in inducing chloroplast movement in *A. thaliana*.

Possible association of actin filaments with chloroplasts was examined in cosedimentation analysis between isolated spinach chloroplasts and skeletal muscle actin filaments (Takagi 2003; Kumatani et al. 2006). Although endogenous actin filaments were not co-purified with chloroplasts during Percoll centrifugation, exogenously added skeletal muscle actin filaments co-sedimented with chloroplasts after incubation for 30 min (Takagi 2003; Kumatani et al. 2006). When chloroplasts were pretreated with trypsin so that outer membrane proteins were destructed, actin–chloroplast interaction was abolished, indicating that actin filaments bound with chloroplasts via outer membrane proteins (Kumatani et al. 2006).

Although the recent development of in vivo imaging of actin filaments using GFP-actin binding protein (talin and fimblin) transgenic plants (Yoneda et al. 2007) provided the opportunity to observe actin organization during chloroplast photorelocation movement, actin dynamics associated with chloroplast movement have not yet been found (Oikawa et al. 2003; DeBlasio et al. 2005; Luesse et al. 2006).

4.2 Myosins

It is thought that actin-mediated movement of mitochondria, peroxisomes, and Golgi bodies in plants is myosin-dependent (Wada and Suetsugu 2004) but the involvement of myosin in chloroplast movement is controversial. The effectiveness of known myosin inhibitors such as *N*-ethylmaleimide (NEM) and 2,3-butanedione monoxime (BDM) on chloroplast movement varied in several studies. They effectively inhibited both chloroplast accumulation and avoidance response (Kadota and Wada 1992a; Malec et al. 1996) or mechanical stress-induced chloroplast movement (Sato et al. 1999), but BDM was ineffective in inhibiting chloroplast movement in the aquatic plant *Elodea* (McCurdy 1999). In *A. thaliana*, BDM, NEM and another myosin inhibitor ML-7 partially inhibited the accumulation response but did not inhibit the avoidance response at all (Paves and Truve 2007). Immunolabeling with myosin antibodies against animal myosins or plant myosin XI or VIII suggested that myosins localized on chloroplast surfaces (Liebe and Menzel 1995; Malec et al. 1996; Wang and Pesacreta 2004; Wojtaszek et al. 2005; Krzeszowiec and

Gabry's 2007). Curiously, in experiments in *A. thaliana* using rabbit muscle myosin antibody, labeling was found on chloroplast surfaces and at cytoplasmic foci in darkness, weak blue light induced uniform labeling on chloroplast surfaces and antibody labeling disappeared from chloroplasts under strong blue light (Krzeszowiec and Gabry's 2007). Delocalization under strong blue light was phot2-dependent (Krzeszowiec and Gabry's 2007). Although it was not shown that rabbit muscle myosin antibody specifically detected *A. thaliana* myosin proteins, at least the localization of proteins that reacted to this antibody may be light- and phot2-regulated.

However, no YFP-tail domain fusion proteins of six of 13 class XI *A. thaliana* myosins (MYA1, MYA2, XI-B, XI-I, XI-J, XI-K) were targeted to chloroplasts but they localized to mitochondria, Golgi, peroxisomes, and small organelles (Li and Nebenführ 2007; Reisen and Hanson 2007). Most importantly, myosin mutant lines in Arabidopsis and tobacco (T-DNA knockout, RNAi-knockdown and dominant negative lines) retain normal chloroplast photorelocation (Aviser et al. 2008; Peremyslov et al. 2008), indicating that myosins are not involved in chloroplast photorelocation movement.

4.3 Microtubules

Microtubule-dependent chloroplast movement is very rare and found only in a few alga and the moss *P. patens*. Microtubule inhibitors but not actin inhibitors suppressed light-induced trapping of streaming chloroplasts at the irradiated area with a microbeam in two Caulerpales alga (Mizukami and Wada 1981; Maekawa et al. 1986). Interestingly, in the centric diatom *P. laevis*, blue light-induced chloroplast assemblage into the nuclear periphery (this is not the directional movement towards light and is dependent on extracellular Ca^{2+}) is microtubule-dependent whereas the green light-induced accumulation response (this is the directional movement towards light and is independent of extracellular Ca^{2+}) is actin-dependent (Shihira-Ishikawa et al. 2007), indicating that blue and green light mediate chloroplast movement via totally distinct mechanisms (Shihira-Ishikawa et al. 2007).

Among land plants examined, *P. patens* alone utilizes microtubules as well as actin filaments to mediate chloroplast movement (Sato et al. 2001a, 2003) (Table 1). In dark-adapted protonemal cells, chloroplasts, which were randomly distributed, repeated short-distance "back and forth" movement along the longitudinal axis and this movement was microtubule-dependent (Sato et al. 2001a). As described above, both red and blue light mediate chloroplast accumulation and avoidance response (Kadota et al. 2000) and mechanical stress induces an accumulation response in this moss (Sato et al. 2003). A microtubule inhibitor Cremart completely inhibited chloroplast movement (both accumulation and avoidance) induced by red light and mechanical stress (Sato et al. 2001a, 2003). However, inhibition of blue light-induced chloroplast movement (both accumulation and avoidance) required simultaneous treatment of Cremart and cytochalasin B, indicating that blue light-induced chloroplast movement utilizes both microtubule- and actin-based systems and that light quality but not light intensity determines the motile systems (Sato et al. 2001a) (Fig. 4). Disruption of microtubules decreased the velocity of blue light-induced chloroplast movement whereas disruption of actin filaments increased those of blue light but not red light-induced chloroplast movement (Sato et al. 2001a), suggesting that the microtubule system was differentially regulated between red and blue light and that microtubule and actin systems function competitively to some extent in blue light-induced chloroplast movement. The mechanism of choice of cytoskeleton systems between red light and blue light-induced responses is difficult to imagine since both are phototropin-dependent (Kasahara et al. 2004). Detailed analysis of knockout lines of phototropin or phytochrome genes using a cytoskeleton inhibitor may clarify this question.

4.4 CHUP1, a Possible Regulator of Actin Filaments

With the white band assay, mutants with no band were isolated (Fig. 2), termed *chloroplast unusual positioning1* (*chup1*) (Oikawa et al. 2003). The *chup1* mutant lacked accumulation and avoidance responses and their chloroplasts aggregated at the cell bottom irrespective of light conditions (Kasahara et al. 2002a; Oikawa et al. 2003), suggesting that CHUP1 is essential for the anchoring of chloroplasts on the plasma membrane. In spite of severe defects in chloroplast movement and positioning, distribution and movement of nuclei, mitochondria, and peroxisomes is normal in the *chup1* mutant although peroxisomes associated with chloroplasts in *chup1* mesophyll cells resided at the cell bottom as a result of a secondary effect of chloroplast aggregation (Oikawa et al. 2003). T-DNA tagging of the *CHUP1* gene revealed that CHUP1 is a novel plant-specific protein consisting of an N-terminal hydrophobic region (N), a coiled-coil region, an actinin-type actin-binding domain (ABD), a proline-rich motif (PRM), and a C-terminal conserved region (C) (Oikawa et al. 2003) (Fig. 6). Three *A. thaliana* proteins have a region similar to the CHUP1 C-terminus but their functions are unknown (Oikawa et al. 2003). Since CHUP1 has an ABD and a PRM, which is known as a profilin-binding motif in many actin regulators (Holt and Koffer 2001), it was suggested that CHUP1 may be a regulator for actin organization. Recombinant CHUP1 proteins could bind to actin filaments and G-actins in vitro (Oikawa et al.

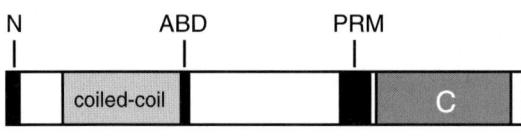

Fig. 6 Structure of *A. thalina* CHUP1 protein. CHUP1 consists of N-terminal hydrophobic region functioning as targeting signal to chloroplast outer envelope (N), a coiled-coil domain, an actinin-like actin-binding domain that can bind to F-actin in vitro (ABD), a proline-rich motif (PRM), C-terminal region similar to those of the other three *A. thaliana* proteins (C). *Bar*: 100 amino acids

2003, 2008; Schmidt von Brown and Schleiff 2008). Significantly, when transiently expressed in epidermal cells and mesophyll cells of *A. thaliana* leaves, CHUP1 GFP fusion proteins localized on the chloroplast periphery (Oikawa et al. 2003, 2008; Schmidt von Brown and Schleiff 2008), similar to the AtOEP7-GFP fusion protein which is a well-characterized marker for chloroplast outer membrane targeting (Lee et al. 2001), suggesting that CHUP1 may localize on the chloroplast outer envelope. These results suggested that CHUP1 may regulate actin dynamics on the chloroplast periphery and are consistent with previous results that binding of actin filaments with chloroplasts is dependent on proteins localized on the chloroplast outer membrane (Kumatani et al. 2006). However, actin dynamics and organization in mesophyll cells is similar in both wild-type and *chup1* leaves (Oikawa et al. 2003). More detailed observation of actin dynamics and biochemical analysis of CHUP1 are required to understand regulation of actin dynamics during chloroplast movement.

5 Ecological Significance of Chloroplast Photorelocation Movement

Although it has not yet been demonstrated, it is reasonable to think that the chloroplast accumulation response under low light conditions facilitates efficient light capture for photosynthesis (Zurzycki 1955). However, there are three hypotheses for ecological advantages of chloroplast movement to anticlinal walls by the avoidance response; (1) protection from photodamage (Zurzycki 1957; Brugnoli and Björkman 1992; Park et al. 1996); (2) promotion of light penetration to deeper layers (Brugnoli and Björkman 1992; Terashima and Hikosaka 1995; Gorton et al. 1999); (3) facilitation of CO_2 uptake (Gorton et al. 2003). The second and third hypotheses can apply only to chloroplast movement in multiple cell-layered leaves of seed plants. Comparison of CO_2 diffusion efficiency between plants whose chloroplasts were positioned on periclinal walls or on anticlinal walls revealed no evidence that chloroplast movement to anticlinal walls by the avoidance response enhanced CO_2 diffusion (Gorton et al. 2003). The first hypothesis is the most plausible since the avoidance response usually is induced by strong light in various plant species from alga to flowering plants. Shade plants showed greater chloroplast avoidance response than nonshade plants and this was positively correlated with tolerance to light stress (Brugnoli and Björkman 1992; Park et al. 1996). Experiments in *A. thaliana* clearly revealed that one of the advantages of the chloroplast avoidance response is the protection from photodamage (Kasahara et al. 2002a). When continuously irradiated with strong white light (1,400 mmol m^{-2} s^{-1}), leaves of *phot2* and *chup1* mutant plants that lack the avoidance response and are normal in other phototropin-mediated responses such as stomatal opening (Kagawa et al. 2001; Oikawa et al. 2003) began to bleach after 10 h and became severely bleached after 22 h whereas wild-type and *phot1* mutant plants did not bleach even after 31 h of strong light irradiation (Kasahara et al. 2002a). The severe bleaching phenotype in *phot2* and *chup1* was accompanied with enhanced photodamage and incomplete recovery after photodamage of photosystem II (Kasahara et al.

2002a). Both *phot2* and *chup1* were normal in chlorophyll content, photochemical and nonphotochemical quenching, the activities of reactive oxygen-scavenging enzymes and the levels of antioxidants, indicating that extensive photodamage in *phot2* and *chup1* resulted mainly from the defect in chloroplast avoidance response (Kasahara et al. 2002a). Note that field experiments revealed that chloroplasts change their positions most of the time, rarely realizing the extreme low-light (periclinal) or high-light (anticlinal) positions (Williams et al. 2003). Therefore, although availability of mutants deficient in chloroplast photorelocation movement provides the opportunity to examine the ecological significance of chloroplast movement in a more definitive way, we should consider natural light conditions to examine this.

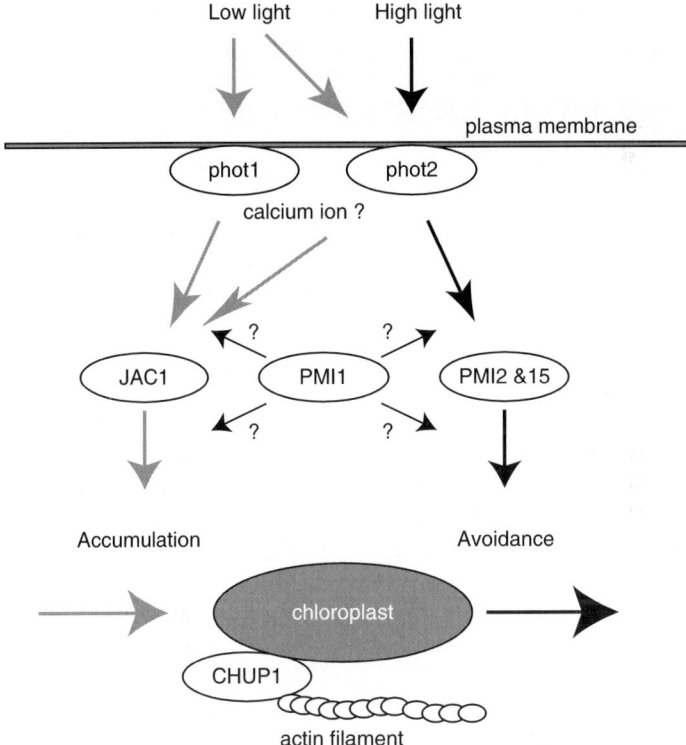

Fig. 7 A model of the mechanism for chloroplast photorelocation movement in *Arabidopsis thaliana*. phot1 and phot2 are blue light receptors localized on the plasma membrane. phot1 and phot2 regulate chloroplast accumulation response and phot2 alone mediates chloroplast avoidance response as shown by *arrows*. Although phototropins mediate blue light-induced calcium ion influx, the involvement of calcium ion in chloroplast movement is controversial. JAC1 specifically mediates chloroplast accumulation response (*gray arrows*). PMI2 and PMI15 regulate chloroplast avoidance response together with other unidentified proteins (not shown in this figure for clarity) (*black arrows*). PMI1 is necessary for both the accumulation and the avoidance responses but it is unknown when PMI1 functions on the signal transduction pathways. CHUP1 is localized on chloroplasts and may regulate actin polymerization

6 Conclusion

For about 100 years since chloroplast photorelocation movement was discovered, researchers have been making every endeavor to identify the photoreceptor molecules, signaling factors, and proteins for the motility system. Therefore, recent identification of photoreceptors for chloroplast movement, phototropins and neochromes is a major breakthrough in research on chloroplast movement. Although the signal transduction pathway and motility system for chloroplast movement have not yet been clearly revealed, thorough analysis of proteins identified from *Arabidopsis* research (CHUP1, JAC1, PMI1, PMI2, etc.) (Fig. 7) and in vivo imaging of signaling molecules and of cytoskeleton will clarify the molecular mechanism of chloroplast photorelocation movement in the near future.

Acknowledgments This work was partly supported by the Japanese Ministry of Education, Sports, Science and Technology (MEXT 13139203, 17084006 to M.W.) and the Japan Society of Promotion of Science (16107002, 13304061 to M.W. and a Research Fellowship for Young Scientists to N.S.).

References

Augustynowicz J, Lekka M, Burda K (2001) Correlation between chloroplast motility and elastic properties of tobacco mesophyll protoplasts. Acta Physiol Plant 23:291–302

Aviser D, Prokhnevsky AI, Makarova KS (2008) Myosin XI-K is required for rapid trafficking of Golgi stacks, peroxisomes, and mitochondria in leaf cells of *Nicotiana benthamiana*. Plant Physiol 146:1098–1108

Babourina O, Newman I, Shabala S (2002) Blue light-induced kinetics of H^+ and Ca^{2+} fluxes in etiolated wild-type and phototropin-mutant *Arabidopsis* seedlings. Proc Natl Acad Sci USA 99:2433–2438

Baum S, Long JC, Jenkins GI (1999) Stimulation of the blue light phototropic receptor NPH1 causes a transient increase in cytosolic Ca^{2+}. Proc Natl Acad Sci USA 96:13554–13559

Blatt MR, Wessells NK, Briggs WR (1980) Actin and cortical fiber reticulation in the siphonaceous alga *Vaucheria sessilis*. Planta 147:363–375

Boccara M, Schwartz W, Guiot E (2007) Early chloroplastic alterations analyzed by optical coherence tomography during a harpin-induced hypersensitive response. Plant J 50:338–346

Bögre L, Ökrész L, Henriques R (2003) Growth signaling pathways in *Arabidopsis* and the AGC protein kinases. Trends Plant Sci 8:424–431

Brett TJ, Traub LM, Fremont DH (2002) Accessory protein recruitment motifs in clathrin-mediated endocytosis. Structure 10:797–809

Brugnoli E, Björkman O (1992) Chloroplast movements in leaves: influence on chlorophyll fluorescence and measurements of light-induced absorbance changes related to ΔpH and zeaxanthin formation. Photosynth Res 32:23–35

Christie JM (2007) Phototropin blue-light receptors. Annu Rev Plant Biol 58:21–45

Christie JM, Salomon M, Nozue K (1999) LOV (light, oxygen, or voltage) domains of the blue-light photoreceptor phototropin (nph1): binding sites for the chromophore flavin mononucleotide. Proc Natl Acad Sci USA 96:8779–8783

Christie JM, Swartz TE, Bogomolni RA (2002) Phototropin LOV domains exhibit distinct roles in regulating photoreceptor function. Plant J 32:205–219

Cove D (2005) The moss *Physcomitrella patens*. Annu Rev Genet 39:339–358

Cox G, Hawes CR, van der Lubbe L (1987) High-voltage electron microscopy of whole, critical-point dried plant cells. 2. Cytoskeletal structures and plastid motility in *Selaginella*. Protoplasma 140:173–186

DeBlasio SL, Mullen JL, Luesse DR (2003) Phytochrome modulation of blue light-induced chloroplast movement in Arabidopsis. Plant Physiol 133:1471–1479

DeBlasio SL, Luesse DR, Hangarter RP (2005) A plant-specific protein essential for blue-light-induced chloroplast movements. Plant Physiol 139:101–114

Dong X-J, Ryu J-H, Takagi S (1996) Dynamic changes in the organization of microfilaments associated with the photocontrolled motility of chloroplasts in epidermal cells of *Vallisneria*. Protoplasma 195:18–24

Dong X-J, Nagai R, Takagi S (1998) Microfilaments anchor chloroplasts along the outer periclinal wall in *Vallisneria* epidermal cells through cooperation of PFR and photosynthesis. Plant Cell Physiol 39:1299–1306

Dreyer EM, Weisenseel MH (1979) Phytochrome-mediated uptake of calcium in *Mougeotia* cells. Planta 146:31–39

Eisenberg E, Greene LE (2007) Multiple roles of auxilin and Hsc70 in clathrin-mediated endocytosis. Traffic 8:640–646

Ezaki B, Kiyohara H, Matsumoto H (2007) Overexpression of an auxilin-like gene (F9E10.5) can suppress Al uptake in roots of *Arabidopsis*. J Exp Bot 58:497–506

Furukawa T, Watanabe M, Shihira-Ishikawa I (1998) Green- and blue-light-mediated chloroplast migration in the centric diatom *Pleurosira laevis*. Protoplasma 203:214–220

Gabry's H, Walczak T, Haupt W (1984) Blue-light-induced chloroplast orientation in *Mougeotia*. Evidence for a separate sensor pigment besides phytochrome. Planta 160:21–24

Gorton HL, William WE, Vogelmann TC (1999) Chloroplast movement in *Alocasia macrorrhiza*. Physiol Plant 106:421–428

Gorton HL, Herbert SK, Vogelmann TC (2003) Photoacoustic analysis indicates that chloroplast movement does not alter liquid-phase CO_2 diffusion in leaves of *Alocasia brisbanensis*. Plant Physiol 132:1529–1539

Harada A, Sakai T, Okada K (2003) phot1 and phot2 mediate blue light-induced transient increases in cytosolic Ca^{2+} differently in *Arabidopsis* leaves. Proc Natl Acad Sci USA 100:8583–8588

Haupt W (1959) Die Chloroplastendrehung bei *Mougeotia*. I. Über den quantitativen und qualitativen Lichtbedarf der Schwachlichtbewegung. Planta 53:484–501

Haupt W (1982) Light-mediated movement of chloroplasts. Annu Rev Plant Physiol 33:205–233

Haupt W (1999) Chloroplast movement: from phenomenology to molecular biology. In: Esser K, Lüttge U, Beyschlag W, Murata J (eds) Progress in Botany, vol 60. Springer, Heidelberg, pp 3–36

Haupt W, Scheuerlein R (1990) Chloroplast movement. Plant Cell Environ 13:595–614

Hetherington AM, Brownlee C (2004) The generation of Ca^{2+} signals in plants. Annu Rev Plant Biol 55:401–427

Holt MR, Koffer A (2001) Cell motility: proline-rich proteins promote protrusions. Trends Cell Biol 11:38–46

Huala E, Oeller PW, Liscum E (1997) *Arabidopsis* NPH1: a protein kinase with a putative redox-sensing domain. Science 278:2120–2123

Izutani Y, Takagi S, Nagai R (1990) Orientation movements of chloroplasts in *Vallisneria* epidermal cells: different effects of light at low- and high-fluence rate. Photochem Photobiol 51:105–111

Jarillo JA, Gabry's H, Capel J (2001) Phototropin-related NPL1 controls chloroplast relocation induced by blue light. Nature 410:952–954

Jiao Y, Yang H, Ma L (2003) A genome-wide analysis of blue-light regulation of Arabidopsis transcription factor gene expression during seedling development. Plant Physiol 133:1480–1493

Kadota A, Wada M (1989) Photoinduction of circular F-actin on chloroplast in a fern protonemal cell. Protoplasma 151:171–174

Kadota A, Wada M (1992a) Photoorientation of chloroplasts in protonemal cells of the fern *Adiantum* as analyzed by use of a video-tracking system. Bot Mag Tokyo 105:265–279

Kadota A, Wada M (1992b) Photoinduction of formation circular structures by microfilaments on chloroplasts during intracellular orientation in protonemal cells of the fern *Adiantum capillus-veneris*. Protoplasma 167:97–107

Kadota A, Wada M (1999) Red light-aphototropic (rap) mutants lack red light-induced chloroplast relocation movement in the fern *Adiantum capillus-veneris*. Plant Cell Physiol 40:238–247

Kadota A, Kohyama I, Wada M (1989) Phototropism and photomovement of chloroplasts in the protonemata of the ferns *Pteris* and *Adiantum*: evidence for the possible lack of dichroic phytochrome in *Pteris*. Plant Cell Physiol 40:523–531

Kadota A, Sato Y, Wada M (2000) Intracellular chloroplast photorelocation in the moss *Physcomitrella patens* is mediated by phytochrome as well as by a blue-light receptor. Planta 210:932–937

Kagawa T, Suetsugu N (2007) Photometrical analysis with photosensory domains of photoreceptors in green algae. FEBS Lett 581:368–374

Kagawa T, Wada M (1993) Light-dependent nuclear positioning in prothallial cells of *Adiantum capillus-veneris*. Protoplasma 177:82–85

Kagawa T, Wada M (1994) Brief irradiation with red or blue light induces orientational movement of chloroplasts in dark-adapted prothallial cells of the fern *Adiantum*. J Plant Res 107:389–398

Kagawa T, Wada M (1996) Phytochrome- and blue-light-absorbing pigment-mediated directional movement of chloroplasts in dark-adapted prothallial cells of fern *Adiantum* as analyzed by microbeam irradiation. Planta 198:488–493

Kagawa T, Wada M (1999) Chloroplast-avoidance response induced by high-fluence blue light in prothallial cells of the fern *Adiantum capillus-veneris* as analyzed by microbeam irradiation. Plant Physiol 119:917–923

Kagawa T, Wada M (2000) Blue light-induced chloroplast relocation in *Arabidopsis thaliana* as analyzed by microbeam irradiation. Plant Cell Physiol 41:84–93

Kagawa T, Wada M (2004) Velocity of chloroplast avoidance movement is fluence rate dependent. Photochem Photobiol Sci 3:592–595

Kagawa T, Kadota A, Wada M (1994) Phytochrome-mediated photoorientation of chloroplasts in protonemal cells of the fern *Adiantum* can be induced by brief irradiation with red light. Plant Cell Physiol 35:371–377

Kagawa T, Sakai T, Suetsugu N (2001) *Arabidopsis*NPL1: a phototropin homolog controlling the chloroplast high-light avoidance response. Science 291:2138–2141

Kagawa T, Kasahara M, Abe T (2004) Functional analysis of phototropin2 using fern mutants deficient in blue light-induced chloroplast avoidance movement. Plant Cell Physiol 45:416–426

Kandasamy MK, Meagher RB (1999) Actin-organelle interaction: association with chloroplast in *Arabidopsis* leaf mesophyll cells. Cell Motil Cytoskel 44:110–118

Kanegae T, Hayashida E, Kuramoto C (2006) A single chromoprotein with triple chromophores acts as both a phytochrome and a phototropin. Proc Natl Acad Sci USA 103:17997–18001

Kasahara M, Kagawa T, Oikawa K (2002a) Chloroplast avoidance movement reduces photodamage in plants. Nature 420:829–832

Kasahara M, Swartz TE, Olney MA (2002b) Photochemical properties of the flavin mononucleotide-binding domains of the phototropins from Arabidopsis, rice and *Chlamydomonas reinhardtii*. Plant Physiol 129:762–773

Kasahara M, Kagawa T, Sato Y (2004) Phototropins mediate blue and red light-induced chloroplast movements in *Physcomitrella patens*. Plant Physiol 135:1388–1397

Kawai H, Kanegae T, Christensen S (2003) Responses of ferns to red light are mediated by an unconventional photoreceptor. Nature 421:287–290

Kelley WL (1998) The J-domain family and the recruitment of chaperone power. Trends Biochem Sci 23:222–227

Kinoshita T, Doi M, Suetsugu N (2001) phot1 and phot2 mediate blue light regulation of stomatal opening. Nature 414:656–660
Knieb E, Salomon M, Rüdiger W (2004) Tissue-specific and subcellular localization of phototropin determined by immuno-blotting. Planta 218:843–851
Kodama Y, Tsuboi H, Kagawa T (2008) Low temperature-induced chloroplast relocation mediated by a blue light receptor, phototropin 2, in fern gametophytes. J Plant Res 121:441–448
Kondo A, Kaikawa J, Funaguma T (2004) Clumping and dispersal of chloroplasts in succulent plants. Planta 219:500–506
Kong S-G, Suzuki T, Tamura K (2006) Blue light-induced association of phototropin 2 with the Golgi apparatus. Plant J 45:994–1005
Kong S-G, Kinoshita T, Shimazaki K (2007) The C-terminal kinase fragment of Arabidopsis phototropin 2 triggers constitutive phototropin responses. Plant J 51:862–873
Krzeszowiec W, Gabry's H (2007) Phototropin mediated relocation of myosins in *Arabidopsis thaliana*. Plant Signal Behav 2:333–336
Krzeszowiec W, Rajwa B, Dobrucki J (2007) Actin cytoskeleton in *Arabidopsis thaliana* under blue and red light. Biol Cell 99:251–260
Kumatani T, Sakurai-Ozato N, Miyawaki N (2006) Possible association of actin filaments with chloroplasts of spinach mesophyll cells in vivo and in vitro. Protoplasma 229:45–52
Lam BC-H, Sage TL, Bianchi F (2001) Role of SH3 domain-containing proteins clathrin-mediated vesicle trafficking in *Arabidopsis*. Plant Cell 13:2499–2512
Lee YJ, Kim DH, Kim Y-W (2001) Identification of a signal that distinguishes between the chloroplast outer envelope membrane and the endomembrane system in vivo. Plant Cell 13:2175–2190
Li J-F, Nebenführ A (2007) Organelle targeting of myosin XI is mediated by two globular tail subdomains with separate cargo binding sites. J Biol Chem 282:20593–20602
Li QH, Yang HQ (2007) Cryptochrome signaling in plants. Photochem Photobiol 83:94–101
Liebe S, Menzel D (1995) Actomyosin-based motility of endoplasmic reticulum and chloroplasts in *Vallisneria* mesophyll cells. Biol Cell 85:207–222
Luesse DR, DeBlasio SL, Hangarter RP (2006) Plastid movement impaired 2, a new gene involved in normal blue-light-induced chloroplast movements in *Arabidopsis*. Plant Physiol 141:1328–1337
Maekawa T, Tsutsui I, Nagai R (1986) Light-regulated translocation of cytoplasm in green alga *Dichotomosiphon*. Plant Cell Physiol 27:837–851
Makita N, Shihira-Ishikawa I (1997) Chloroplast assemblage by mechanical stimulation and its intercellular transmission in diatom cells. Protoplasma 197:87–95
Malec P, Rinaldi RA, Gabry's H (1996) Light-induced chloroplast movements in *Lemna trisulca*. Identification of the motile system. Plant Sci 120:127–137
Marcotte EM, Pellegrini M, Ng H-L (1999) Detecting protein function and protein-protein interactions from genome sequences. Science 285:751–753
Matsuoka D, Tokutomi S (2005) Blue light-regulated molecular switch of Ser/Thr kinase in phototropin. Proc Natl Acad Sci USA 102:13337–13342
Matsushita T, Mochizuki N, Nagatani A (2003) Dimers of the N-terminal domain of phytochrome B are functional in the nucleus. Nature 424:571–574
McCurdy DW (1999) Is 2,3-butanedione monoxime an effective inhibitor of myosin-based activities in plant cells? Protoplasma 209:120–125
Mineyuki Y, Kataoka H, Masuda R (1995) Dynamic changes in the actin cytoskeleton during the high-fluence rate response of the *Mougeotia* chloroplast. Protoplasma 185:222–229
Mittmann F, Brückner G, Zeidler (2004) Targeted knockout in *Physcomitrella* reveals direct actions of phytochrome in the cytoplasm. Proc Natl Acad Sci USA 101:13939–13944
Mizukami M, Wada S (1981) Action spectrum for light-induced chloroplast accumulation in a marine coenocytic green alga, *Bryopsis plumosa*. Plant Cell Physiol 22:1245–1255
Mommer L, Pons TL, Wolters-Arts M (2005) Submergence-induced morphological, anatomical, and biochemical responses in a terrestrial species affect gas diffusion resistance and photosynthetic performance. Plant Physiol 139:497–508

Nozue K, Kanegae T, Imaizumi T (1998) A phytochrome from the fern *Adiantum* with features of the putative photoreceptor NPH1. Proc Natl Acad Sci USA 95:15826–15830

Ohgishi M, Saji K, Okada K (2004) Functional analysis of each blue light receptor, cry1, cry2, phot1, and phot2, by using combinatorial multiple mutants in *Arabidopsis*. Proc Natl Acad Sci USA 101:2223–2228

Oikawa K, Kasahara M, Kiyosue T (2003) CHLOROPLAST UNUSUAL POSITIONNING is essential for proper chloroplast positioning. Plant Cell 15:2805–2815

Oikawa K, Yamasato A, Kong SG (2008) Chloroplast outer envelope protein CHUP1 is essential for chloroplast anchorage to the plasma membrane and chloroplast movement. Plant Physiol 10.1104/pp.108.123075

Onodera A, Kong S-G, Doi M (2005) Phototropin from *Chlamydomonas reinhardtii* is functional in *Arabidopsis thaliana*. Plant Cell Physiol 46:367–374

Park Y-L, Chow WS, Anderson JM (1996) Chloroplast movement in the shade plant *Tradescantia albiflora* helps protect photosystem II against light stress. Plant Physiol 111:867–875

Paves H, Truve E (2007) Myosin inhibitors block accumulation movement of chloroplasts in *Arabidopsis thaliana*. Protoplasma 230:165–169

Peremyslov VV, Prokhnevsky AI, Aviser D (2008) Two class XI myosins function in organelle trafficking and root hair development in Arabidopsis. Plant Physiol 146:1109–1116

Reisen D, Hanson MR (2007) Association of six YFP-myosin XI-tail fusions with mobile plant cell organelles. BMC Plant Biol 7:6

Rensing SA, Lang D, Zimmer AD (2008) The *Physcomitrella* genome reveals evolutionary insights into the conquest of land by plants. Science 319:64–69

Rockwell NC, Su Y-S, Lagarias JC (2006) Phytochromes structure and signaling mechanisms. Annu Rev Plant Biol 57:837–858

Russ U, Grolig F, Wagner G (1991) Changes of cytoplasmic free Ca^{2+} in the green alga *Mougeotia scalaris* as monitored with indo-1, and their effect on the velocity of chloroplast movements. Plant Physiol 121:37–44

Russell AJ, Cove DJ, Trewavas AJ (1998) Blue light but not red light induces a calcium transient in the moss *Physcomitrella patens* (Hedw.) B., S. & G. Planta 206:278–283

Sakai Y, Takagi S (2005) Reorganized actin filaments anchor chloroplasts along the anticlinal walls of *Vallisneria* epidermal cells under high-intensity blue light. Planta 221:823–830

Sakai T, Kagawa T, Kasahara M (2001) *Arabidopsis* nph1 and npl1: blue light receptors that mediate both phototropism and chloroplast relocation. Proc Natl Acad Sci USA 98:6969–6974

Sakamoto K, Briggs WR (2002) Cellular and subcellular localization of phototropin 1. Plant Cell 14:1723–1735

Sakurai N, Domoto K, Takagi S (2005) Blue-light-induced reorganization of the actin cytoskeleton and the avoidance response of chloroplasts in epidermal cells of *Vallisneria gigantea*. Planta 221:66–74

Salomon M, Christie JM, Knieb E (2000) Photochemical and mutational analysis of the FMN-binding domain of the plant blue light receptor, phototropin. Biochemistry 39:9401–9410

Sato Y, Kadota A, Wada M (1999) Mechanically induced avoidance response of chloroplasts in fern protonemal cells. Plant Physiol 121:37–44

Sato Y, Wada M, Kadota A (2001a) Choice of tracks, microtubules and/or actin filaments for chloroplast photo-movement is differentially controlled by phytochrome and a blue light receptor. J Cell Sci 114:269–279

Sato Y, Wada M, Kadota A (2001b) External Ca^{2+} is essential for chloroplast movement induced by mechanical stimulation but not by light stimulation. Plant Physiol 127:497–504

Sato Y, Wada M, Kadota A (2003) Accumulation response of chloroplasts induced by mechanical stimulation in bryophyte cells. Planta 216:772–777

Schmidt von Braun S, Schleiff E (2007) Movement of endosymbiotic organelles. Curr Protein Pept Sci 8:426–438

Schmidt von Braun S, Schleiff E (2008) The chloroplast outer membrane protein CHUP1 interacts with actin and profilin. Planta 227:1151–1159

Schneider H, Schuettpelz E, Pryer KM (2004) Ferns diversified in the shadow of angiosperms. Nature 428:553–557
Schönbohm E, Meyer-Wegner J, Schönbohm E (1990) No evidence for Ca^{2+} influx as an essential link in the signal transduction chains of either light-oriented chloroplast movements or Pfr-mediated chloroplast anchorage in *Mougeotia*. J Photochem Photobiol 5:331–341
Senn G (1908) Die Gestalts- und Lageveränderung der Pflanzen-Chromatophoren. Engelmann, Stuttgart
Serlin BS, Roux SJ (1984) Modulation of chloroplast movement in the green alga *Mougeotia* by the Ca^{2+} ionophore A23187 and by calmodulin antagonists. Proc Natl Acad Sci USA 81:6368–6372
Shihira-Ishikawa I, Nakamura T, Higashi S (2007) Distinct responses of chloroplasts to blue and green laser microbeam irradiations in the centric diatom *Pleurosira laevis*. Photochem Photobiol 83:1101–1109
Sinclair J, Hall CE (1995) Photosynthetic energy storage in aquatic leaves measured by photothermal deflection. Photosynth Res 45:157–168
Stoelzle S, Kagawa T, Wada M (2003) Blue light activates calcium-permeable channels in *Arabidopsis* mesophyll cells via the phototropin signaling pathway. Proc Natl Acad Sci USA 100:1456–1461
Suetsugu N, Wada M (2005) Photoreceptor gene families in lower plants. In: Briggs WR, Spudich JL (eds) Handbook of Photosensory Receptors. Wiley-VCH, Weinheim, pp 349–369
Suetsugu N, Wada M (2007a) Phytochrome-dependent photomovement responses mediated by phototropin family proteins in cryptogam plants. Photochem Photobiol 83:87–93
Suetsugu N, Wada M (2007b) Chloroplast photorelocation movement mediated by phototropin family proteins in green plants. Biol Chem 83:87–93
Suetsugu N, Kagawa T, Wada M (2005a) An auxilin-like J-domain protein, JAC1, regulates phototropin-mediated chloroplast movement in *Arabidopsis*. Plant Physiol 139:151–162
Suetsugu N, Mittmann F, Wagner G (2005b) A chimeric photoreceptor gene, *NEOCHROME*, has arisen twice during plant evolution. Proc Natl Acad Sci USA 102:13705–13709
Sullivan S, Thomson CE, Lamont DJ (2008) In vivo phosphorylation site mapping and functional characterization of Arabidopsis phototropin 1. Mol Plant 1:178–194
Takagi S (2003) Actin-based photo-orientation movement of chloroplasts in plant cells. J Exp Biol 206:1963–1969
Taylor BL, Zhulin IB (1999) PAS domains: internal sensors of oxygen, redox potential, and light. Microbiol Mol Biol Rev 63:479–506
Tazawa M, Kurosawa S, Amino S (1991) Induction of cytoplasmic streaming and movement of chloroplast induced by l-Histidine and its derivatives in leaves of *Egeria densa*. Plant Cell Physiol 32:253–260
Terashima I, Hikosaka K (1995) Comparative ecophysiology of leaf and canopy photosynthesis. Plant Cell Environ 18:1111–1128
Tlalka M, Fricker M (1999) The role of calcium in blue-light-dependent chloroplast movement in *Lemna trisulca* L. Plant J 20:461–473
Tlalka M, Gabry's H (1993) Influence of calcium on blue-light-induced chloroplast movement in *Lemna trisulca* L. Planta 189:491–498
Trojan A, Gabry's H (1996) Chloroplast distribution in *Arabidopsis thaliana* (L.) depends on light conditions during growth. Plant Physiol 111:419–425
Tsuboi H, Suetsugu N, Wada M (2006) Negative phototropic response of rhizoid cells in the fern *Adiantum capillus-veneris*. J Plant Res 119:505–512
Tsuboi H, Suetsugu N, Toyooka-Kawai H (2007) Phototropins and neochrome1 mediate nuclear movement in the fern *Adiantum capillus-veneris*. Plant Cell Physiol 48:892–896
Tucker EB, Lee M, Alli S (2005) UV-A induces two calcium waves in *Physcomitrella patens*. Plant Cell Physiol 46:1226–1236
Uenaka H, Kadota A (2007) Functional analyses of the *Physcomitrella patens* phytochromes in regulating chloroplast avoidance movement. Plant J 51:1050–1061

Wada M (1988) Chloroplast photoorientation in enucleated fern protonemata. Plant Cell Physiol 29:1227–1232
Wada M, Kadota A (1989) Photomorphogenesis in lower green plants. Annu Rev Plant Physiol Plant Mol Biol 40:169–191
Wada M, Suetsugu N (2004) Plant organelle positioning. Curr Opin Plant Biol 7:626–631
Wada M, Grolig F, Haupt W (1993) Light-oriented chloroplast positioning. Contribution to progress in photobiology. J Photochem Photobiol B Biol 17:3–25
Wada M, Kagawa T, Sato Y (2003) Chloroplast movement. Annu Rev Plant Biol 54:455–468
Wagner G, Haupt W, Laux A (1972) Reversible inhibition of chloroplast movement by cytochalasin B in the green alga *Mougeotia*. Science 176:808–809
Walsh P, Bursac D, Law YC (2004) The J-domain family: modulating protein assembly, disassembly and translocation. EMBO Rep 5:567–571
Wan Y-L, Eisinger W, Ehrhardt D (2008) The subcellular localization and blue-light-induced movement of phototropin 1-GFP in etiolated seedlings of *Arabidopsis thaliana*. Mol Plant 1:103–117
Wang Z, Pesacreta TC (2004) A subclass of myosin XI is associated with mitochondria, plastids, and the molecular chaperone subunit TCP-1α in maize. Cell Motil Cytoskel 57:218–232
Weidinger M, Ruppel HG (1985) Ca^{2+}-requirement for a blue-light-induced chloroplast translocation in *Eremosphaera viridis*. Protoplasma 124:184–187
Williams WE, Gorton HL, Witiak SM (2003) Chloroplast movements in the field. Plant Cell Environ 26:2005–2014
Wojtaszek P, Anielska-Mazur A, Gabry's H (2005) Recruitment of myosin VIII towards plastid surfaces is root-cap specific and provides the evidence for actomyosin involvement in root osmosensing. Funct Plant Biol 32:721–736
Yatsuhashi H (1996) Photoregulation systems for light-oriented chloroplast movement. J Plant Res 109:139–146
Yatsuhashi H, Kobayashi H (1993) Dual involvement of phytochrome in light-oriented chloroplast movement in *Dryopteris sparsa* protonemata. J Photochem Photobiol B 19:25–31
Yatsuhashi H, Wada M (1990) High-fluence rate responses in the light-oriented chloroplast movement in *Adiantum* protonemata. Plant Sci 68:87–94
Yatsuhashi H, Kadota A, Wada M (1985) Blue- and red-light action in photoorientation of chloroplasts in *Adiantum* protonemata. Planta 165:43–50
Yatsuhashi H, Hashimoto T, Wada M (1987a) Dichroic orientation of photoreceptors for chloroplast movement in *Adiantum* protonemata. Non-helical orientation. Plant Sci 51:165–170
Yatsuhashi H, Wada M, Hashimoto T (1987b) Dichroic orientation of phytochrome and blue-light photoreceptor in *Adiantum* protonemata as determined by chloroplast movement. Acta Physiol Plant 9:163–173
Yoneda A, Kutsuna N, Higaki T (2007) Recent progress in living cell imaging of plant cytoskeleton and vacuole using fluorescent-protein transgenic lines and three-dimensional imaging. Protoplasma 230:129–139
Zurzycki J (1955) Chloroplasts arrangement as a factor in photosynthesis. Acta Soc Bot Pol 24:27–63
Zurzycki J (1957) The destructive effect of intense light on the photosynthetic apparatus. Acta Soc Bot Pol 26:157–175
Zurzycki J (1980) Blue light-induced intracellular movement. In: Senger H (ed) Blue Light Syndrome. Springer, Heidelberg, pp 50–68

A Sentinel Role for Plastids

F. Bouvier, A.S. Mialoundama, and B. Camara (✉)

Abstract Plastids are best known for their role in photosynthesis and metabolism. However, accumulating evidence indicates that plastids could be important environmental sensors and executors of the adaptive responses of plants. The mechanisms involved are diverse, ranging from the recognition of pathogen invaders to the execution of signaling cascades that trigger their destruction. Thus, several pathogens and parasites exploit or hijack plastid functions to evade host defenses. Plastids are also implicated in the adaptation of plants to low nutrient environments. Finally, at the end of the life cycle or during the reproduction stage, the adequate implementation of the programs leading to senescence or to the increased pigmentation necessary for the attraction of pollinators and seed dispersal are also among the key roles assigned to plastids.

1 Introduction

One of the most important aspects of plant evolution was the acquisition of plastids which resulted in massive reduction of the genome of the primary cyanobacterial endosymbiont and its transfer into the nucleus (Kim and Archibald 2008). For instance from the 5,366 genes in the cyanobacterium *Anabaena* sp. PCC 7120 (Kaneko et al. 2001a; Kaneko et al. 2001b), we arrive at 251 plastid genes in the less depleted plastid genome of the red algae *Porphyra purpurea* (Reith and Munholland 1995). Thus, most plastid proteins are nucleus-encoded (Abdallah et al. 2000; Leister 2003). The logic of the association is often considered to be the acquisition of photosynthesis. However, it is important to realize that before the primary endosymbiosis, the free prokaryotic endosymbiont organism possessed diverse adaptive responses to

---

B. Camara
Institut de Biologie Moléculaire des Plantes
Centre National de la Recherche Scientifique and Université Louis Pasteur,
67084, Strasbourg Cedex, France
e-mail: bilal.camara@ibmp-ulp.u-strasbg.fr

environmental stresses that were essential to ensure its survival. Thus, one may conjecture that in addition to the acquisition of new functions, remnants or the memory of these early functions predating the endosymbiotic events as well as the propensity of individual prokaryotes to communicate are now conserved in plastids hosted in plants. Low molecular weight products and macromolecules derived from plastid metabolism play a decisive role in the adaptation of plants to the environment.

2 Plastid Low Molecular Weight Sensors

2.1 Tocopherol-Dependent Signaling

Tocopherols are synthesized in plastids from homogentisic acid and isopentenyl diphosphate (Bouvier et al. 2005b). It is assumed that they play a major role in photoprotection and antioxidant protection by scavenging lipid peroxy radicals. α-Tocopherol increases membrane rigidity like phytosterols (Leiken and Brenner 1989), thus its concentration, together with that of the other membrane lipid components, may be regulated to afford adequate fluidity for membrane function. Recent studies using tocopherol-deficient plants revealed that tocopherol may have nonantioxidant roles. The observation that callose deposition is upregulated (Provencher et al. 2001), concomitantly to the deficiency of tocopherol cyclase (Porfirova et al. 2002) suggests a functional link between the synthesis of tocopherol in plastids and the signaling of the stress response leading to callose. Callose is a β-1,3-glucan polymer that is induced during pathogen attack below or near the entry zone. Because tocopherols are recognized as efficient scavengers of lipid peroxy radicals (Kamal-Eldin and Appelqvist 1996), one may suggest that in the absence of functional tocopherol, peroxy radical-derived products activate defense reactions involving callose formation and deposition. This contention is further supported by the induction of the wound-induced *proteinase inhibitor Pin2* gene and by the increased accumulation of the proline biosynthesis enzyme Δ-1-pyrroline-5-carboxylate synthase in potato plants where the expression of the tocopherol cyclase gene has been down-regulated by a RNAi approach (Hofius et al. 2004). Studies using a tocopherol-deficient mutant of *Arabidopsis thaliana* (*vte1*) that accumulates the pathway intermediate 2,3-dimethyl-5-phytyl-1,4-benzoquinone and *vte2*, which lacks all tocopherols and pathway intermediates, point to the same conclusion.

Indeed, it has been postulated that in addition to its antioxidant role, tocopherols are implicated in signaling mechanisms in mammals (Azzi 2007; Rimbach et al. 2002; Traber and Atkinson 2007) and in plants (Munne-Bosch 2005). The observation that the *allene oxide cyclase* mRNA is upregulated in tocopherol-deficient potato suggests a functional relationship between tocopherol and jasmonate signaling in plants (Hofius et al. 2004). The *Arabidopsis vte2* mutant in contrast to *vte1* accumulates massive amounts of hydroxy fatty acids, malondialdehyde, and phytoprostanes that derive from the non enzymic oxidation of membrane lipids. This phenomenon is paralleled by the induction of several defense genes and the phytoalexin camalexin up to 100-fold in *vte2* compared to *vt1* and the wild-type (Cheng et al. 2003).

The signaling role of tocopherol has also been observed in cyanobacteria and does not appear to be linked to the redox homeostasis. For example, the photoautotrophic growth rates of cyanobacterium *Synechocystis* sp. PCC 6803 are not affected in an α-tocopherol-deficient mutant strain. On the other hand, when the α-tocopherol-deficient mutant was grown photomixotrophically with glucose, the photosystem II activity and the carboxysome gene transcripts were lost within 24 h. These changes take place via a signaling mechanism involving the *pmgA* gene, which encodes a putative serine-threonine kinase (Sakuragi et al. 2006).

The signaling role played by tocopherol is further supported by the fact that α-tocopherol and γ-tocopherol may have different functions. The knockdown of the γ-tocopherol methyl transferase (γ-TMT) gene resulted in an up to 95% reduction of α-tocopherol in tobacco leaves paralleled by an increased accumulation of γ-tocopherol. The resulting γ-TMT knockdown plants accumulated less lipid peroxy radicals and the membrane damage was limited under stress conditions (Abbasi et al. 2007). The role played by tocopherol is exacerbated when plants are grown at low temperatures. Studies using the *Arabidopsis* mutant *vte2* revealed no significant modification when plants are grown at 22°C, but significant differences became apparent after 14 days at 7°C. Under these conditions, the *vte2* mutant accumulated lower levels of linolenic acid (18:3) and higher levels of linoleic acid (18:2) compared with the wild-type. Lipid profiling revealed that oxidative reactions affecting C18:3 could not account for the decrease, rather the endoplasmic reticulum pathway of C18:3 synthesis was affected (Maeda et al. 2008). This unexpected effect further indicates how plastid tocopherol contributes to the acclimation of plants to low temperatures.

Tocopherol could exert a surveillance role, i.e., in the absence of lipid peroxidation they restrain the premature transcription of defense genes. However, under biotic or abiotic stress conditions, lipid peroxidation products may induce the expression of genes involved in stress responses and defense mechanisms through TGA (TGACG-sequence-specific binding-protein) transcription factors (Mueller et al. 2008).

2.2 Chlorophyll Precursors and Metabolites as Cellular Signaling Mediators of Cell Defense

2.2.1 Role of Tetrapyrrole Precursors

During plant–pathogen interactions the hypersensitive reaction, characterized among other features by the death of cells surrounding the infection zone and the induction of defense responses, is triggered. Further analysis of this phenomenon, using plant mutants displaying a spontaneous cell death phenotype revealed the key role played by several metabolites involved in the synthesis or the catabolism of plastid tetrapyrroles (Fig. 1). In maize, the *lesion mimic* mutant *Les22* is characterized by the formation of necrotic spots on leaves in a light-dependent manner (Hu et al. 1998). The phenotype is very similar to the symptoms of diseases affecting maize, collectively termed disease *lesion mimic*s (Neuffer and Calvert 1975). *Les22* encodes uroporphyrinogen III decarboxylase (UROD), an enzyme of the plastid

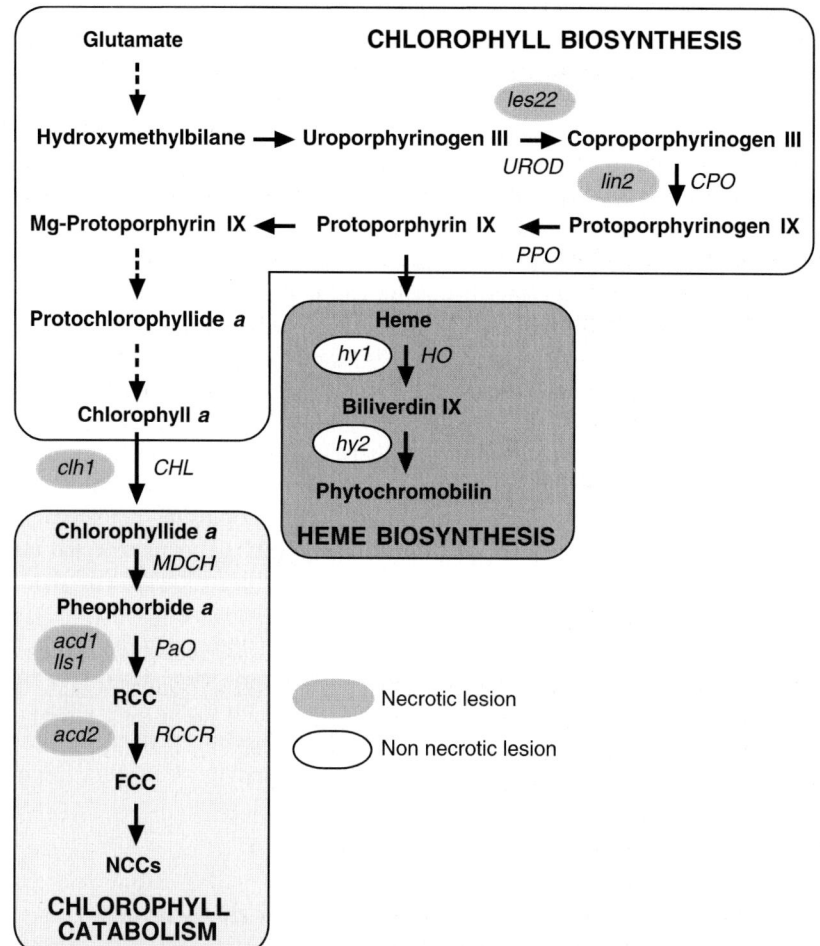

Fig. 1 Pathways of tetrapyrrole biosynthesis and chlorophyll catabolism. The steps inducing the apparition of necrotic lesions and the corresponding mutants are indicated. Enzyme abbreviations refer to: *CLH* chlorophyllase; *CPO* coproporphyrinogen III oxidase; *MDCH* magnesium dechelatase; *HO* heme oxygenase; *PaO* pheophorbide *a* oxygenase; *PPO* protoporphyrinogen IX oxidase; *RCCR* red chlorophyll catabolite reductase; *UROD* uroporphyrinogen III decarboxylase. Substrate abbreviations refer to: *FCC* fluorescent chlorophyll catabolite; *NCCs*, non-fluorescent chlorophyll catabolites; *RCC*, red chlorophyll catabolite. Pathway-specific mutants are indicated: *Acd1* accelerated cell death 1; *Acd2* accelerated cell death 2; *clh1* chlorophyllase 1; *Hy1* heme oxygenase 1; *Hy2* phytochromobilin synthase 2; *Les22* lesion mimic 22; *Lin2* lesion initiation 2; *lls1* lethal leaf spot 1

porphyrin pathway leading to heme and chlorophylls (Hu et al. 1998) (Fig. 1). In a similar vein, the down-regulation of tobacco, through an antisense technology led to the formation of necrotic lesions that were correlated with the accumulation of uroporphyrinogen due to the formation of reactive oxygen species under light conditions (Mock and Grimm 1997). A very similar phenomenon was observed in transgenic tobacco plants, when the conversion of coproporphyrinogen III to protoporphyrinogen IX was inhibited through the antisense expression of coproporphyrinogen oxidase III gene (CPO) (Mock et al. 1999) (Fig. 1). A characteristic feature of tobacco expressing antisense UROD and CPO is the accumulation of the fluorescent coumarin phytoalexin scopolin and the pathogenesis-related protein PR-1, which normally participate in defense responses (Mock et al. 1999). This trend was reinforced by the fact that the level of salicylic acid was concomitantly increased. The same trend was observed in the *Arabidopsis lesion initiation mutant 2* (lin2) which results from a T-DNA insertion in the *LIN2* gene which encodes CPO (Ishikawa et al. 2001) (Fig. 1). Further studies indicate that in *Arabidopsis* the down-regulation of the expression of protoporphyrinogen oxidase (PPO) which encodes the enzyme converting protoporphyrinogen IX into protoporphyrin IX causes the apparition of necrotic spots on the leaves associated with a systemic acquired resistance (SAR) (Molina et al. 1999). PPO represents the last enzymes shared by the plastid pathway leading to chlorophylls and the heme pathway (Fig. 1). It is interesting to note that *Arabidopsis* mutants affected in the heme oxygenase (*hy1*) and phytochromobilin synthase (*hy2*) genes which are specific of the branch leading to heme do not apparently form necrotic lesions (Ishikawa 2005) (Fig. 1).

2.2.2 Chlorophyll Catabolite Mediators

The downstream steps involved in the breakdown of chlorophylls are also implicated in the signaling process. The down-regulation of *Arabidopsis* chlorophyllase *AtCLH1* (Fig. 1), which converts chlorophyll into chlorophyllide + phytol, results in an increased resistance to the bacterial pathogen *Erwinia carotovora* and an increased susceptibility to the fungal pathogen *Alternaria brassicicola* (Kariola et al. 2005). The increased resistance of *Arabidopsis* to *E. carotovora* observed when *AtCLH1* was silenced is light-dependent (Kariola et al. 2005). The *Arabidopsis accelerated cell death*1 (*acd1*) mutant (Greenberg and Ausubel 1993), like the maize *lethal leaf spot1* (*lls1*) mutant (Gray et al. 1997), is affected in the *pheophorbide a oxygenase* (*PaO*) which encodes a plastid Rieske-type iron–sulfur protein that cleaves the macrocycle of pheophorbide *a* into the red chlorophyll catabolite(RCC) (Pruzinska et al. 2003) (Fig. 1). The *ACCELERATED CELL DEATH 2* (*ACD2*) gene has been cloned and shown to encode the red chlorophyll catabolite reductase which functions in connection to PaO as a chaperone-like protein in a later step of the chlorophyll breakdown pathway leading to the conversion of RCC into fluorescent chlorophyll catabolite (Pruzinska et al. 2007) (Fig. 1). *Arabidopsis acd1* and *acd2* mutants accumulate, respectively, pheophorbide *a* and RCC in a light-dependent manner. The formation of singlet oxygen, measured by fluorescence quenching, was elicited when dark-adapted *acd2* mutant accumulating

RCC was illuminated (Pruzinska et al. 2007). Both mutants (*acd1* and *acd2*) exhibit spontaneously necrotic lesions in the absence of pathogens (Fig. 1). This phenomenon is accompanied by the accumulation of defense-related gene transcripts, camalexin, phytoalexin, and salicylic acid. These mutants display an increased resistance to phytopathogenic bacteria (Greenberg et al. 1994).

2.2.3 Signaling Mechanisms via Plastid Tetrapyrroles

How can changes affecting the plastid tetrapyrrole biosynthetic pathway translate into a global induction of defense response? Because of their photoreactivity, it is usually suggested that the mechanism proceeds via reactive oxygen species (ROS). Usually ROS are produced as singlet oxygen ($^1O_2$) from the photosystem II and the superoxide anion generated by the photosystem I and from hydrogen peroxide generated by plastids under stress conditions. In the *Arabidopsis flu* mutant, protochlorophyllide accumulates in dark conditions and behaves as a photosensitizer upon illumination to trigger the formation of $^1O_2$ that subsequently activates up to 5% of the total genome of *Arabidopsis* (Meskauskiene et al. 2001; op den Camp et al. 2003). The signaling effect of $^1O_2$ is modulated by two plastidial proteins named Executer 1 and 2 (Lee et al. 2007). It is not known how $^1O_2$ produced in the plastid elicits a signal that can regulate the expression of nuclear genes involved in the adaptative response. Whatever the answer, one has to consider that the half-life time of $^1O_2$ is approximately 200 ns in cells (Gorman and Rodgers 1992) and its diffuses up to 10 nm under biological conditions (Sies and Menck 1992). Furthermore, inside the plastid, β, β-xanthophylls (Dall'Osto et al. 2007), tocopherols (Trebst et al. 2002), and the D1 protein of photosystem II (Aro et al. 1993) are potential quenchers of $^1O_2$. As an alternate possibility, one could invoke the chloroplast-to-nucleus retrograde signaling for which Mg-protoporphyrin (Strand et al. 2003) or heme (von Gromoff et al. 2008) is required to coordinate the functional assembly of chloroplasts through a recently identified plastid pentatricopeptide-repeat protein and the AP2 (APETALA 2)-type transcription factor ABI4 (ABSCISIC ACID-INSENSITIVE 4) (Koussevitzky et al. 2007). This mechanism operates in a limited scale (Richly et al. 2003). For instance, when carotenoid biosynthesis is inhibited the resulting ROSs induce the expression of genes that are different from those induced when the prochlorophyllide-accumulating mutant *flu* is illuminated (op den Camp et al. 2003; Strand et al. 2003). To account for the global response triggered by the products of the tetrapyrrole pathway, one could consider that protoporphyrinogen IX could diffuse from the plastid to the cytosol where it could be enzymatically or non enzymatically converted into protoporphyrin IX (Yamamoto et al. 1995). This could implicate a plastid transporter similar to the ABC transporter required for the possible transport of protophorhyrin IX (Moller et al. 2001). Alternatively, a fatty acid-derived signal could be involved in the signaling pathway due to their reactivity with $^1O_2$ (Weber 2002). In this context, it is interesting to note that in *Arabidopsis*, the chloroplast lipoxygenase gene At*lox2* is induced upon pathogen infection (Moran and Thompson 2001) and its expression is modulated by redox signals (Gadjev et al. 2006). For chloroplast-nucleus signaling, also see Dietzel et al. (2008).

2.3 Plastid Guanosine-3', 5'-(bis)Pyrophosphate Nucleotides(ppGpp) as Plant Alarmones

The adaptative response of *E. coli* subjected to amino acid starvation is usually qualified as the stringent response. It was shown that under these conditions, the synthesis of ribosomal RNA (rRNA) and transfer RNA (tRNA) is rapidly shutdown (Stent and Brenner 1961) concomitantly to the accumulation of hyperphosphorylated derivatives of guanosine, ppGpp and pppGpp, termed alarmone (Cashel and Kalbacher 1970; Sy and Lipmann 1973). It has been suggested that the stress conditions favor an increase in the ratio of deacylated to acylated tRNA. The translation machinery recruits deacylated tRNAs which blocks the translation and trigger the enzymic synthesis of guanosine nucleotides catalyzed by the bacterial enzyme RelA (Chatterji and Ojha 2001; Wendrich et al. 2002) (Fig. 2). The resulting guanosine nucleotides behave like key sensors of the adaptative response through the regulation of RNA polymerase. Once ppGpp is no longer required, a second bacterial enzyme SpoT catalyzes their degradation (Chatterji and Ojha 2001) (Fig. 2).

Plant genes homologous to the bacterial *RelA/SpoT* genes have been identified in several plants. These include *Clamydomonas* (Kasai et al. 2004), *Arabidopsis*

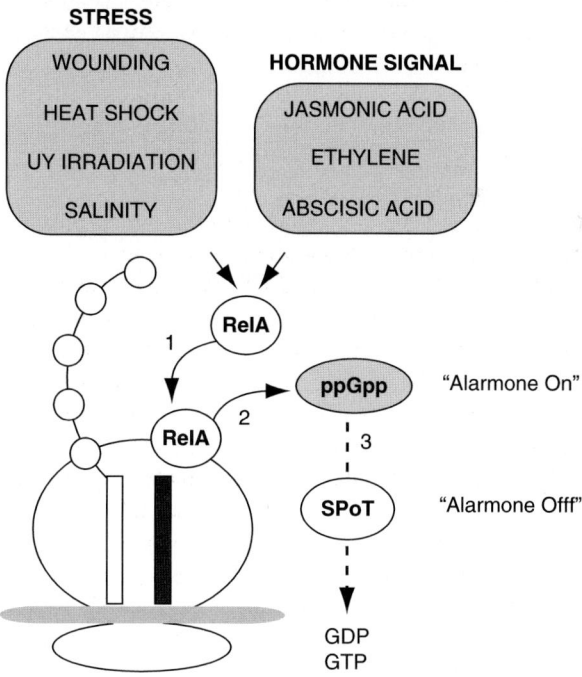

Fig. 2 Plastid plant alarmone (ppGPP) synthesis. The ppGpp level is modulated by the opposing activities of RelA and SpoT which, respectively, catalyze the synthesis of ppGpp and its hydrolysis. Abbreviations refer to: *GDP* guanosine 5'-diphosphate; *GTP*, guanosine 5'-triphosphate; *ppGpp* guanosine 5'-diphosphate 3'-diphosphate; *RelA*, guanosine 5'-triphosphate 3'-diphosphate synthase; *SpoT* guanosine 5'-triphosphate 3'-diphosphate phosphohydrolase

(van der Biezen et al. 2000), tobacco (Givens et al. 2004), and rice (Xiong et al. 2001). Interestingly, it has been shown that plant RelA is localized in the plastid and catalyzes the synthesis of ppGpp (Givens et al. 2004; Takahashi et al. 2004). Induced expression of *RelA* and increased synthesis of plastid ppGpp have been observed when plants are subjected to different stress including wounding, heat shock, heavy metals, UV irradiation, salinity and abrupt changes in light and also during microbial infection (Givens et al. 2004; Takahashi et al. 2004). In addition, the level of plastid ppGpp is increased when plants are treated with any of the plant hormones jasmonic acid, ethylene and abscisic acid, which are all known to induce defense responses in plants (Givens et al. 2004; Takahashi et al. 2004). Collectively, these characteristics indicate that the bacterial stringent response is conserved by plastids and is integrated into various aspects of plant stress responses.

2.4 Invading Plant Pathogens Avoid or Hijack Plastid-Derived Signals

The implication of the metabolism of plastid-derived signals during plant–pathogen interactions could be inferred from the strategies that pathogens have developed to suppress plant defense responses and perform their cell cycles within their hosts. Indeed, several subplastidial pathways leading to the synthesis of signaling molecules are exploited by plant pathogens (Fig. 3).

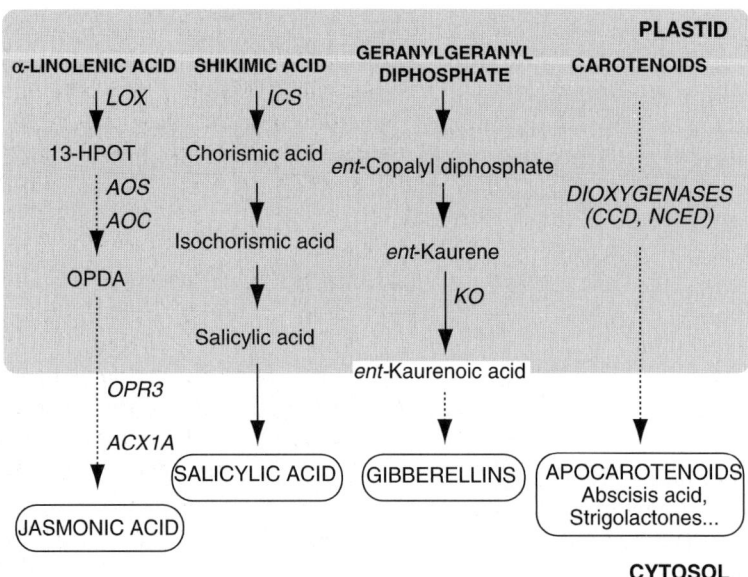

Fig. 3 Plastid pathways leading to the formation of signaling molecules. Abbreviations refer to enzymes catalyzing the different steps: *ACX* Acyl-CoA oxidase; *AOC* allene oxide cyclase; *AOS* allene oxide synthase; *CCD* carotenoid cleavage dioxygenase; *HPOT* hydroperoxyoctadecatrienoic acid; *ICS* isochorismate synthase; *KO ent*-kaurene oxidase; *LOX* lipoxygenase; *NCED* nine-*cis*-epoxycarotenoid dioxygenase; *OPDA cis*(+)-12-oxophytodienoic acid; *OPR* OPDA reductase

2.4.1 Jasmonic Acid

The expression of the gene encoding chloroplast lipoxygenase AtlOX2 is induced following pathogen attacks or under oxidative stress (Gadjev et al. 2006; Moran and Thompson 2001). This change reflects the activation of the initial steps of the jasmonic acid pathway which operate inside the plastids (Wasternack 2007) (Fig. 3). The last plastidial step involves the formation of $cis(+)$-12-oxophytodienoic acid (OPDA) which is often incorporated into plastid monogalactosyldiacylglycerol and digalactosyldiacylglycerol (DGDG) to yield the oxylipins called arabidopsides (Andersson et al. 2006; Hisamatsu et al. 2003, 2005, Ribot et al 2008). Some molecular species of arabidopsides accumulate after wounding (Buseman et al 2006) and also form up to 7–8% of the total lipid content of *Arabidopsis* upon recognition of the *P. syringae* avirulence protein AvrRpm1 and could inhibit the bacterial growth in vitro (Andersson et al. 2006). A subset of genes responds to OPDA but not to jasmonic acid. These include the genes encoding the synthesis of signaling components, transcription factors, and stress response factors (Taki et al. 2005). Some *P. syringae* strains under infectious conditions produce plastid destined toxins such as coronatine, a jasmonic acid analog, and tagetitoxin to cause stomata closure and to produce pigment-deficient leaves (Lukens et al. 1987; Melotto et al. 2006).

2.4.2 Salicylic Acid

Upon infection by fungal biotroph *Erysiphe orontii* or bacteria such as *Pseudomonas syringae* pv. *Maculicola*, the pathway leading to salicylic acid is triggered and this phenomenon is associated with a localized or systemic acquired resistance response. For instance, a salicylic acid-regulated defense pathway is activated in callose synthase-deficient plants as a compensatory defense mechanism (Nishimura et al. 2003). Initially, a phenylpropanoid pathway of salicylic acid was suggested. However, through the isolation of *Arabidopsis* mutants it was shown that salicylic acid is mainly synthesized in plastids. The pathway proceeds via a monofunctional isochorismate synthase 1 (ICS1) that converts chorismate to isochorismate (Strawn et al. 2007; Wildermuth et al. 2001) (Fig. 3). Mutations of the *ICS1* gene reduce salicylic acid accumulation after infection to only 5–10% of wild-type levels and compromise SAR induction (Wildermuth et al. 2001).

When *P. syringae* infects plants, a virulence effector (HOPL1) is injected inside plant cells and localizes to chloroplasts to suppress the synthesis of salicylic acid (Jelenska et al. 2007) (Fig. 4). Salicylic acid is also implicated in the acclimation of plants to water drought stress, which predisposes plants to pathogen attacks. Apparently, drought stress increases the level of salicylic acid which probably changes the antioxidant status of the plastids (Munné-Bosch et al. 2007). Recent data suggest that during *Arabidopsis–Pseudomonas aeruginosa* interactions, salicylic acid directly down-regulates the production of the bacterial virulence factor and the fitness of the bacteria (Prithiviraj et al. 2005). It has also been shown that salicylic acid directly shuts down the expression of *vir* genes required for *Agrobacterium* T-DNA transfer (Anand et al. 2008; Yuan et al. 2007). Thus, in addition to its signaling role, salicylic acid has a direct role on the infectivity of bacterial pathogens (Prithiviraj et al. 2005). The promoter

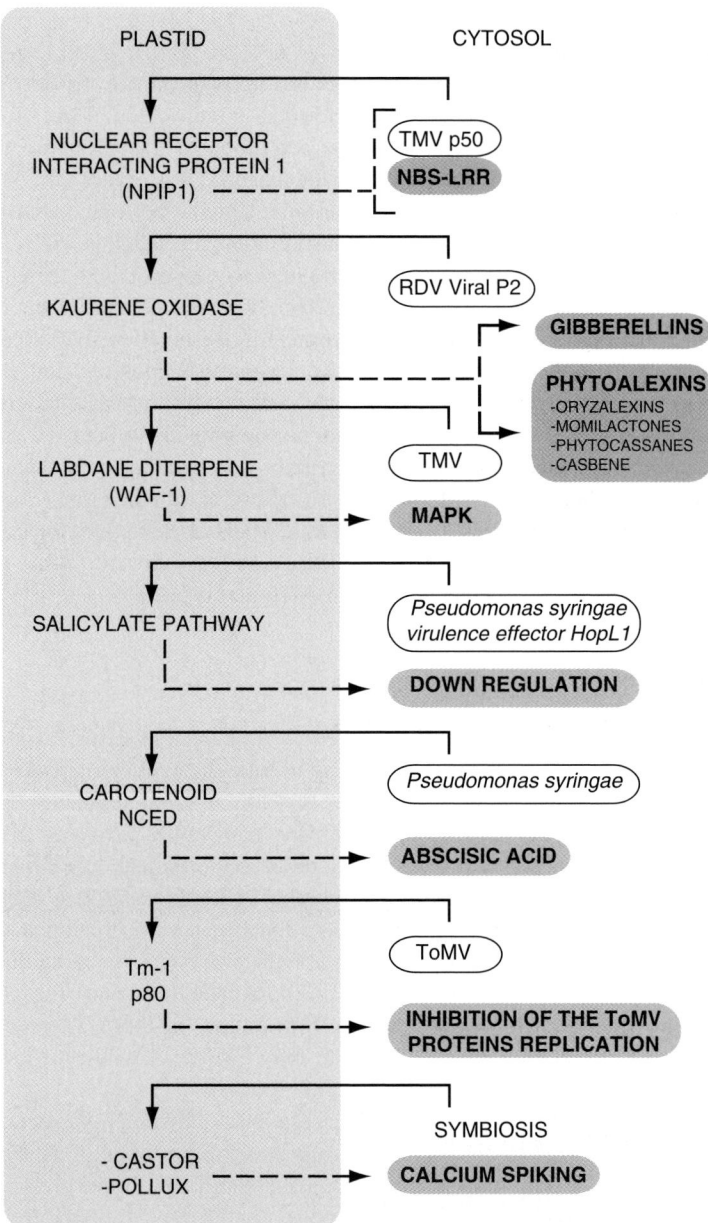

Fig. 4 Defense responses implicating plastids. The *unbroken arrows* symbolize pathogen attack or symbiotic interaction and the *dotted arrows* indicate the diverse responses triggered from the plastids. Abbreviations refer to: CASTOR and POLLUX, plastid proteins; MAPK, mitogen-activated protein kinase; NBS-LRR, nucleotide-binding site leucine-rich-repeat protein family; NCED nine-*cis*-epoxycarotenoid dioxygenase; *NRIP1* N receptor-interacting protein 1; *RDV Viral P2* rice dwarf virus capsid protein P2; *Tm-1 p80* protein conferring resistance to ToMV (tomato mosaic virus); *TMV* tobacco mosaic virus; *TMV p50* tobacco mosaic virus helicase protein p50; *ToMV* (tomato mosaic virus); *WAF-1* labdane diterpene elicitor

of *ICS1* displays *cis*-elements which bind WRKY and Myb transcription factors (Metraux 2002). These transcription factors are known to regulate defense genes and associated secondary metabolism and senescence (Eulgem and Somssich 2007; Robatzek and Somssich 2002).

2.4.3 Abscisic Acid

Previous data showed that exogenous abscisic acid (ABA) treatment could regulate the susceptibility of plants to bacterial and fungal pathogens (Thaler and Bostock 2004). A microarray analysis reveals that in *Arabidopsis* infested by *P. syringae* 42% of up-regulated genes correspond to ABA responsive genes (de Torres-Zabala et al. 2007). The genes encoding the enzymes catalyzing the plastid steps of ABA synthesis (Figs. 3 and 4) belong to the cluster of the most induced genes. In plants, the regulation exerted by ABA is mediated through an increase of intracellular calcium via cyclic ADP-ribose (cADPR) (Wu et al. 1997). It is interesting to note that the opportunistic protozoan pathogen *Toxoplasma gondii*, exploits a signaling pathway regulated by ABA via cADPR. Apparently, *T. gondii* produces ABA using a biosynthetic pathway operating in its apicoplast, a parasite organelle derived from an ancient algal endosymbiont. The similarity between the ABA pathway operating in *T. gondii* and the plant pathway is reinforced by the fact that the herbicide fluridone, which blocks the desaturation of phytoene into colored carotenoids, inhibits ABA synthesis in *T. gondii*. A tight correlation has been observed between the level of endogenous ABA produced in *T. gondii* and its capacity to replicate. Fluridone and probably other carotenoid desaturase inhibitors specifically block further dissemination of the parasite (Nagamune et al. 2008). These data extend and further illustrate the importance of plastids in the biosynthesis of molecules that control the invasiveness or the survival of pathogens. In this context, it is interesting to note that in metazoans, ABA could also modulate calcium-mediated signaling pathways regulated by cADPR (Bruzzone et al. 2007).

2.5 Roles of Plastids During Phosphate Limitation

Phosphorus is an essential element required for plant growth but is present in soils as insoluble inorganic phosphate or organic phosphate. As such, phosphorus is often limiting for plant development and crop yield (Vance et al. 2003) in particular in alkaline soils. Two aspects illustrate how plastids are implicated in the adaptation of plants to phosphate deprivation.

2.5.1 Down-Regulation of Plastid PNPase

Chloroplast polynucleotide phosphorylase (PNPase) catalyzes polyadenylation-mediated RNA degradation (Kudla et al. 1996). Because of its bifunctional activity, the reaction catalyzed by PNPases consumes nucleotide diphosphates (NDPs) and either generates or liberates inorganic phosphate (Yehudai-Resheff et al. 2001). During phosphate starvation, the chloroplast PNPase is down-regulated to preserve the level of chloroplast RNA. On the other hand, the content of chloroplast DNA declines (Yehudai-Resheff et al. 2007). On the basis of this behavior, it has been suggested that under phosphate deprivation, chloroplast polyploidy could be exploited at least as a repository of phosphate through a signaling mechanism that leads to the down-regulation of PNPase expression (Yehudai-Resheff et al. 2007). Interestingly, in the *Arabidopsis* mutant *rif* (resistant to inhibition with fosmidomycin), the down-regulation of PNPase expression correlates with the posttranscriptional regulation of several enzymes of the chloroplast prokaryotic pathway of isopentenyl diphosphate synthesis (Sauret-Gueto et al. 2006).

2.5.2 Remodeling Galactolipids

It has been shown that the composition of membrane glycerolipids of bacterial cell and prokaryotic photosynthetic organisms is drastically changed during phosphate deprivation (for more details, see Rolland et al. 2008; Andersson and Dörmann 2008). One of the prominent modifications concerns the replacement of phospholipids by non phosphorous lipids such as glycolipids (Benning et al. 1993; Minnikin et al. 1974). The corresponding modification occurs also in higher plants, where following phosphate starvation, DGDG, which exclusively is compartmentalized in plastids under normal conditions, is transported to the plasma membrane, the tonoplast and the mitochondria due to the deficit in polar phospholipids (Andersson et al. 2005; Jouhet et al. 2004). In *Arabidopsis*, the plastidial DGDG synthases 1 and 2 (DGDG1 and 2) account for the increased DGDG incorporated into the extraplastidial membranes (Kelly and Dormann 2002; Kelly et al. 2003). Recent data indicate that the signal triggering this alternative galactoglycerolipid pathway is linked to reactive oxygen originating from the mitochondria (Xu et al. 2008). The contribution of the plastid galactolipids is not limited to inorganic phosphate deprivation because the expressions of DGDG 1 and 2 are also induced upon nitrogen deficiency (Gaude et al. 2007).

2.5.3 Apocarotenoid Signaling

Mycorrhizal associations between the majority of terrestrial plants and symbiotic fungi facilitate phosphate acquisition by plants. The association involves a complex signaling mechanism and a branching factor released from the plant roots has been implicated for the initial step. Recent data reveal that strigolactones exuded from the roots of the host plants trigger the fungus–plant root association (Akiyama et al. 2005). Strigolactones are formed from apocarotenoids via the oxidative cleavage of carotenoids (Bouvier et al. 2005a; Matusova et al. 2005) (Fig. 3). Strigolactones were previously recognized as seed germination stimulants for parasitic weeds *Striga* and *Orobanche* (Akiyama and Hayashi 2006). Interestingly, during nitrogen and phosphate starvation the synthesis of strigolactones is stimulated (Yoneyama et al. 2007). Little is known about the signal transduction induced by apocarotenoids, but the plastid proteins CASTOR and POLLUX could be implicated. These proteins were recently characterized from the roots of the wild perennial legume *Lotus japonicus* and they represent two indispensable components of the signal transduction cascade leading to the symbiotic association between the Rhizobia symbiont and *L. japonicum* (Imaizumi-Anraku et al. 2005) (Fig. 4).

2.6 Plastidial Production of Phytoalexins

Cereal crop plants often produce plastid-derived phytoalexins when challenged with biotic or abiotic elicitors (Peters 2006). For instance in rice cell suspension cultures treated with a chitin elicitor to mimic pathogen attack, the production of plastidial labdane-related diterpene phytoalexins (momilactones, oryzalexins, and phytocassanes) via geranylgeranyl diphosphate is associated with an increased expression of the genes encoding the different steps of the plastid enzymes of isopentenyl diphosphate synthesis (Okada et al. 2007) (Fig. 4). Similarly the antifungal plastid diterpene casbene is produced in infected castor bean (Dudley et al. 1986) (Fig. 4). In tobacco infected with tobacco mosaic virus, the defense response involves a plastid labdane-type diterpene named WAF-1 that activates a mitogen-activated protein kinase (MAPK) signaling cascade (Seo et al. 2003) (Fig. 4). One could note that antifungal compounds produced from maize roots contain apocarotenoids derived from the oxidative cleavage of carotenoids (Park et al. 2004). Further involvement of the plastid pathway of IPP synthesis in plant defense against pathogens is provided by the *Arabidopsis constitutive subtilisin3* (*csb*3) mutant which has a low 1-hydroxy-2-methyl-2-butenyl 4-diphosphate synthase (HDS) activity and displays an increased resistance to biotrophic pathogens (Gil et al. 2005).

3 Plastid High Molecular Weight Sensors

3.1 Plastid NAD Kinase and Epoxy Xanthophyll Modulate the Dissipation of Excess Excitation Energy

In addition to their roles in energy metabolism, NAD^+ and its phosphorylated counterpart NADP have signaling roles. For instance they serve as a substrate for ADP-ribosylation reactions and for the synthesis of cADPR. The *Arabidopsis* mutant *nadk2* has been identified recently and shown to be hypersensitive to oxidative stress, drought, salinity, heat shock and UVB (Chai et al. 2005). *NADK*2 encodes chloroplast NAD kinase that catalyzes the phosphorylation of NAD^+ (Chai et al. 2005; Takahashi et al. 2006). The *nadk2* mutant accumulates zeaxanthin under low and high light conditions. This represents a diagnostic feature of an altered functioning of the de-epoxidation state of the epoxy-xanthophyll cycle (Demming-Adams et al. 1996;

Takahashi et al. 2006). Although the accumulation of zeaxanthin could be explained by the depletion of NADPH required to sustain the epoxidation reactions according to zeaxanthin -> antheraxanthin -> violaxanthin, the data suggest that NADK2 is probably a mediator of more global responses. This contention is supported by the fact that in bacteria or mitochondria the increased concentration of NAD^+ provides resistance to diverse stresses (Foster et al. 1990; Yang et al. 2007b).

3.2 A Plastid Surveillance System Interacting with Pathogen Recognition Receptors

During plant evolution, plastids have been integrated into a sophisticated strategy to perceive attack by pathogens or symbiotic partners and to translate the perceived information into appropriate adaptive responses. During their replication positive-strand RNA viruses, which include the plant viruses tobacco mosaic virus (TMV) and tomato mosaic virus (ToMV), recruit genomic RNA templates to the intracellular endomembrane systems. It has been shown recently that the recognition of the 50-kDa helicase (p50) domain of TMV by the plant N immune receptor is mediated by the host protein termed N-receptor-interacting protein 1 (NRIP1) before the activation of defense responses through a host protein belonging to the nucleotide-binding site–leucine-rich repeat (NBS-LRR) family (Caplan et al. 2008; DeYoung and Innes 2006). NRIP1 is localized in the chloroplast stroma and functions as true plastid sulfur transferase (Caplan et al. 2008) (Fig. 4). Interestingly, transcriptome analysis of virus-infected leaf tissues from *Arabidopsis*, revealed that several genes involved in sulfur assimilation in the chloroplasts are down-regulated by the infection. Furthermore, a productive infection correlates with a decreased expression of genes along the whole sulfur assimilation pathway, from sulfur transport to its incorporation into complex molecules, such as the defense compounds glucosinolates (Yang et al. 2007a). In a similar vein, the tomato Tm-1 gene that confers resistance to ToMV encodes a plastid-targeted protein that functions as an inhibitor of viral RNA replication (Ishibashi et al. 2007) (Fig. 4). In rice plants infected with rice dwarf virus (RDV), plastid *ent*-kaurene oxidase isoforms involved in the biosynthesis of gibberellin and different diterpene phytoalexins interact with the P2 capsid protein of RDV (Itoh et al. 2004; Zhu et al. 2005) (Fig. 4). Because of its capacity to induce membrane fusion in insect cells (Zhou et al. 2007), one could suggest that in plants P2 induces fusion with the plastid envelope membrane, where *ent*-kaurene oxidase has been located (Helliwell et al. 2001).

3.3 Role of Plastid Resident Proteins Containing Nuclear Transcription Motifs

Continuous integration of endosymbiont DNA into the nucleus occurred during plant evolution. This phenomenon was apparently much more intense for higher plants compared to unicellular algae (Lister et al. 2003). Five plastid proteins displaying nuclear transcription factor motifs are located in plastids. These include the tobacco

CDN41 and NtWIN14 and the pea PD1, PD3 and PEND (Kodama 2007). CDN41, is a multifunctional protein displaying a DNA binding activity due to its zinc finger motif, in addition to its proteolytic activity (Kato et al. 2004; Nakano et al. 1997). Increased expression of the *CDN4* gene is negatively correlated with the decrease of several plastid transcripts and the degradation of Rubisco (Kato et al. 2004; Nakano et al. 1997). PEND possesses a bZIP motif and overexpression of the *PEND* gene induces chlorosis and blocks the differentiation of palisade tissue (Sato and Ohta 2001; Wycliffe et al. 2005). Pea *PD1* and *PD3* possess AT-hook motifs but their functions deserve further studies (Kodama 2007). NtWIN14 represents a basic helix-loop-helix protein (bHLH) (Kodama 2007). The expression of the tobacco *NtWIN*14 is increased following wound stress and pathogen attack and is up-regulated by jasmonate and ROS (Kodama and Sano 2006).

3.4 Requisite Role of Plastid During Senescence and Signaling for Reproduction

During plant senescence degenerative processes as well as nutrient remobilization take place in an orderly fashion. Several regulatory factors associated to plastids play a decisive role. For instance, it has been observed that angiosperms displaying developmental leaf senescence possess plastid genes that code for a NADH-specific dehydrogenase complex (ndh). The *ndh* genes are apparently absent from several evergreen gymnosperms (Sabater et al. 2002). This observation is reinforced by the fact that transgenic tobacco plants in which the plastid *ndhF* has been knocked-down exhibit delayed senescence (Zapata et al. 2005).

3.4.1 Sensing the Carbon to Nitrogen Status

The sensing of the ratio between nitrogen and carbon (sugar) contents plays a key role in the signaling process that leads to the dismantling of the chloroplast structures. High sucrose and low nitrogen concentrations usually trigger the dismantling of chloroplast and modulate the chloroplast to chromoplast transition (Hörtensteiner and Feller 2002; Huff 1983; Iglesias et al. 2001; Masclaux et al. 2000; Ono et al. 1999). Thus, the major chloroplast protein Rubisco is actively degraded by the CND41 (41-kD chloroplast nucleoid DNA binding protein) protease in naturally senescing tobacco leaves or in leaves incubated in the presence of a high sucrose to nitrogen ratio (Kato et al. 2005).

In prokaryotic organisms, the PII proteins represent key sensors of the carbon to nitrogen status that govern the adaptation of their metabolism (Commichau et al. 2006). In eukaryotes, PII proteins have been characterized only in plants and red algae and are located inside plastids (Moorhead and Smith 2003). Under nitrogen-replete conditions, plant PII proteins, like their bacterial homologs bind 2-oxoglutarate and interact with the plastid enzyme *N*-acetylglutamate kinase (NAGK), the second enzyme of the pathway leading arginine from glutamate (Chen et al. 2006). This relieves the feed back inhibition exerted by the pathway product arginine (Chen et al. 2006).

During nitrogen starvation, PII dissociates from NAGK thus leading to decreased NAGK (Osanai and Tanaka 2007).

3.5 Plastid Stay-GreenProtein (SGR) Ensures Adequate Plastid Reconversion

Photosynthetic organisms have evolved pathways to ensure the destruction of chlorophylls and the disposal of its byproducts throughout the life cycle. Chlorophyll breakdown *in planta* is tightly connected to the extensive dismantling of thylakoid proteins and ultrastructural changes of plastid membranes. A characteristic aspect of this catabolic process occurs in senescing plant tissues when chloroplasts transform into gerontoplasts (Parthier 1988), and during flower development and fruit ripening when chloroplasts differentiate into chromoplasts to attract pollinators or seed dispersal vectors (Camara et al. 1995) (Fig. 5). Although key enzymes catalyzing the different steps of chlorophyll breakdown have been identified (Hörtensteiner 2006), several additional proteins are required for the proper functioning of the catabolic pathway of chlorophylls. The multifaceted aspect of chlorophyll breakdown is uncovered by several genetic alterations not caused by defective chlorophyll catabolic enzymes that affect leaf color and are referred to as *stay-green* or *nonyellowing* mutants (Thomas and Howarth 2000). These include the NON-YELLOW COLORING1 (NYC1) protein (Kusaba et al. 2007), the NONYELLOWING protein (Ren et al. 2007), and the

Fig. 5 Induction of senescence and chloroplast to chromoplast differentiation. Analysis through the pepper chlorophyll retainer mutant (*cl*). (**A, B**) Phenotype of the pepper chlorophyll retainer (*cl*) mutant compared to a red-fruited wild-type *CL*. Color phenotype of basal leaves. (**A**) *cl* mutant; (**B**) wild-type (*CL*) plant homozygous for the *CL* allele. The two 124-day-old plants were grown concomitantly under the same greenhouse conditions. *Square brackets* indicate the senescing zone which is more pronounced in the wild-type (*CL*) compared to the *cl* mutant. (**C, D**) Color phenotype of ripe fruits from *cl* (C) and *CL* (D) plants. (**E, F**) Electron microscope images of chromoplasts from pepper *cl* mutant (E) and wild-type (*CL*) (F) fruits. The retention of thylakoid membranes in the chromoplast is indicated by *asterisks*. (A.S. Mialoundama, B. Camara and F. Bouvier, unpublished)

STAY-GREEN (SGR) protein whose function is largely unknown (Armstead et al. 2007; Jiang et al. 2007; Park et al. 2007; Sato et al. 2007).

Concerning reproductive organs and fruits, a single recessive mutation designated *chlorophyll retainer* (*cl*) in pepper (Smith 1950), and *green flesh* (*gf*) in tomato (Kerr 1958), blocks the breakdown of chlorophylls during the chloroplast to chromoplast differentiation in ripening fruits (Akhtar et al. 1999; Cheung et al. 1993; Roca and Minguez-Mosquera 2006). Both mutations have been mapped to chromosome 1 for pepper (Efrati et al. 2005) and chromosome 8 for tomato (Kerr 1958), respectively, and do not correspond to putative *CHLOROPHYLLASEs* (Schenk et al. 2007) and *PHEOPHORBIDE a OXYGENASE* (*PaO*) loci (Efrati et al. 2005). In pepper, under normal culture conditions, the *cl* mutants and wild-type plants grow and flower with no significant developmental differences. Phenotypic color differences between *cl* mutants and wild-type plants exist and are particularly pronounced during the reproductive stage. The mutant exhibits a delay in chlorophyll breakdown compared to wild-type (Fig. 5a–d). Electron micrographs of ripe fruits revealed that the chlorophyllous thylakoid membranes were completely dismantled in wild-type plants but were retained in the *cl* mutants (Fig. 5e, f). SDS-PAGE followed by immunoblot analysis revealed that the chromoplast-specific carotenogenic enzyme (Bouvier et al. 1994), capsanthin-capsorubin synthase (CCS) accumulated specifically in the *cl* mutant and wild-type chromoplasts. The chloroplast-specific light-harvesting chlorophyll-a/b protein (LHCII) was retained during the chloroplast to chromoplast transition in *cl* mutant plants.

The senescence characteristics of *cl* mutants described above suggest the existence of similarity to *Arabidopsis* (Armstead et al. 2006), rice (Park et al. 2007; Sato et al. 2007), and pea (Sato et al. 2007) *SGR* mutants. This assertion is further reinforced by the fact that a screen of plastid proteomics (http://ppdb.tc.cornell.edu; http://www.plprot.ethz.ch) to search for senescence-specific proteins accumulating in chromoplasts, identified one candidate meeting this criterion: SGR1-type protein in pepper chromoplasts (Siddique et al. 2006). The mechanism by which plastid SGR induces chlorophyll breakdown is not known. Transient overexpression of a rice *SGR* (Os*SGR*) homolog in *N. benthamiana* followed by an in vitro pull-down assay revealed that OsSGR interacts with LHCII (Park et al. 2007). Furthermore, it is possible that other proteins interact with SGR given the complexity and the compartmentalization of chlorophyll catabolism. Heterocomplexes comprising SGR, could include the FTSH6-type protease which specifically acts on LHCII (Zelisko et al. 2005). Whether SGR and chlorophyll catabolic enzymes directly cooperate is still unclear.

4 Conclusion

The impact of plastids in plant physiology is not restricted to photosynthesis and associated metabolism. Plastids play key roles during the adaptation of plants to environmental changes. There are possibly overlaps and cooperation between the sentinel function of plastids and mitochondria which have also been integrated through an endosymbiotic process. This is suggested by the fact that the impaired chloroplast activity in the barley mutant *albostrians*, is associated with the amplification and the increased expression of the mitochondrial genome (Hedtke et al. 1999). In the same vein, in the *Nicotiana sylvestris* mutant CMSII (cytoplasmic male-sterile) having a deficient mitochondrial NADH dehydrogenase, the photosynthetic efficiency is decreased (Sabar et al. 2000). Further developments in the field are likely to unravel at the molecular level the mechanisms involved in the integration of the diverse sets of signals that lead to appropriate responses. The fact that NRIP1, a protein localized in plastids helps plants to detect the presence of a virus could make it a promising target candidate for the search of transducer proteins (Caplan et al. 2008). In a similar vein, one could exploit the fact that functional screens for effector proteins injected in plant cells by the plant pathogen *P. syringae* reveal that several effector proteins are plastid-targeted (Guttman et al. 2002).

References

Abbasi AR, Hajirezaei M, Hofius D, Sonnewald U, Voll LM (2007) Specific roles of alpha- and gamma-tocopherol in abiotic stress responses of transgenic tobacco. Plant Physiol 143:1720–1738
Abdallah F, Salamini F, Leister D (2000) A prediction of the size and evolutionary origin of the proteome of chloroplasts of *Arabidopsis*. Trends Plant Sci 5:141–142
Andersson MX, Dörmann P (2008) Chloroplast membrane lipid biosynthesis and transport. Plant Cell Monogr., doi:10.1007/7089_2008_18
Akhtar MS, Goldschmidt EE, John I, Rodoni S, Matile P, Grierson D (1999) Altered patterns of senescence and ripening in *gf*, a stay-green mutant of tomato (*Lycopersicon esculentum* Mill.). J Exp Bot 50:1115–1122
Akiyama K, Hayashi H (2006) Strigolactones: chemical signals for fungal symbionts and parasitic weeds in plant roots. Ann Bot (Lond) 97:925–931
Akiyama K, Matsuzaki K, Hayashi H (2005) Plant sesquiterpenes induce hyphal branching in arbuscular mycorrhizal fungi. Nature 435:824–827
Anand A, Uppalapati SR, Ryu CM, Allen SN, Kang L, Tang Y, Mysore KS (2008) Salicylic acid and systemic acquired resistance play a role in attenuating crown gall disease caused by *Agrobacterium tumefaciens*. Plant Physiol 146:703–715
Andersson MX, Larsson KE, Tjellstrom H, Liljenberg C, Sandelius AS (2005) Phosphate-limited oat. The plasma membrane and the tonoplast as major targets for phospholipid-to-glycolipid replacement and stimulation of phospholipases in the plasma membrane. J Biol Chem 280:27578–27586
Andersson MX, Hamberg M, Kourtchenko O, Brunnstrom A, McPhail KL, Gerwick WH, Gobel C, Feussner I, Ellerstrom M (2006) Oxylipin profiling of the hypersensitive response in *Arabidopsis thaliana*. Formation of a novel oxo-phytodienoic acid-containing galactolipid, arabidopside E. J Biol Chem 281:31528–31537
Armstead I, Donnison I, Aubry S, Harper J, Hörtensteiner S, James C, Mani J, Moffet M, Ougham H, Roberts L et al. (2006) From crop to model to crop: identifying the genetic basis of the staygreen mutation in the *Lolium/Festuca* forage and amenity grasses. New Phytol 172:592–597
Armstead I, Donnison I, Aubry S, Harper J, Hörtensteiner S, James C, Mani J, Moffet M, Ougham H, Roberts L et al. (2007) Cross-species identification of Mendel's I locus. Science 315:73

Aro EM, Virgin I, Andersson B (1993) Photoinhibition of Photosystem II. Inactivation, protein damage and turnover. Biochim Biophys Acta 1143:113–134

Azzi A (2007) Molecular mechanism of alpha-tocopherol action. Free Radic Biol Med 43:16–21

Benning C, Beatty JT, Prince RC, Somerville CR (1993) The sulfolipid sulfoquinovosyldiacylglycerol is not required for photosynthetic electron transport in *Rhodobacter sphaeroides* but enhances growth under phosphate limitation. Proc Natl Acad Sci U S A 90:1561–1565

Bouvier F, Hugueney P, d'Harlingue A, Kuntz M, Camara B (1994) Xanthophyll biosynthesis in chromoplast: isolation and molecular cloning of an enzyme catalyzing the conversion of 5,6-epoxycarotenoid into ketocarotenoid. Plant J 6:45–54

Bouvier F, Isner JC, Dogbo O, Camara B (2005a) Oxidative tailoring of carotenoids: a prospect towards novel functions in plants. Trends Plant Sci 10:187–194

Bouvier F, Rahier A, Camara B (2005b) Biogenesis, molecular regulation and function of plant isoprenoids. Prog Lipid Res 44:357–429

Bruzzone S, Moreschi I, Usai C, Guida L, Damonte G, Salis A, Scarfi S, Millo E, De Flora A, Zocchi E (2007) Abscisic acid is an endogenous cytokine in human granulocytes with cyclic ADP-ribose as second messenger. Proc Natl Acad Sci U S A 104:5759–5764

Buseman CM, Tamura P, Sparks AA, Baughman EJ, Maatta S, Zhao J, Roth MR, Esch SW, Shah J, Williams TD, Welti R (2006) Wounding stimulates the accumulation of glycerolipids containing oxophytodienoic acid and dinor-oxophytodienoic acid in *Arabidopsis* leaves. Plant Physiol 142:28–39

Camara B, Hugueney P, Bouvier F, Kuntz M, Monéger R (1995) Biochemistry and molecular biology of chromoplast development. Inter Rev Cytol 163:175–247

Caplan JL, Mamillapalli P, Burch-Smith TM, Czymmek K, Dinesh-Kumar SP (2008) Chloroplastic protein NRIP1 mediates innate immune receptor recognition of a viral effector. Cell 132:449–462

Cashel M, Kalbacher B (1970) The control of ribonucleic acid synthesis in *Escherichia coli*. V. Characterization of a nucleotide associated with the stringent response. J Biol Chem 245:2309–2318

Chai MF, Chen QJ, An R, Chen YM, Chen J Wang XC (2005) NADK2, an *Arabidopsis* chloroplastic NAD kinase, plays a vital role in both chlorophyll synthesis and chloroplast protection. Plant Mol Biol 59:553–564

Chatterji D, Ojha AK (2001) Revisiting the stringent response, ppGpp and starvation signaling. Curr Opin Microbiol 4:160–165

Cheng Z, Sattler S, Maeda H, Sakuragi Y, Bryant DA, DellaPenna D (2003) Highly divergent methyltransferases catalyze a conserved reaction in tocopherol and plastoquinone synthesis in cyanobacteria and photosynthetic eukaryotes. Plant Cell 15:2343–2356

Chen YM, Ferrar TS, Lohmeier-Vogel EM, Morrice N, Mizuno Y, Berenger B, Ng KK, Muench DG, Moorhead GB (2006) The PII signal transduction protein of *Arabidopsis thaliana* forms an arginine-regulated complex with plastid N-acetyl glutamate kinase. J Biol Chem 281:5726–5733

Cheung AY, McNellis T, Piekos B (1993) Maintenance of chloroplast components during chromoplast differentiation in the tomato mutant *green flesh*. Plant Physiol 101:1223–1229

Commichau FM, Forchhammer K, Stulke J (2006) Regulatory links between carbon and nitrogen metabolism. Curr Opin Microbiol 9:167–172

Dall'Osto L, Fiore A, Cazzaniga S, Giuliano G, Bassi R (2007) Different roles of alpha- and beta-branch xanthophylls in photosystem assembly and photoprotection. J Biol Chem 282:35056–35068

Demming-Adams B, Gilmore AM, Adam WW III (1996) In vivo functions of carotenoids in higher plants. FASEB 10:403–412

de Torres-Zabala M, Truman W, Bennett MH, Lafforgue G, Mansfield JW, Rodriguez Egea P, Bogre L, Grant M (2007) *Pseudomonas syringae* pv. tomato hijacks the *Arabidopsis* abscisic acid signalling pathway to cause disease. Embo J 26:1434–1443

DeYoung BJ, Innes RW (2006) Plant NBS-LRR proteins in pathogen sensing and host defense. Nat Immunol 7:1243–1249

Dietzel L, Steiner S, Schröter Y, Pfannschmidt T (2008) Retrograde signalling. Plant Cell Monogr., doi:10.1007/7089_2008_41

Dudley MW, Dueber MT, West CA (1986) Biosynthesis of the macrocyclic diterpene casbene in Castor bean (*Ricinus communis* L.) seedlings. Changes in enzyme levels induced by fungal infection and intracellular localization of the pathway. Plant Physiol 81:335–342

Efrati A, Eyal Y, Paran I (2005) Molecular mapping of the chlorophyll retainer (cl) mutation in pepper (*Capsicum* spp.) and screening for candidate genes using tomato ESTs homologous to structural genes of the chlorophyll catabolism pathway. Genome 48:347–351

Eulgem T, Somssich IE (2007) Networks of WRKY transcription factors in defense signaling. Curr Opin Plant Biol 10:366–371

Foster JW, Park YK, Penfound T, Fenger T, Spector MP (1990) Regulation of NAD metabolism in *Salmonella typhimurium*: molecular sequence analysis of the bifunctional nadR regulator and the nadA-pnuC operon. J Bacteriol 172:4187–4196

Gadjev I, Vanderauwera S, Gechev TS, Laloi C, Minkov IN, Shulaev V, Apel K, Inze D, Mittler R, Van Breusegem F (2006) Transcriptomic footprints disclose specificity of reactive oxygen species signaling in *Arabidopsis*. Plant Physiol 141:436–445

Gaude N, Brehelin C, Tischendorf G, Kessler F, Dormann P (2007) Nitrogen deficiency in *Arabidopsis* affects galactolipid composition and gene expression and results in accumulation of fatty acid phytyl esters. Plant J 49:729–739

Gil MJ, Coego A, Mauch-Mani B, Jorda L, Vera P (2005) The *Arabidopsis csb3* mutant reveals a regulatory link between salicylic acid-mediated disease resistance and the methyl-erythritol 4-phosphate pathway. Plant J 44:155–166

Givens RM, Lin MH, Taylor DJ, Mechold U, Berry JO, Hernandez VJ (2004) Inducible expression, enzymatic activity, and origin of higher plant homologues of bacterial RelA/SpoT stress proteins in *Nicotiana tabacum*. J Biol Chem 279:7495–7504

Gorman AA, Rodgers MA (1992) Current perspectives of singlet oxygen detection in biological environments. J Photochem Photobiol B 14:159–176

Gray J, Close PS, Briggs SP, Johal GS (1997) A novel suppressor of cell death in plants encoded by the *Lls1* gene of maize. Cell 89:25–31

Greenberg JT, Ausubel FM (1993) *Arabidopsis* mutants compromised for the control of cellular damage during pathogenesis and aging. Plant J 4:327–341

Greenberg JT, Guo A, Klessig DF, Ausubel FM (1994) Programmed cell death in plants: a pathogen-triggered response activated coordinately with multiple defense functions. Cell 77:551–563

Guttman DS, Vinatzer BA, Sarkar SF, Ranall MV, Kettler G, Greenberg JT (2002) A functional screen for the type III (Hrp) secretome of the plant pathogen *Pseudomonas syringae*. Science 295:1722–1726

Hedtke B, Wagner I, Borner T, Hess WR (1999) Inter-organellar crosstalk in higher plants: impaired chloroplast development affects mitochondrial gene and transcript levels. Plant J 19:635–643

Helliwell CA, Sullivan JA, Mould RM, Gray JC, Peacock WJ, Dennis ES (2001) A plastid envelope location of *Arabidopsis ent*-kaurene oxidase links the plastid and endoplasmic reticulum steps of the gibberellin biosynthesis pathway. Plant J 28:201–208

Hisamatsu Y, Goto N, Hasegawa K, Shigemori H (2003) Arabidopsides A and B, two new oxylipins from *Arabidopsis thaliana*. Tetrahedron Lett 44:5553–5556

Hisamatsu Y, Goto N, Sekiguchi M, Hasegawa K, Shigemori H (2005) Oxylipins arabidopsides C and D from *Arabidopsis thaliana*. J Nat Prod 68:600–603

Hofius D, Hajirezaei MR, Geiger M, Tschiersch H, Melzer M, Sonnewald U (2004) RNAi-mediated tocopherol deficiency impairs photoassimilate export in transgenic potato plants. Plant Physiol 135:1256–1268

Hörtensteiner S (2006) Chlorophyll degradation during senescence. Annu Rev Plant Biol 57:55–77

Hörtensteiner S, Feller U (2002) Nitrogen metabolism and remobilization during senescence. J Exp Bot 53:927–937

Huff A (1983) Nutritional control of regreening and degreening in *Citrus* peel segments. Plant Physiol 73:243–249

Hu G, Yalpani N, Briggs SP, Johal GS (1998) A porphyrin pathway impairment is responsible for the phenotype of a dominant disease lesion mimic mutant of maize. Plant Cell 10:1095–1105

Iglesias DJ, Tadeo FR, Legaz F, Primo-Millo E, Talon M (2001) In vivo sucrose stimulation of colour change in citrus fruit epicarps: interactions between nutritional and hormonal signals. Physiol Plant 112:244–250

Imaizumi-Anraku H, Takeda N, Charpentier M, Perry J, Miwa H, Umehara Y, Kouchi H, Murakami Y, Mulder L, Vickers K et al. (2005) Plastid proteins crucial for symbiotic fungal and bacterial entry into plant roots. Nature 433:527–531

Ishibashi K, Masuda K, Naito S, Meshi T, Ishikawa M (2007) An inhibitor of viral RNA replication is encoded by a plant resistance gene. Proc Natl Acad Sci U S A 104:13833–13838

Ishikawa A (2005) Tetrapyrrole metabolism is involved in lesion formation, cell death, in the *Arabidopsis lesion initiation 1* mutant. Biosci Biotechnol Biochem 69:1929–1934

Ishikawa A, Okamoto H, Iwasaki Y, Asahi T (2001) A deficiency of coproporphyrinogen III oxidase causes lesion formation in *Arabidopsis*. Plant J 27:89–99

Itoh H, Tatsumi T, Sakamoto T, Otomo K, Toyomasu T, Kitano H, Ashikari M, Ichihara S, Matsuoka M (2004) A rice semi-dwarf gene, Tan-Ginbozu (D35), encodes the gibberellin biosynthesis enzyme, *ent*-kaurene oxidase. Plant Mol Biol 54:533–547

Jelenska J, Yao N, Vinatzer BA, Wright CM, Brodsky JL, Greenberg JT (2007) A J domain virulence effector of *Pseudomonas syringae* remodels host chloroplasts and suppresses defenses. Curr Biol 17:499–508

Jiang H, Li M, Liang N, Yan H, Wei Y, Xu X, Liu J, Xu Z, Chen F, Wu G (2007) Molecular cloning and function analysis of the *stay green* gene in rice. Plant J 52:197–209

Jouhet J, Marechal E, Baldan B, Bligny R, Joyard J, Block MA (2004) Phosphate deprivation induces transfer of DGDG galactolipid from chloroplast to mitochondria. J Cell Biol 167:863–874

Kamal-Eldin A, Appelqvist LA (1996) The chemistry and antioxidant properties of tocopherols and tocotrienols. Lipids 31:671–701

Kaneko K, Nakamura Y, Wolkers CP, Kuritz T, Sasamoto S, Watanabe A, Iriguchi M, Ishikawa A, Kawashima K, Kimura T et al. (2001a) Complete genomic sequence of the filamentous nitrogen-fixing *Cyanobacterium Anabaena* sp. Strain PCC 7120. DNA Res 8:227–253

Kaneko K, Nakamura Y, Wolkers CP, Kuritz T, Sasamoto S, Watanabe A, Iriguchi M, Ishikawa A, Kawashima K, Kimura T et al. (2001b) Complete genomic sequence of the filamentous nitrogen-fixing *Cyanobacterium Anabaena* sp. Strain PCC 7120. DNA Res 8:205–213

Kariola T, Brader G, Li J, Palva ET (2005) Chlorophyllase 1, a damage control enzyme, affects the balance between defense pathways in plants. Plant Cell 17:282–294

Kasai K, Kanno T, Endo Y, Wakasa K, Tozawa Y (2004) Guanosine tetra- and pentaphosphate synthase activity in chloroplasts of a higher plant: association with 70S ribosomes and inhibition by tetracycline. Nucleic Acids Res 32:5732–5741

Kato Y, Murakami S, Yamamoto Y, Chatani H, Kondo Y, Nakano T, Yokota A, Sato F (2004) The DNA-binding protease, CND41, and the degradation of ribulose-1,5-bisphosphate carboxylase/oxygenase in senescent leaves of tobacco. Planta 220:97–104

Kato Y, Yamamoto Y, Murakami S, Sato F (2005) Post-translational regulation of CND41 protease activity in senescent tobacco leaves. Planta 222:643–651

Kelly AA, Dormann P (2002) DGD2, an *Arabidopsis* gene encoding a UDP-galactose-dependent digalactosyldiacylglycerol synthase is expressed during growth under phosphate-limiting conditions. J Biol Chem 277:1166–1173

Kelly AA, Froehlich JE, Dormann P (2003) Disruption of the two digalactosyldiacylglycerol synthase genes *DGD1* and *DGD2* in *Arabidopsis* reveals the existence of an additional enzyme of galactolipid synthesis. Plant Cell 15:2694–2706

Kerr EA (1958) Linkage relations of *gf*. Tomato Genet Coop Rep 8:21

Kim E, Archibald JM (2008) Diversity and evolution of plastids and their genomes. Plant Cell Monogr., doi:10.1007/7089_2008_17

Kodama Y (2007) Plastidic proteins containing motifs of nuclear transcription factors. Plant Biotechnol 24:165–170

Kodama Y, Sano H (2006) Evolution of a basic helix-loop-helix protein from a transcriptional repressor to a plastid-resident regulatory factor: involvement in hypersensitive cell death in tobacco plants. J Biol Chem 281:35369–35380

Koussevitzky S, Nott A, Mockler TC, Hong F, Sachetto-Martins G, Surpin M, Lim J, Mittler R, Chory J (2007) Signals from chloroplasts converge to regulate nuclear gene expression. Science 316:715–719

Kudla J, Hayes R, Gruissem W (1996) Polyadenylation accelerates degradation of chloroplast mRNA. Embo J 15:7137–7146

Kusaba M, Ito H, Morita R, Iida S, Sato Y, Fujimoto M, Kawasaki S, Tanaka R, Hirochika H, Nishimura M, Tanaka A (2007) Rice *NON-YELLOW COLORING1* is involved in light-harvesting complex II and grana degradation during leaf senescence. Plant Cell 19:1362–1375

Lee KP, Kim C, Landgraf F, Apel K (2007) *EXECUTER1-* and *EXECUTER2*-dependent transfer of stress-related signals from the plastid to the nucleus of *Arabidopsis thaliana*. Proc Natl Acad Sci U S A 104:10270–10275

Leiken AL, Brenner R (1989) Fatty acid desaturase activities are modulated by phytosterol incorporation in microsomes. Biochem Biophys Acta 1005:1187–1191

Leister D (2003) Chloroplast research in the genomic age. Trends Genet 19:47–56

Lister DL, Bateman JM, Purton S, Howe CJ (2003) DNA transfer from chloroplast to nucleus is much rarer in *Chlamydomonas* than in tobacco. Gene 316:33–38

Lukens JH, Mathews DE, Durbin RD (1987) Effect of tagetitoxin on the levels of ribulose 1,5-bisphosphate carboxylase, ribosomes, and RNA in plastids of wheat leaves. Plant Physiol 84:808–813

Maeda H, Sage TL, Isaac G, Welti R, Dellapenna D (2008) Tocopherols modulate extraplastidic polyunsaturated fatty acid metabolism in *Arabidopsis* at low temperature. Plant Cell 20:452–470

Masclaux C, Valadier MH, Brugiere N, Morot-Gaudry JF, Hirel B (2000) Characterization of the sink/source transition in tobacco (*Nicotiana tabacum* L.) shoots in relation to nitrogen management and leaf senescence. Planta 211:510–518

Matusova R, Rani K, Verstappen FW, Franssen MC, Beale MH, Bouwmeester HJ (2005) The strigolactone germination stimulants of the plant-parasitic *Striga* and *Orobanche* spp. are derived from the carotenoid pathway. Plant Physiol 139:920–934

Melotto M, Underwood W, Koczan J, Nomura K, He SY (2006) Plant stomata function in innate immunity against bacterial invasion. Cell 126:969–980

Meskauskiene R, Nater M, Goslings D, Kessler F, op den Camp R, Apel K (2001) FLU: a negative regulator of chlorophyll biosynthesis in *Arabidopsis thaliana*. Proc Natl Acad Sci U S A 98:12826–12831

Metraux JP (2002) Recent breakthroughs in the study of salicylic acid biosynthesis. Trends Plant Sci 7:332–334

Minnikin DE, Abdolrahimzadeh H, Baddiley J (1974) Replacement of acidic phosphates by acidic glycolipids in *Pseudomonas diminuta*. Nature 249:268–269

Mock HP, Grimm B (1997) Reduction of Uroporphyrinogen decarboxylase by antisense RNA expression affects activities of other enzymes involved in tetrapyrrole biosynthesis and leads to light-dependent necrosis. Plant Physiol 113:1101–1112

Mock HP, Heller W, Molina A, Neubohn B, Sandermann H, Jr, Grimm B (1999) Expression of uroporphyrinogen decarboxylase or coproporphyrinogen oxidase antisense RNA in tobacco induces pathogen defense responses conferring increased resistance to tobacco mosaic virus. J Biol Chem 274:4231–4238

Molina A, Volrath S, Guyer D, Maleck K, Ryals J, Ward E (1999) Inhibition of protoporphyrinogen oxidase expression in *Arabidopsis* causes a lesion-mimic phenotype that induces systemic acquired resistance. Plant J 17:667–678

Moller SG, Kunkel T, Chua NH (2001) A plastidic ABC protein involved in intercompartmental communication of light signaling. Genes Dev 15:90–103

Moorhead GB, Smith CS (2003) Interpreting the plastid carbon, nitrogen, and energy status. A Role for PII? Plant Physiol 133:492–498

Moran PJ, Thompson GA (2001) Molecular responses to aphid feeding in *Arabidopsis* in relation to plant defense pathways. Plant Physiol 125:1074–1085

Mueller S, Hilbert B, Dueckershoff K, Roitsch T, Krischke M, Mueller MJ, Berger S (2008) General detoxification and stress responses are mediated by oxidized lipids through TGA transcription factors in *Arabidopsis*. Plant Cell doi: 10.1105/tpc.107.054809

Munne-Bosch S (2005) Linking tocopherols with cellular signaling in plants. New Phytol 166:363–366

Munné-Bosch S, Penuelas J, Llusià J (2007) A deficiency in salicylic acid alters isoprenoid accumulation in water-stressed NahG transgenic *Arabidopsis* plants. Plant Sci 172:756–762

Nagamune K, Hicks LM, Fux B, Brossier F, Chini EN, Sibley LD (2008) Abscisic acid controls calcium-dependent egress and development in *Toxoplasma gondii*. Nature 451:207–210

Nakano T, Murakami S, Shoji T, Yoshida S, Yamada Y, Sato F (1997) A novel protein with DNA binding activity from tobacco chloroplast nucleoids. Plant Cell 9:1673–1682

Neuffer MG, Calvert OH (1975) Dominant disease lesion mimics in maize. *J Hered* 66:265–270

Nishimura MT, Stein M, Hou BH, Vogel JP, Edwards H, Somerville SC (2003) Loss of a callose synthase results in salicylic acid-dependent disease resistance. Science 301:969–972

Okada A, Shimizu T, Okada K, Kuzuyama T, Koga J, Shibuya N, Nojiri H, Yamane H (2007) Elicitor induced activation of the methylerythritol phosphate pathway toward phytoalexins biosynthesis in rice. Plant Mol Biol 65:177–187

Ono K, Ishimaru K, Aoki N, Ohsugi R (1999) Transgenic rice with low sucrose-phosphate synthase activities retain more soluble protein and chlorophyll during flag leaf senescence. Plant Physiol Biochem 37:949–953

op den Camp RG, Przybyla D, Ochsenbein C, Laloi C, Kim C, Danon A, Wagner D, Hideg E, Gobel C, Feussner Iet-al. (2003) Rapid induction of distinct stress responses after the release of singlet oxygen in *Arabidopsis*. Plant Cell 15:2320–2332

Osanai T, Tanaka K (2007) Keeping in touch with PII: PII-interacting proteins in unicellular cyanobacteria. Plant Cell Physiol 48:908–914

Park S, Takano Y, Matsuura H, Yoshihara T (2004) Antifungal compounds from the root and root exudate of *Zea mays*. Biosci Biotechnol Biochem 68:1366–1368

Park SY, Yu JW, Park JS, Li J, Yoo SC, Lee NY, Lee SK, Jeong SW, Seo HS, Koh HJ et al. (2007) The senescence-induced staygreen protein regulates chlorophyll degradation. Plant Cell 19:1649–1664

Parthier B (1988) Gerontoplasts – the yellow end in the ontogenesis of chloroplasts. Endocytobiosis Cell Res 5:163–190

Peters RJ (2006) Uncovering the complex metabolic network underlying diterpenoid phytoalexin biosynthesis in rice and other cereal crop plants. Phytochemistry 67:2307–2317

Porfirova S, Bergmuller E, Tropf S, Lemke R, Dormann P (2002) Isolation of an *Arabidopsis* mutant lacking vitamin E and identification of a cyclase essential for all tocopherol biosynthesis. Proc Natl Acad Sci U S A 99:12495–12500

Prithiviraj B, Bais HP, Weir T, Suresh B, Najarro EH, Dayakar BV, Schweizer HP, Vivanco JM (2005) Down regulation of virulence factors of *Pseudomonas aeruginosa* by salicylic acid attenuates its virulence on *Arabidopsis thaliana* and *Caenorhabditis elegans*. Infect Immun 73:5319–5328

Provencher LM, Miao L, Sinha N, Lucas WJ (2001) *Sucrose export defective1* encodes a novel protein implicated in chloroplast-to-nucleus signaling. Plant Cell 13:1127–1141

Pruzinska A, Tanner G, Anders I, Roca M, Hörtensteiner S (2003) Chlorophyll breakdown: pheophorbide *a* oxygenase is a Rieske-type iron-sulfur protein, encoded by the *accelerated cell death 1* gene. Proc Natl Acad Sci U S A 100:15259–15264

Pruzinska A, Anders I, Aubry S, Schenk N, Tapernoux-Luthi E, Muller T, Krautler B, Hörtensteiner S (2007) In vivo participation of red chlorophyll catabolite reductase in chlorophyll breakdown. Plant Cell 19:369–387

Reith ME, Munholland J (1995) Complete nucleotide sequence of the *Porphyra purpurea* chloroplast genome. Plant Mol Biol Rep 13:333–335

Ren G, An K, Liao Y, Zhou X, Cao Y, Zhao H, Ge X, Kuai B (2007) Identification of a novel chloroplast protein AtNYE1 regulating chlorophyll degradation during leaf senescence in *Arabidopsis*. Plant Physiol 144:1429–1441

Ribot C, Zimmerli C, Farmer EE, Reymond P, Poirier Y (2008) Induction of the *Arabidopsis* PHO1;H10 gene by 12-oxo-phytodienoic acid but not jasmonic acid via a CORONATINE INSENSITIVE1-dependent pathway. Plant Physiol 147:696–706

Richly E, Dietzmann A, Biehl A, Kurth J, Laloi C, Apel K, Salamini F, Leister D (2003) Covariations in the nuclear chloroplast transcriptome reveal a regulatory master-switch. EMBO Rep 4:491–498

Rimbach G, Minihane AM, Majewicz J, Fischer A, Pallauf J, Virgli F, Weinberg PD (2002) Regulation of cell signalling by vitamin E. Proc Nutr Soc 61:415–425

Robatzek S, Somssich IE (2002) Targets of *AtWRKY6* regulation during plant senescence and pathogen defense. Genes Dev 16:1139–1149

Roca M, Minguez-Mosquera MI (2006) Chlorophyll catabolism pathway in fruits of *Capsicum annuum* (L.): stay-green versus red fruits. J Agric Food Chem 54:4035–4040

Rolland N, Ferro M, Seigneurin-Berny D, Garin J, Block M, Joyard J (2008) The chloroplast envelope proteome and lipidome. Plant Cell Monogr., doi:10.1007/7089_2008_33

Sabar M, De Paepe R, de Kouchkovsky Y (2000) Complex I impairment, respiratory compensations, and photosynthetic decrease in nuclear and mitochondrial male sterile mutants of *Nicotiana sylvestris*. Plant Physiol 124:1239–1250

Sabater B, Martin M, Schmitz-Linneweber KC, Maier RM (2002) Is clustering of plastid RNA editing sites a consequence of transitory loss of gene function? Implications for past environmental events. Perspect Plant Ecol Evol Systemat 5:81–90

Sakuragi Y, Maeda H, Dellapenna D, Bryant DA (2006) alpha-Tocopherol plays a role in photosynthesis and macronutrient homeostasis of the *Cyanobacterium Synechocystis* sp. PCC 6803 that is independent of its antioxidant function. Plant Physiol 141:508–521

Sato N, Ohta N (2001) DNA-binding specificity and dimerization of the DNA-binding domain of the PEND protein in the chloroplast envelope membrane. Nucleic Acids Res 29:2244–2250

Sato Y, Morita R, Nishimura M, Yamaguchi H, Kusaba M (2007) Mendel's green cotyledon gene encodes a positive regulator of the chlorophyll-degrading pathway. Proc Natl Acad Sci U S A 104:14169–14174

Sauret-Gueto S, Botella-Pavia P, Flores-Perez U, Martinez-Garcia JF, San Roman C, Leon P, Boronat A, Rodriguez-Concepcion M (2006) Plastid cues posttranscriptionally regulate the accumulation of key enzymes of the methylerythritol phosphate pathway in *Arabidopsis*. Plant Physiol 141:75–84

Schenk N, Schelbert S, Kanwischer M, Goldschmidt EE, Dormann P, Hörtensteiner S (2007) The chlorophyllases *AtCLH1* and *AtCLH2* are not essential for senescence-related chlorophyll breakdown in *Arabidopsis thaliana*. FEBS Lett 581:5517–5525

Seo S, Seto H, Koshino H, Yoshida S, Ohashi Y (2003) A diterpene as an endogenous signal for the activation of defense responses to infection with tobacco mosaic virus and wounding in tobacco. Plant Cell 15:863–873

Siddique MA, Grossmann J, Gruissem W, Baginsky S (2006) Proteome analysis of bell pepper (*Capsicum annuum* L.) chromoplasts. Plant Cell Physiol 47:1663–1673

Sies H, Menck CF (1992) Singlet oxygen induced DNA damage. *Mutat Res* 275:367–375

Smith PG (1950) Inheritance of brown and green mature fruit color in peppers. J Hered 41:138–140

Stent GS, Brenner S (1961) A genetic locus for the regulation of ribonucleic acid synthesis. Proc Natl Acad Sci *U S A* 47:2005–2014

Strand A, Asami T, Alonso J, Ecker JR, Chory J (2003) Chloroplast to nucleus communication triggered by accumulation of Mg-protoporphyrinIX. Nature 421:79–83

Strawn MA, Marr SK, Inoue K, Inada N, Zubieta C, Wildermuth MC (2007) *Arabidopsis* isochorismate synthase functional in pathogen-induced salicylate biosynthesis exhibits properties consistent with a role in diverse stress responses. J Biol Chem 282:5919–5933

Sy J, Lipmann F (1973) Identification of the synthesis of guanosine tetraphosphate (MS I) as insertion of a pyrophosphoryl group into the 3'-position in guanosine 5'-diphosphate. Proc Natl Acad Sci U S A 70:306–309

Takahashi H, Watanabe A, Tanaka A, Hashida SN, Kawai-Yamada M, Sonoike K, Uchimiya H (2006) Chloroplast NAD kinase is essential for energy transduction through the xanthophyll cycle in photosynthesis. Plant Cell Physiol 47:1678–1682

Takahashi K, Kasai K, Ochi K (2004) Identification of the bacterial alarmone guanosine 5'-diphosphate 3'-diphosphate (ppGpp) in plants. Proc Natl Acad Sci U S A 101:4320–4324

Taki N, Sasaki-Sekimoto Y, Obayashi T, Kikuta A, Kobayashi K, Ainai T, Yagi K, Sakurai N, Suzuki H, Masuda T et al. (2005) 12-oxo-phytodienoic acid triggers expression of a distinct set of genes and plays a role in wound-induced gene expression in *Arabidopsis*. Plant Physiol 139:1268–1283

Thaler JS, Bostock RM (2004) Interactions between abscisic-acid-mediated responses and plant resistance to pathogens and insects. Ecology 85:48–58

Thomas H, Howarth CJ (2000) Five ways to stay green. J Exp Bot 51:329–337

Traber MG, Atkinson J (2007) Vitamin E, antioxidant and nothing more. Free Radic Biol Med 43:4–15

Trebst A, Depka B, Hollander-Czytko H (2002) A specific role for tocopherol and of chemical singlet oxygen quenchers in the maintenance of photosystem II structure and function in *Chlamydomonas reinhardtii*. FEBS Lett 516:156–160

Vance CP, Uhde-Stone C, Allan DL (2003) Phosphorus acquisition and use: critical adaptations by plants for securing a norenewable resource. New Phytol 157:423–447

van der Biezen EA, Sun J, Coleman MJ, Bibb MJ, Jones JD (2000) *Arabidopsis* RelA/SpoT homologs implicate (p)ppGpp in plant signaling. Proc Natl Acad Sci U S A 97:3747–3752

von Gromoff ED, Alawady A, Meinecke L, Grimm B, Beck CF (2008) Heme, a plastid-derived regulator of nuclear gene expression in *Chlamydomonas*. Plant Cell doi: 10.1105/tpc.107.054650

Wasternack C (2007) Jasmonates: an update on biosynthesis, signal transduction and action in plant stress response, growth and development. Ann Bot (Lond) 100:681–697

Weber H (2002) Fatty acid-derived signals in plants. Trends Plant Sci 7:217–224

Wendrich TM, Blaha G, Wilson DN, Marahiel MA, Nierhaus KH (2002) Dissection of the mechanism for the stringent factor RelA. Mol Cell 10:779–788

Wildermuth MC, Dewdney J, Wu G, Ausubel FM (2001) Isochorismate synthase is required to synthesize salicylic acid for plant defence. Nature 414:562–565

Wu Y, Kuzma J, Marechal E, Graeff R, Lee HC, Foster RChua NH (1997) Abscisic acid signaling through cyclic ADP-ribose in plants. Science 278:2126–2130

Wycliffe P, Sitbon F, Wernersson J, Ezcurra I, Ellerstrom M, Rask L (2005) Continuous expression in tobacco leaves of a *Brassica napus* PEND homologue blocks differentiation of plastids and development of palisade cells. Plant J 44:1–15

Xiong L, Lee MW, Qi M, Yang Y (2001) Identification of defense-related rice genes by suppression subtractive hybridization and differential screening. Mol Plant Microbe Interact 14:685–692

Xu C, Moellering ER, Fan J, Benning C (2008) Mutation of a mitochondrial outer membrane protein affects chloroplast lipid biosynthesis. Plant J 54:163–175

Yamamoto S, Suzuki Y, Katagiri M, Ohkawa H (1995) Protoporphyrinogen-oxidizing enzymes of tobacco cells with respect to light-dependent herbicide mode of action. Pest Sci 43:357–358

Yang C, Guo R, Jie F, Nettleton D, Peng J, Carr T, Yeakley JM, Fan JB, Whitham SA (2007a) Spatial analysis of *Arabidopsis thaliana* gene expression in response to Turnip mosaic virus infection. Mol Plant Microbe Interact 20:358–370

Yang H, Yang T, Baur JA, Perez E, Matsui T, Carmona JJ, Lamming DW, Souza-Pinto NC, Bohr VA, Rosenzweig Aet-al. (2007b) Nutrient-sensitive mitochondrial NAD+ levels dictate cell survival. Cell 130:1095–1107

Yehudai-Resheff S, Hirsh M, Schuster G (2001) Polynucleotide phosphorylase functions as both an exonuclease and a poly(A) polymerase in spinach chloroplasts. Mol Cell Biol 21:5408–5416

Yehudai-Resheff S, Zimmer SL, Komine Y, Stern DB (2007) Integration of chloroplast nucleic acid metabolism into the phosphate deprivation response in *Chlamydomonas reinhardtii*. Plant Cell 19:1023–1038

Yoneyama K, Xie X, Kusumoto D, Sekimoto H, Sugimoto Y, Takeuchi Y (2007) Nitrogen deficiency as well as phosphorus deficiency in *Sorghum* promotes the production and exudation of 5-deoxystrigol, the host recognition signal for arbuscular mycorrhizal fungi and root parasites. Planta 227:125–132

Yuan ZC, Edlind MP, Liu P, Saenkham P, Banta LM, Wise AA, Ronzone E, Binns AN, Kerr K, Nester EW (2007) The plant signal salicylic acid shuts down expression of the *vir* regulon and activates quormone-quenching genes in *Agrobacterium*. Proc Natl Acad Sci U S A 104:11790–11795

Zapata JM, Guera A, Esteban-Carrasco A, Martin M, Sabater B (2005) Chloroplasts regulate leaf senescence: delayed senescence in transgenic *ndhF*-defective tobacco. Cell Death Differ 12:1277–1284

Zelisko A, Garcia-Lorenzo M, Jackowski G, Jansson S, Funk C (2005) *AtFtsH6* is involved in the degradation of the light-harvesting complex II during high-light acclimation and senescence. Proc Natl Acad Sci U S A 102:13699–13704

Zhou F, Pu Y, Wei T, Liu H, Deng W, Wei C, Ding B, Omura T, Li Y (2007) The P2 capsid protein of the nonenveloped rice dwarf phytoreovirus induces membrane fusion in insect host cells. Proc Natl Acad Sci U S A 104:19547–19552

Zhu S, Gao F, Cao X, Chen M, Ye G, Wei C, Li Y (2005) The rice dwarf virus P2 protein interacts with *ent*-kaurene oxidases in vivo, leading to reduced biosynthesis of gibberellins and rice dwarf symptoms. Plant Physiol 139:1935–1945

Index

A

ABC transporter protein complex, 135, 136, 149, 272
Abscisic acid, (ABA), 47, 73, 79, 189–191, 195, 225, 272, 274, 277
Actin, 242, 253–259
Acyl carrier protein, (ACP), 132–134, 137, 148
Acyl chain, 23, 42, 43, 46, 47, 49, 70–72, 77, 79, 127–132, 134, 137, 138, 141, 146–148, 166–168, 268, 272.
  See also Fatty acid
Acyl-CoA, 72, 77, 133, 134, 138, 139, 148, 274
Acyltransferases, 133, 134, 137
α-helix, 106
α-tocopherol, tocopherol, 48, 73, 74, 79, 142, 148, 268, 269, 272
Alarmone, 5 –(bis)pyrophosphate nucleotides, 273–274. *See also* Guanosine–3, ppGpp
Alga, algae
  amino acid, 23, 28, 42, 68, 74, 76, 101, 111, 161, 167, 218, 223, 240, 244, 246, 252, 253, 257, 273
  diatom, 9, 17, 27, 163, 238, 241, 256
  dinoflagellate, 3, 4, 8, 10–12, 14–18, 23, 28, 163
  green, 2, 10–13, 17, 18, 28, 109, 160, 161, 163, 213, 218
  red, 2, 4, 10, 12, 13, 23, 28, 163–165, 172, 173, 218, 219, 267, 281
Amylopectin, 12
Amyloplast, 169, 171, 174, 207
Amylose, 12
Annexin, 49
Anterograde, 182, 209
Antioxidant, 79, 80, 197, 199, 220, 259, 268, 275

Antipathogenic, 132
Antiporter, 75, 162, 163, 173, 174
Apicomplexan parasites, 3, 4, 8, 11, 12, 15, 17, 23, 27
Apicoplast, 8, 15–17, 22, 23, 164–167, 277
Apocarotenoid, 278–279
Arabidopside, 275
ARC6, 78, 212, 213, 219
Archaeplastida, 160, 161
Ascorbate, ascorbic acid, 79, 80, 191, 195, 197, 199
ATP, ADP, 9, 13, 75–78, 91, 100, 102, 105–108, 111, 125, 131, 135, 144, 163, 167, 169–172, 174, 214, 252, 277
ATPase, metal transporting, 76, 80, 100, 102, 103, 106, 107, 136, 168, 214, 215, 222, 250
Auxilin, 252, 253
Avoidance response, 236–239, 241–244, 246–249, 251–259

B

β-strand, 101
Blue light
  induced, 237, 241–244, 246, 247, 249–251, 254–257, 259
  receptor, 237–242, 245, 246, 259

C

Calcium, calcium ion, Ca2+, 110, 212, 215, 222, 250, 251, 259, 277
Callose, 268, 275
Carbohydrate, 12, 13, 42, 76, 89, 161, 222
Carotenoid, 9, 10, 17, 47, 72, 73, 79, 129, 130, 142, 182, 185, 187, 190, 191, 272, 274, 276–279

293

CASTOR and POLLUX, 75, 79, 276, 279
Cell cycle, 208, 210, 274
Cell death, 79, 109, 168, 198, 217, 223, 269–271
Channel
    Ca2+, 222, 250, 251
    cation, 104, 105, 170, 215, 222, 250
    cation-selective, 76
    import, 94
    ion, 71, 75, 76, 101, 105, 170, 217, 221, 222, 250
    mechanosensitive, 71, 212, 217
Chaperone, 51, 68, 76, 77, 80, 91, 99, 100, 102, 103, 105–108, 111, 271
Chlorarachniophytes, 2, 3, 6, 9, 12, 14
Chlorophyll
    breakdown, 271, 282, 283
    catabolite, 270, 271
    Chl, 9–11, 189, 191, 192, 195
    red, 270, 271
Chlorophyllase, 148, 270, 271, 283
Chlorophyllide, 48, 112, 148, 192, 271
Chlorophyta, 182, 237, 240
Chloroplast
    anchoring, 254, 257
    avoidance movement, 80, 239, 243, 249, 252, 254
    division, chloroplast constriction, 78, 208, 210, 213, 215, 217–225
    envelope, 41–44, 47–52, 69, 70, 72–78, 80, 81, 113, 129, 131, 134–139, 141, 142, 146, 148, 165, 168, 172, 173, 198, 217, 222
    movement, 236–242, 244–249, 251, 253–260
    outer membrane, 258
    unusual positioning protein 1 (CHUP1), 78–80, 235, 238, 257–260
Chromalveolates, 165
Chromophore, 188, 240, 245, 246, 248
Chromoplast, differentiation, 282, 283
Circadian, 220, 221, 238
Clathrin, 252
Clp, 78, 106
Colchicine, 254
Co-translational, 97
Cotransporter, 174
Cryptochrome, cry, 193, 238,
Cyanobacteria, cyano bacteria, 2, 9–13, 15, 17, 18, 22, 23, 27, 28, 90, 101, 109, 126, 137, 144, 145, 160–162, 182, 188, 208, 209, 213, 216, 225, 237, 241, 267, 269
Cyclic ADP-ribose, 277

Cytochalasin, 254, 256
Cytoplasm, 11–15, 23, 42, 162, 165, 183, 243, 246, 248, 256, 284
Cytoskeleton, cytoskeletal, 239, 249, 250, 257, 260
Cytosol, 15, 16, 41, 42, 44, 67, 73–75, 91, 94, 97, 99, 100, 110, 126, 132–134, 139, 143, 160, 162, 164, 165, 167, 169, 171–174, 183, 184, 187–199, 208, 212, 214, 215, 218, 219, 222, 224, 250, 251, 253, 272, 274

D

Dark positioning, 236, 239, 241, 252, 253
δ-aminolevulinic acid, (ALA), 189, 192
Desaturase, 72, 73, 130, 132, 141, 142, 146, 187, 191, 277
Development, gametophyte, 168
Diacylglycerol, 43, 46, 72, 134
Diatom, 9, 17, 27, 163, 238, 241, 256
Dicarboxylate translocator, (DiT), 168, 173
Digalactosyl diacylglycerol, DGDG, DGD, 43, 44, 46–50, 92, 93, 126–131, 139–143, 145–147, 149, 225, 275, 278
Dinoflagellate, 3, 4, 8, 10–12, 14, 15–18, 23, 28, 163
Diterpene, labdane-type, 279
DNA
    damage, 215, 216, 222, 225
    rDNA+A201,
    recombinant+A56,
    replication, 27, 78, 166, 183, 213
Domain
    chromophore-binding (CBD), 235, 248
    histidine kinase-related (HKRD), 248
    light oxygen voltage (LOV), 235, 240, 245, 246
    N-terminal extension (NTE), 248
    PAS-related (PRD), 248
    photosensory, 240, 245, 246
Dynamin, 144, 218, 219

E

Endocytosis, 218, 252
Endomembrane, 15, 42, 77, 110, 112, 113, 160, 162, 163, 165, 168, 280
Endoplasmic reticulum (ER), 15, 49, 50, 97, 110, 113, 132–139, 141–143, 146–149, 163, 174, 250, 269
Endosymbiont
    secondary, 4, 14, 15, 165
    tertiary, 161

Envelope
  inner, 45, 48, 49, 51, 70–74, 77, 78, 91, 92, 100, 111, 129, 137–139, 142–144, 149, 160, 163, 169, 192, 212, 219, 222
  intermembrane space, 51, 92, 100, 111, 219
  outer, 42, 44, 45, 48–51, 68, 74, 76, 77, 91, 92, 102, 110, 129–131, 134, 135, 139, 141, 160, 218, 219, 257, 258
Epoxy-xantophyll cycle, 279
EST database, 241
Ethylene, 198, 225, 250, 274
Etioplast, 41, 102, 112, 197
Eukaryot, eukaryotic, 1–3, 6, 8–10, 13–16, 22, 46, 47, 49, 50, 77, 80, 89, 109, 132–135, 137–139, 142–144, 159–164, 168–170, 181, 182, 208, 209, 211, 212, 218, 226, 281
Eukaryotic pathway, eukaryotic lipids, 132, 134, 142
Evolution, 1, 2, 8, 10, 11, 14–16, 18, 22, 27, 28, 89, 90, 134, 160–163, 165, 168, 169, 173, 174, 182, 199, 208–215, 222, 241, 243, 246, 267, 280

F

Far red light, far-red light, 237, 242, 244, 247
Fatty acid
  free, 148
  oxygenated, 10, 131, 132
  polyunsaturated, 79, 129, 131, 147
  synthesis, fatty acid biosynthesis, 23, 71, 72, 132, 134, 147, 166–168
Fern, 235, 237, 241, 244–246, 251
Ferredoxin, 93, 108, 109, 142, 172, 194, 195, 197
Fission, 209, 218
Fucoxanthin, 10, 17
Fungus, fungi, 11, 13, 219, 225, 278

G

Galactolipid
  galactosyltransferase, 44, 140, 141
  prokaryotic-type, 50
Galactosidase, 147
Gametophyte, 167, 168
Gated pore-forming protein, 74, 76

Gene expression, 73, 75, 80, 108, 145, 182–187, 189–199, 209, 212, 220, 224, 226, 238, 247, 249, 252
Gene transfer, lateral, 23, 173
Genome, 2, 7, 18, 19, 22, 23, 27, 28, 42, 50, 70, 71, 135–137, 143, 146, 148, 160, 162–166, 168, 169, 172, 173, 182, 183, 188, 208, 210, 211, 214, 215, 217, 241, 245, 247, 267, 272, 284
Gerontoplast, 282
GFP-fusion protein, 240, 243
Gibberellin, 47, 280
Glucose phosphate transporter (GPT), 164, 165, 167, 168, 174
Glutamine synthetase (GS), 172, 173
Glutathione, 79, 80, 194, 197, 199
Glyceraldehyde-3-phosphate dehydrogenase (GAPDH), 15–17, 183
Glycolipase, 147
Glycosidase, 147
Golgi, 113, 126, 134, 143, 163, 173, 243, 254, 255, 256
GPT. *See* Glucose phosphate transporter
Green alga, 2, 10–13, 17, 18, 28, 109, 160, 161, 163, 213, 218
Green fluorescent protein (GFP), 79, 92, 98, 135, 136, 240, 243, 253, 255, 258
GS. *See* Glutamine synthetase
GTP, GDP, 78, 91, 95, 97, 98, 102, 171, 211, 273
Guanosine-3,5 -(bis)pyrophosphate nucleotides, ppGpp, 273, 274

H

Heat shock protein (Hsp), 144, 252
Heme, 73, 166, 185, 186, 270–272
Heterotroph, 2, 8, 12, 161, 181
Hexadecatrienoic acid, 128
High-fluence-rate response (HFR), 235–237
Hypersensitive response, 132, 223, 269, 279

I

Import, 13, 23, 48, 49, 51, 71, 74, 76, 77, 89–113, 134, 137, 138, 167, 168, 170, 174, 182, 184, 192, 208, 209, 221, 222, 226
Intermembrane space, 51, 92, 100, 111
Isopentylpyrophosphate (IPP), 279
Isoprenoid, 47, 72, 148

J
JAC1 protein, 252, 253
Jasmonic acid, 72, 130, 198, 225, 274, 275

K
Kinase
  domain, 240, 243–245
  NAD, 109, 132, 279, 280
  serine/threonine domain, 91, 95, 240, 243, 245, 269
Kinesin-like protein, 78

L
Latrunculin, 254
Lesion, 170, 186, 188, 189, 215, 270–272
Lethal, 96, 100, 102, 104, 106–108, 110, 136, 137, 166, 168, 185, 270, 271
Leucoplast, 207
Light-harvesting
  complex, 10
  proteins, 184
Light-induced, 48, 186, 193, 195, 222, 236–238, 241–251, 254–257, 259
Linoleic acid, 18:2, 142, 269
Linoleic acid, 18:3, 269
Lipid
  bilayer, 91, 129
  eukaryotic, 47, 134, 135, 138, 143
  geometry, 129
  metabolism, 50, 70, 72, 139, 144, 145
  phase
    cubic, 129
    cylindrical, 129
    inverted hexagonal, 129
  prokaryotic, 134, 137
  trafficking, 71, 77, 148, 149
  transfer, transport, 50, 134, 135, 142, 144
Lipoxygenase, 51, 72, 130, 225, 272, 274, 275
Low-fluence-rate response, (LFR), 235–237
Lutein, 48, 73
Lyso-PC, 133, 134, 136, 138, 139

M
MAPK, 198, 276, 279
Mechanosensitive ion channel, 217
Membrane
  contact site, 100, 103, 134, 135, 142, 143, 148
  invagination, 49, 142–144, 149
  photosynthetic, 48, 49, 129, 142, 148
Membrane-localized, 244, 247
Meristem, meristematic, 183, 207, 208, 213
Mesophyll, 208, 213, 221, 249, 250, 255, 257, 258
Metabolic, 23, 47, 80, 90, 138, 159–163, 172–174, 181, 207, 215, 220, 224
Metabolism, 13, 42, 50, 51, 68, 70–74, 79, 80, 135, 137, 139, 144, 145, 148, 159, 160, 164, 166, 168, 169, 171–174, 185, 196, 220, 268, 274, 277, 279, 281, 283
Metabolite, 51, 70, 71, 74, 79, 133, 159–164, 166, 170, 173, 174, 222, 269
Mg-chelatase, 186, 188, 189
Mg-Proto-IX, 186, 188–190, 192
Microtubule, 78, 79, 144, 242, 254, 256, 257
Mitochondrial carrier family (MCF), 75, 169, 170, 171
Mitochondria, mitochondrial, 17, 23, 42, 45, 50, 51, 72, 74–76, 79, 99, 103, 104, 108, 132, 135–137, 143, 145, 146, 160, 162, 169, 170, 173, 183, 185, 188, 218, 219, 225, 254, 256, 257, 278, 280, 283, 284
Mitogen-activated protein kinase, 198, 279
Monogalactosyl diacylglycerol, MGDG (MGD), 43, 44, 46–49, 68, 72, 92, 93, 111, 126–131, 134, 140–142, 145, 147, 225
Moss, 96, 99, 148, 211, 225, 236, 237, 241, 242, 247, 250, 251, 256
MSL protein, 75, 78
Myosin, 255–256

N
NAD kinase, 279–280
NADPH:protochlorophyllide oxidoreductase, 112
NADP, NADPH, 109, 110, 112, 167, 168, 183, 279, 280
Neochrome, 235, 240, 244–248, 260
Neoxanthin, 47, 73
N-ethylmaleimide (NEM), 139, 255
Nitrogen, 9, 148, 171–173, 183, 278, 281, 282
Nitrogen sensing, 281
Non-photosynthetic, 96, 108, 112
Norflurazon, 187, 196
N-receptor-interacting protein, 280
Nuclear encoded, 14, 17, 42, 76, 182, 183, 187, 189, 199, 208, 212
Nucleomorph, 8, 9

Nucleotide-binding site-leucine-rich repeat, 276, 280
Nucleotide transporters (NTTs), 163, 168–170, 174
Nucleus, 3, 8, 13, 16, 17, 22, 23, 28, 42, 71, 73, 75, 79, 90, 91, 160, 162, 165, 182–185, 187, 189, 192–198, 200, 208, 209, 212, 213, 224, 226, 236, 253, 267, 272, 280

O

Oleic acid, 132, 142, 269
Oligogalactolipids, 131, 141
Oryzalin, 254
Oxidative pentose phosphate pathway (OPPP), 164, 166, 167
Oxo-phytodienoic acid, 72, 131, 132, 274, 275
Oxygen evolution, 222
Oxylipins, 132, 147, 275

P

Pathogen, 199, 210, 212, 213, 220, 223–225, 268, 269, 271, 272, 274–277, 279–281, 284
Permease-like protein, 136
Peroxisome, 45, 75, 148, 173, 254–257
Phagocytosis, 162
Phosphate translocator, phosphate transporter, 68, 162–167, 174, 195
Phosphatidic acid posphatase (PAP), 136, 138, 139
Phosphatidyl choline (PC), 43–45, 47, 50, 127–130, 133, 134, 136, 138, 139, 146, 147, 195
Phosphatidylethanolamine (PE), 43, 45, 128, 146
Phosphatidylglycerol (PG), 43, 44, 46, 47, 49, 50, 72, 126–128, 130, 137, 138, 142
Phosphatidylinositol (PI), 43, 128, 131
Phospholipase, 44, 45, 72, 138, 139, 146, 250
Phospholipase C (PLC), 44, 45, 138, 139, 146, 250
Phospholipase D (PLD), 138, 139, 146
Phospholipid, 43, 44, 80, 127, 129, 134, 137, 145–147
Phosphorylate, phosphorylation, 91, 92, 95, 131, 133, 169, 170, 196–198, 220, 240, 244, 245, 279
Photobiological, 238, 240, 242, 246, 248
Photodamage, 80, 236, 258, 259
Photooxidation, 170
Photoperception, 237–248
Photoprotection, 10, 80, 190, 268
Photoreceptor, 190, 200, 237–239, 242, 245, 247, 248, 260
Photorelocation, 78, 236–238, 243, 248–251, 255, 256, 258–260
Photosensory domain, 240, 245, 246
Photosynthesis, photosynthetic, 1–9, 11–15, 17, 22, 23, 42, 47–49, 73, 89, 96, 98, 108, 109, 112, 125, 126, 129, 130, 132, 142, 144, 148, 160, 161, 163–165, 168–170, 181–185, 187, 191, 193–196, 207, 208, 211, 212, 220–225, 258, 267, 278, 282–284
Photosynthetic electron transport, 126, 163, 182, 191, 194, 195, 223, 224
Photosystem II (PSII), 10, 28, 184, 195, 198, 223, 224, 258, 269, 272
Photosystem I (PSI), 195–197, 199, 224, 272
Phototropin (phot), 238–251, 257–260
Phototropism, 238–241, 244, 245, 247
Phycobilin, 9, 11
Phycocyanobilin, 11, 245
Phylloquinone, 47, 48, 142, 148
Phylogenetic, 2, 10, 17, 18, 27, 28, 101, 163–165, 168, 171, 211, 218, 246
Phylogenomic, 163, 168, 173
Phytoalexin, 269, 271, 272, 279, 280
Phytochrome (phy), 185, 188, 190, 193, 237, 238, 240, 242, 244–248, 250, 257
Phytoene, 73, 187, 190, 277
Phytohormone, 47, 135
Phytol, 10, 148, 271
Phytyl esters, 148
PII protein, 281
PLAM, 113, 135, 136
16:3 plant, 128, 132, 134, 137, 278. *See also* Prokaryotic pathway
18:3 plant, 132, 142. *See also* Eukaryotic pathway
Plasma membrane, 49, 50, 75, 77, 126, 135, 139, 143, 145, 146, 160, 162, 163, 165, 174, 238, 240, 243, 246, 250, 251, 254, 257, 259, 278
Plastid
  division, 71, 78–79, 207–226
  less, 6, 8, 12, 15
  phosphate translocator family (pPTs), 163, 164, 166–168
  primary, 2, 13–14, 23, 160
  secondary, 14–16, 161, 163, 165
  stay-green protein (SGR), 282–283
  tertiary, 161
Plastome, plastid genome, 89, 90, 182, 185

Plastoquinone, 47, 48, 73, 74, 142, 190, 191, 194–197, 222, 224
PLC, 138, 139, 146
PLD, 138, 139, 146
PNPase, 277, 278
Polarotropism, 247
Polynucleotide phosphorylase (PNPase), 277, 278
Porin, 68, 74
POR. *See* NADPH:protochlorophyllide oxidoreductase
Prenylquinone, 48, 72–74
Prokaryotic
   division machinery, 78, 209, 210, 212, 216, 218, 219, 226
   pathway, procaryotic lipids, 128, 132, 134, 137, 278
Proplastid, 41, 187, 199, 207, 208
Protease, 44, 51, 68, 77, 78, 80, 96, 101, 106, 108, 111, 138, 218, 281, 283
Protease, FTSH6-type, 283
Protein
   binding motif, 252
   GFP fusion, 98, 136, 240, 243, 258
   import, 13, 49, 51, 76, 90–91, 93, 94, 96, 99, 102, 109, 160, 192, 221, 226
   kinesin-like, 78
   N-receptor-interacting, 280
   permease-like, 136
   phosphorylation, 220
   protein interaction, protein-protein interaction, 95, 100, 106, 135, 213, 240
   proteolytic, 78
   translocation, 78, 92, 100, 105
Proteome, proteomic, 2, 23–28, 41–81, 96, 170, 283
Protist, 165
Protoalga, 160, 163, 169
Protochlorophyllide, Pchlide, 10, 48, 73, 112, 198, 272
Protonema, protonemal, 241, 247, 249–251, 254, 256
Protoporphyrin IX, 73, 170, 188, 212, 271, 272

Q
Quinone, 48–49, 72

R
Ratchet, 100, 102, 107
Reactive oxygen species (ROS), 182, 194, 197–199, 272, 281

Receptor
   blue light, 237–242, 245, 246, 259
   far red light, 237, 242, 244, 247
Redox
   metabolism, 71, 79–80
   sensitive, 196, 197, 199
   state, 71, 103, 108, 182, 191, 194–196, 212, 220, 222, 224
RelA, 273, 274
Retrograde, 181–200, 209, 212, 221, 222, 224, 272
Rhodophyta, rhodophyte, 2, 3, 5, 11, 13–15, 27
RNA
   messenger (mRNA), 185–187, 225, 252, 268
   ribosomal (rRNA), 8, 16–18, 22, 23, 27, 185, 223, 224, 273
   transfer+A315 (tRNA), 16, 18, 22, 23, 27, 185, 192, 224, 225, 273
   transfer-messenger (tmRNA), 23
Rubisco, 17, 23, 27, 92, 183, 220, 281

S
S-adenosylmethionine (SAM), 75, 170
Salicylic acid, 225, 271, 272, 275–277
Senescence, senescing, 78, 132, 148, 217, 224, 277, 281–283
Shikimic pathway, 164
Signalling, 130, 131, 139, 181–200, 209, 221, 222, 224–226
Signal recognition particle (SRP), 23, 97
Signal transduction, 48, 75, 79, 189, 190, 196, 218, 237, 248–254, 259, 260, 279
Singlet oxygen, 194, 198, 271, 272
SpoT, 273
Starch, 11–13, 164, 165, 167–171, 174, 222–224
Stay-green, 282, 283
Stomata, 132, 220, 239–241, 258, 275
Stress
   mechanical, 255, 256
   oxidative, 48, 76, 79, 80, 185, 187, 222, 275, 279
   water, 236
Strigolactones, 278
Stromule, 135
Sugar, 76, 77, 135, 139, 141, 163, 168, 169, 190, 194, 195, 212, 222, 281
Sulfoquinovosyl diacylglycerol (SQDG), sulfolipid, 43, 44, 46, 49, 72, 126–128, 130, 140–142, 145, 147
Superoxide dismutase (SOD), 79, 80, 194, 197

Symbiosis, 89, 172, 208, 209, 237, 276, 278–280
   secondary, 237
Systemic acquired resistance (SAR), 223, 225, 271, 275

T
Tetragalactosyl diacylglycerol (TeGDG), 43, 44, 46
Tetrapyrrole, 73, 182, 185–189, 196, 199, 200, 269, 270, 272
Tetratricopeptide (TPR), 99, 100, 106, 107
THF1, 77, 144
Thylakoid
   formation, 77, 144
   lumen, 129, 130, 170
   membrane, 3, 10, 11, 48, 70, 125, 126, 129, 130, 142, 169–171, 187, 197, 219, 222, 223, 225, 282, 283
   stacking, 222
   targeting, 97
Thylakoid ATP carrier (TAAC), 170, 171
TIC, Tic, 76, 77, 90–92, 99–113, 192
TOC, Toc, 76, 77, 90–92, 99–113
Trafficking
   lipid, 71, 77, 148, 149
   vesicular, 77
Trans-3-hexadecenoic acid, 129
Translocation, 78, 90–92, 96–98, 100–103, 105, 107, 108, 110–112, 130
Transmembrane, 42, 47, 66, 68–70, 80, 96, 99, 104, 106, 108, 110, 111, 136, 218
Transport
   lipid, 77, 134–136, 141–144
   vesicular, 113, 134, 142–144
Transporter
   amino acid, 74
   transporting, 161, 168
Trienoic fatty acid, 130
Trifluoperazine, 254
Trigalactosyl diacylglycerol (TGDG), 43, 44, 46, 131, 136, 140, 141
Triose phosphate, 76, 162–165, 167, 173
Triosephosphate transporter (TPP), triosephosphate translocator, 162–165
Tubulin, 211, 212

U
Ubiquinone, 47
Ultrastructure, ultrastructural, 14, 130, 220, 225, 226, 282
Uniporter, 171, 174
UV light, UV irradiation, 215, 274

V
Vesicle, vesicular, 12, 41, 77, 93, 113, 134, 135, 142–144, 149, 223, 252
Violaxanthin, 47, 48, 191, 225, 280
VIPP1, 77, 144

W
Wounding, 132, 225, 274, 275

X
Xanthophyll, 10, 73, 79, 130, 191, 272, 279, 280
Xanthophyll cycle, 130, 279

Z
Zeaxanthin, 48, 73, 191, 225, 279, 280
Z-ring, 210–215, 219

Printing: Krips bv, Meppel, The Netherlands
Binding: Stürtz, Würzburg, Germany

DATE DUE